CTS®

Certified Technology Specialist

EXAM GUIDE

Second Edition

Brad Grimes

New York • Chicago • San Francisco • Lisbon
London • Madrid • Mexico City • Milan • New Delhi
San Juan • Seoul • Singapore • Sydney • Toronto

Library of Congress Cataloging-in-Publication Data

Grimes, Brad.
　　CTS certified technology specialist exam guide / Brad Grimes.—Second
edition.
　　　　pages cm
　　ISBN 978-0-07-180794-4 (book)—ISBN 0-07-180794-2 (book)—ISBN
978-0-07-180793-7 (CD)—ISBN 0-07-180793-4 (CD)—ISBN
978-0-07-180796-8 (set)—ISBN 0-07-180796-9 (set)　1. Audio-visual
equipment industry—Employees—Certification.　2. Audio-visual
equipment—Examinations—Study guides.　I. Laurik, Sven. CTS certified
technology specialist exam guide　II. Title.
　　TS2301.A7L38　2013
　　621.38—dc23　　　　　　　　　　　　　　2013022661

McGraw-Hill Education books are available at special quantity discounts to use as premiums and sales promotions, or for use in corporate training programs. To contact a representative, please e-mail us at bulksales@mcgraw-hill.com.

CTS® Certified Technology Specialist Exam Guide, Second Edition

10　LCR　21 20 19 18

ISBN: Book p/n 978-0-07-180794-4 and CD p/n 978-0-07-180793-7
of set 978-0-07-180796-8

MHID: Book p/n 0-07-180794-2 and CD p/n 0-07-180793-4
of set 0-07-180796-9

Sponsoring Editor *Timothy Green*	**Technical Editors** *Amanda Beckner, Michelle Streffon*	**Composition** *Cenveo® Publisher Services*
Editorial Supervisor *Jody McKenzie*	**Copy Editor** *Marilyn Smith*	**Illustration** *Cenveo Publisher Services*
Project Editor *Howie Severson,* *Fortuitous Publishing Services*	**Proofreader** *Paul Tyler*	**Art Director, Cover** *Jeff Weeks*
Acquisitions Coordinator *Stephanie Evans*	**Indexer** *Jack Lewis*	**Cover Designer** *Pehrsson Design*
	Production Supervisor *James Kussow*	

ABOUT THE AUTHOR

Brad Grimes is Senior Writer/Editor for InfoComm International, an Adjunct Faculty member of InfoComm University, and the former Editor of *Pro AV* magazine. He has been writing about technology for more than 20 years, including positions covering information technology for Ziff Davis, International Data Group, and Post Newsweek Tech Media (now 1105 Government Information Group). Brad continues to author articles and white papers on everything from video communications to cloud computing. His work has been recognized by the American Business Media, the American Society of Business Publication Editors, and the American Society of Magazine Editors.

About the Technical Editors

Amanda Beckner, CTS, is the Director of Education for InfoComm International. Since joining InfoComm in 2000, Amanda has researched, designed, authored, or revised more than 40 online courses, classroom courses, publications, and assessments for the audiovisual industry. Amanda holds a master's degree in Communication, Culture, and Technology from Georgetown University, and a bachelor's degree in Education Studies from The Catholic University of America.

Michelle Streffon, CTS, is a Training Developer for InfoComm International. She has worked for InfoComm International since 2011, and contributed to the development, revision, and integration of online audiovisual courses and assessments for InfoComm and InfoComm University's learning content management system. Michelle graduated from Hillsdale College in 2012 with a bachelor's degree in English.

CONTENTS AT A GLANCE

CONTENTS

FOREWORD

You've heard the phrase, "The only constant is change." When it comes to working in the audiovisual (AV) industry, those are words to live by.

For years, AV technology has been critical to helping people communicate, whether through simple presentation systems or videoconferencing networks that span the globe, emergency notification solutions, or public address systems in high school auditoriums. That hasn't changed.

What has changed is the breadth and variety of the AV systems our industry delivers, spurred in large part by the digitization of audio and video and the integration of AV systems with that other prominent technology that has come to permeate people's lives: information technology (IT). Today's video walls, unified communications systems, command-and-control rooms, digital signage networks, and other systems represent the perfect blend of twenty-first century innovation—innovation that customers of the $78-billion commercial AV industry want for their offices, boardrooms, theaters, stores, classrooms, studios, churches, lobbies, and hospital rooms. The professionals who design, install, and manage AV systems work in a dynamic field, and they need to demonstrate a level of expertise that reflects current technology and assures customers of a job done right.

In 2011, InfoComm International and McGraw-Hill Education developed the first-ever *CTS Certified Technology Specialist Exam Guide*. InfoComm is the trade association that has represented the AV industry since 1939. Its commitment to professionalism spans education and training, standards development, and accredited industry certification. InfoComm's CTS, CTS-D, and CTS-I certifications are the only AV credentials to achieve accreditation through the International Organization of Standardization (ISO) and the International Electrotechnical Commission (IEC) ISO/IEC 17024 certification of personnel, as administered in the United States by the American National Standards Institute (ANSI).

But since 2011, as the refrain goes, the industry has changed, and it will continue to change. As a result, the CTS exam has changed. InfoComm volunteers, professional instructional designers, and psychometricians have revised the exam to more accurately reflect the job of an AV professional in today's marketplace.

Much of what a CTS-certified professional needs to know to succeed is the same. But as you might expect, skills and knowledge related to IT, networking, and network security have been deemed necessary for delivering the types of advanced AV systems required today. And increasingly, AV professionals must understand how to deliver AV functionality to mobile users, whether because those users can't attend an important meeting in person or they just want to enjoy a live performance on their favorite smartphone or tablet.

What's more, the universe of skilled technologists who would benefit from the knowledge that comes from earning a CTS credential has grown. IT professionals, electrical engineers, building operators, and others will find that the collective skills reflected in the CTS exam will serve them well now and in the future. For a more detailed explanation of how the CTS exam changes over time, see Chapter 2.

Note that InfoComm, in keeping with the accreditation requirements of the ISO/IEC 17024 standard, does not require any training or education to obtain a CTS credential. You don't even need to read this book to take the exam, although it has been developed by industry experts and designed to reflect much of the content you'll find on the actual CTS exam. That said, the *CTS Exam Guide* is not the only way to help you prepare—and for some, it shouldn't be.

Although many prospective CTS holders will rely on this book to earn their CTS certification, others will benefit from collaborating with instructors and peers through classroom training such as InfoComm University's CTS Prep course. InfoComm University also now offers CTS Prep Online, a comprehensive course that is free to InfoComm members and available to everyone. You may also benefit from taking the Essentials of AV Technology Online course, and/or completing the self-assessment tests that Info-Comm provides.

Again, none of this training is required to become CTS-certified, but it's in your and the AV industry's best interest to develop the most highly educated workforce possible. InfoComm certifies more qualified AV professionals than anyone. And because of the commitment of people like you, the CTS certification is now recognized across multiple industries and market segments for its credibility and integrity. When customers or employers hire a CTS, they know their AV systems will be designed correctly, include the latest technology, and work as promised.

Congratulations on pursuing your CTS certification. We hope to be there with you as you pursue your career in this exciting field, maybe even when you decide to attain an advanced CTS certification—either the CTS-D (design) or CTS-I (installation). For information about these programs and everything else InfoComm offers, visit www.infocomm.org.

David Labuskes, CTS, RCDD

Executive Director and CEO
InfoComm International

ACKNOWLEDGMENTS

First of all, thanks to everyone in the AV industry for making this such a growing, dynamic, and successful field. It's a clear indication of how far and fast professional AV has come that thousands of the best and brightest recognize a need to certify their skills to each other and to the wider world. Hopefully, you're next in line to be a Certified Technology Specialist (CTS).

Thanks to all the volunteer AV experts who have worked with InfoComm International over the years to define what it means to be a CTS-certified professional and share their knowledge with the next generation of great technology minds. Special thanks to InfoComm's Mandy Beckner, Rachel Bradshaw, and Michelle Streffon. Collectively, these smart, dedicated individuals put eyes on every word in this book to ensure it delivered on its value proposition: to give anyone who picks it up the foundational knowledge required to take their careers to a new level. In addition, Tim and Stephanie at McGraw-Hill Education, along with their cast of supporting characters, kept level heads and made sure this guide came out on time. My wife Kim and sons Ben and Matthew are still clueless about Ohm's law, bit rates, and networked AV systems, but I thank them with all my heart for letting me hide in a corner of the house and work this manuscript into shape.

To all of you, good luck on your CTS exam!

—Brad Grimes, 2013

2013 InfoComm Board of Directors

Greg Jeffreys, Paradigm Audio Visual Ltd., Leadership Development Committee Chair
Tony Warner, CTS-D, CDT, LEED AP, RTKL Inc., President
Johanne Bélanger, AVW-TELAV Audio Visual Solutions, President-Elect
Matt Emerson, CTS, CEAVCO Audio Visual Co. Inc., Secretary-Treasurer
Jeff Faber, Sharp's Audio-Visual Ltd.
Gary Hall, CVE, CTS-D, CTS-I, Cisco Systems
Craig Janssen, LEED AP, Acoustic Dimensions
Andrew Milne, Ph.D., Tidebreak Inc.
Thierry Ollivier, projectiondesign AS
Julian Phillips, Whitlock
Janice Sandri, FSR Inc.
Jeff Stoebner, AVI Systems Inc.

InfoComm Staff

David Labuskes, CTS, RCDD, Executive Director and CEO
Alex Damico, Chief Operating Officer

Melissa Taggart, Senior Vice President, Education and Certification
Amanda Beckner, CTS, Director of Education
Michelle Streffon, CTS, Training Developer
Rachel Bradshaw, M. Ed., Curriculum Development Manager
Rod Brown, CTS-D, CTS-I, Staff Instructor
Andrew Buskey, CTS-D, Technical Writing and Publications Coordinator
Tom Kehr, CTS-D, CTS-I, Senior Staff Instructor
Andre LeJeune, CTS, Staff Instructor
Paul Streffon, CTS-D, CTS-I, Senior Staff Instructor
Pamela Taggart, CTS, Manager International Education Programs
Scott Wills, CTS-D, CTS-I, Director of International Education
Shawn Walters, Vice President, Marketing and Communications
DoriAnn Gedris, Director of Marketing Services
Scott Hansbarger, Senior Designer
Betsy Jaffe, Public Relations Director
Nicole Verardi, Director, Education and Certification Marketing
Hank Wieland, Marketing Writer/Editor

PART I

The Certified Technology Specialist and the CTS Exam

What Is a Certified Technology Specialist?

In this chapter, you will learn about
- What it means to be a Certified Technology Specialist (CTS)
- Why you might choose to become a CTS
- The purpose of the CTS exam
- Eligibility criteria for taking the CTS exam
- The CTS exam application process

Before you prepare for the CTS exam, you will have arrived at the conclusion that you actually want to be a CTS. Earning the CTS designation can help advance your career in the professional audiovisual (AV) communications industry—whether you're just starting in professional AV or you've been working on AV projects for years.

This chapter starts out by describing what your decision to study for the CTS exam can mean for your career. It then covers the background and purpose of the exam itself, from eligibility criteria to the application process. It is important to understand that there are reasons why the CTS exam is designed the way it is, and each reason contributes to the strength of the CTS credential.

What Does a CTS Do?

A CTS designs, installs, integrates, and operates AV systems to support advanced communications and exceptional multimedia experiences. Whether for high-definition telepresence rooms or live corporate events, conference rooms or digital signage networks, distance-learning presentation or telemedicine applications, houses of worship or school auditoriums, a CTS provides the best, most cutting-edge AV solutions to meet the client's needs—both on time and within budget.

InfoComm International created and administers the CTS program. Founded in 1939, InfoComm is the leading nonprofit association serving the professional information communications industry worldwide. It has offered certification programs for more than 30 years, as well as industry-specific and general business training and education for people seeking careers in professional AV.

InfoComm developed the CTS program as part of a process for certifying industry professionals. Individuals who hold the CTS certification demonstrate a commitment to excellence in the AV industry, as measured through an objective assessment of that person's AV knowledge and/or skills. Every year, InfoComm certifies more qualified AV professionals than anyone else in the industry.

CTS certification also demonstrates a commitment to ongoing professional growth in the AV industry. After you have earned your CTS, you must maintain certified status through continual education and training. This is achieved by acquiring renewal units for taking classes online, at InfoComm University, or at various other in-person events and trade shows.

There are currently two levels of CTS certification—general and specialized—and they include the following designations:

- **Certified Technology Specialist (CTS)** A general certification that covers common AV industry knowledge and skills required for people active in system design, installation, and operation.

- **Certified Technology Specialist—Design (CTS-D)** A specialized certification that demonstrates the holder's more in-depth knowledge of AV system design skills.

- **Certified Technology Specialist—Installation (CTS-I)** A specialized certification that demonstrates the holder's more in-depth knowledge of AV system installation and integration skills.

InfoComm certifications are currently the only AV credentials to achieve accreditation through the International Organization of Standardization (ISO) and the International Electrotechnical Commission (IEC) ISO/IEC 17024 certification, as administered in the United States by the American National Standards Institute (ANSI). ISO/IEC 17024 is a standard designed to assure the public and the CTS certification holders that InfoComm applies best practices to the administration of the CTS program. The ANSI accreditation provides a higher level of confidence in the CTS certification program. ANSI serves as an unbiased third-party source to verify the competence of certification bodies to develop, manage, and maintain their programs.

Why Earn Your CTS?

Industry certifications are common across many of the trades that AV professionals encounter on the job. Like the CTS, these certifications demonstrate that holders have achieved a certain level of knowledge and expertise in their chosen fields. As AV professionals increasingly work more with architects, contractors, and others who hold industry certifications, their ability to present recognized CTS credentials to partners on the job can instill confidence and boost the value of their work.

Furthermore, CTS certification can instill confidence in clients, providing additional credibility that the work the CTS holder performs will meet an accepted level of professionalism and reliability. Increasingly, requests for proposals (RFPs) for AV projects include requirements that the winning bidder employ CTS-certified professionals. Moreover, an increasing number of technology managers—individuals who might initiate an RFP for an AV project—are themselves CTS holders, increasing the likelihood they would prefer to hire CTS-certified consultants, integrators, and installers.

Finally, InfoComm studies suggest that CTS holders earn more than their peers. According to InfoComm's 2012 *Compensation and Benefits Survey*, CTS holders earn 8 percent more on average than non-CTS holders with similar backgrounds in the AV industry. The CTS designation is also highly recognized and valued among those who earn it. According to InfoComm's most recent *CTS Surveillance Survey*, conducted in 2012 among existing CTS holders, eight of ten respondents say they consider the CTS a sign of professional achievement; nearly 66 percent say their CTS conveys higher value and credibility to their customers. And many who have earned the CTS remain committed to the credentialing process. The study showed that 95 percent say renewing their CTS is important, and 75 percent say they are considering pursuing the more advanced CTS-D and CTS-I designations.

The Purpose of the CTS Exam

As mentioned earlier, AV professionals who hold the CTS certification demonstrate their AV industry knowledge and skill through an objective assessment—the CTS exam. In following the standards set by international bodies, and in support of a professional certification recognized for credibility and integrity, InfoComm's independent Certification Committee oversees a CTS exam that tests individuals against peer-developed standards and competencies.

The exam is designed independently of any course or curriculum, and it is developed and administered to be valid, reliable, defensible, and psychometrically sound. All exams are developed, reviewed, and maintained by a series of AV industry subject matter experts (SMEs) guided by professional test development experts (psychometricians).

NOTE Psychometrics is the study of psychological measurement, including the measurement of knowledge, abilities, attitudes, personality traits, and educational assessment/analysis. In the context of the CTS exam, psychometric principles are applied to exam questions to ensure they correctly evaluate what they are intended to evaluate.

The CTS assessment criteria are based on an analysis of the job tasks performed by professionals in the AV industry as determined by AV industry SMEs guided by professional testing experts. They are not based on any training or education programs, or any educational provider's programs. The specific knowledge areas addressed by the CTS exam, plus the scope of the knowledge and skills in each area, are described in Chapter 2.

The majority of the CTS exam addresses technical skills. However, the Certification Committee and SMEs from various AV job functions determined that the general requirements for a CTS holder should also include a basic understanding of the fundamental concepts of project management, estimating, purchasing, sales, and job costing. This is based on the assumption that during the course of their work, CTS holders may be called upon to oversee projects, assist in estimating expenses, make purchasing decisions, perform basic job-costing, conduct sales, and handle other nontechnical tasks in support of their employer.

The inclusion of this type of content and the assessment of nontechnical skills through the CTS exam helps ensure that a CTS holder has not only the technical aptitude, but also some basic business knowledge to help a project move successfully toward completion. Exposure to such nontechnical skill areas also helps CTS holders determine whether they would like to pursue specialty designations (CTS-D or CTS-I) in order to advance their careers and benefit their employers and clients.

NOTE While the majority of the CTS exam deals with technical skills and knowledge, SMEs from within the AV industry determined that so-called "soft skills" are also important to a successful certified AV professional. Therefore, questions about project management, sales, marketing, and other subjects not strictly related to AV technology are also part of the CTS exam.

Are You Eligible for the CTS Exam?

More often than not, the answer to this question is, yes. It is strongly recommended that you have the skills and/or experience indicated in InfoComm's *CTS Exam Content Outline*. (See Chapter 2 for an outline of the CTS exam content.)

CTS candidates are not required to complete any particular course in order to take the CTS exam or qualify for certification, although there are many ways you can prepare for the exam and determine whether you are ready to take it. However, no single course contains all the information needed to pass the exam.

InfoComm itself offers a number of general courses that can help individuals prepare for each level of CTS certification. The exam guide you are currently reading has been specifically developed to address the content of the general CTS exam. The InfoComm Certification Committee also offers a *CTS Candidate Handbook*, an exam content outline with job-task analysis information, a listing of primary reference materials, a glossary, additional practice exams, and other materials. You can find all the material online at the InfoComm website (www.infocomm.org/cts).

NOTE A job task analysis (JTA) is a study conducted to identify the knowledge, skills, and abilities necessary for professional competence in a particular field. Such an analysis is often conducted to determine the content and competencies that should be included in a certification or exam. InfoComm's independent Certification Committee conducts periodic JTAs to make sure the CTS exam and certification process aligns with the real-world skills required of certified AV professionals.

The online materials can help you prepare for the CTS exam, but they are not intended to include all potentially useful sources of information. Inclusion on the InfoComm list of preparatory materials does not constitute an endorsement by the Certification Committee. The Certification Committee does not endorse any particular reference as being completely accurate and encompassing, and it recommends that applicants use multiple resources in the process of preparing for the exam.

The CTS Exam Application Process

To apply for the CTS exam, visit the InfoComm website (www.infocomm.org/cts). There, you can obtain the free *CTS Candidate Handbook* and current CTS candidate application.

CTS applications may be submitted in one of several ways:

- Completed online at the InfoComm website
- Printed and mailed
- Scanned and e-mailed
- Faxed

Candidates must provide required documentation and payment as noted on the application. All signatures must be submitted by the applicant as provided in the CTS application. This includes agreeing to the CTS Code of Ethics and Conduct, which is printed in the *CTS Candidate Handbook.*

After you send the required documentation, meet eligibility requirements, and have your payment approved, InfoComm will e-mail you an eligibility acceptance letter within ten business days of receiving the application. This message includes a unique candidate ID number and testing instructions. If an application is not accepted, you will receive an e-mail letter of explanation. Applications that are incomplete, or for which payments have been denied, will delay approval.

NOTE Be sure to include all required documentation with your application to avoid denial and/or delay.

Once approved for eligibility, you will be notified by e-mail within one business day of InfoComm submitting the approval information to the third-party exam company, Pearson VUE. After another 24 hours, you may contact Pearson VUE to make an appointment to take the CTS exam. Visit the Pearson VUE website (www.pearsonvue .com/infocomm) and click Locate a Test Center to find an exam location near you.

Any questions about the CTS exam should be directed to the following:

InfoComm International, Attn. Certification Office
11242 Waples Mill Rd., Suite 200, Fairfax, VA 22030
Phone: 1.800.659.7469 or +1.703.273.7200
Fax: +1.703.691.2756
certification@infocomm.org
www.infocomm.org

Chapter Review

A CTS performs general technology solutions tasks in the process of designing, building, operating, and servicing AV communications systems. The CTS certification demonstrates to individuals, employers, trade partners, and customers that the CTS holder meets an established, peer-reviewed benchmark of expertise in AV-specific and business-general topics.

The CTS exam is the vehicle by which CTS holders assess their expertise, through a method of documenting their knowledge within the AV industry. The CTS exam is based on an analysis of the tasks that AV professionals perform, focusing on "best practices" within the AV industry.

This chapter briefly described the background and purpose of the CTS exam and how to register to take it. Candidates should now have a better understanding of the purpose and objectives of the exam, and whether obtaining a CTS certification could help advance their career in the AV information communications industry.

The CTS Exam

In this chapter, you will learn about
- The scope of the CTS exam
- How the CTS exam changes over time
- How to study effectively for the exam
- What types of exam questions to expect
- Question-evaluation strategies for successful test-taking
- What to expect on exam day

As you learned in Chapter 1, the InfoComm CTS exam is part of a professional certification process that is designed to evaluate a person's level of knowledge of general AV-related job tasks. The exam covers a wide range of AV-related topics, and it is based on an analysis of typical job tasks performed by people currently working in the AV industry.

This chapter describes the content addressed within the CTS exam, so that you can better prepare for taking the exam. It also covers how the exam is conducted, so you will know what to expect when you arrive to take the exam.

Keep in mind that InfoComm regularly updates the CTS exam content and procedures for taking the exam, so be sure to visit InfoComm's website (www.infocomm .org/CTS) to obtain the latest information and requirements.

The Scope of the CTS Exam

To create the CTS exam, a group of volunteer AV SMEs, guided by professional test development experts, participated in an AV job task analysis (JTA) study. The results of this study form the basis of a valid, reliable, fair, and realistic assessment of the skills, knowledge, and abilities required for competent job performance by AV professionals.

The original JTA study for creating the ANSI-accredited CTS exam was conducted in December 2006. A follow-up study, as required by ANSI, was conducted in 2012 to revalidate the exam and identify areas where it might be updated to reflect current AV jobs, skills, knowledge bases, and best practices (see the next section, "The CTS Exam Over Time").

In creating the CTS exam based on the JTA, the volunteer SMEs identified major categories (*domains*) to be covered by the certification examination, as well as topics that should be addressed within each domain, based on the jobs and tasks that a certified individual might perform. The exam-development team examined the importance, criticality, and frequency of typical AV-related job tasks, and used the data to determine the number of CTS exam questions related to each domain and task. Table 2-1 lists the domains and tasks, as well as how many questions there are for each.

 NOTE The *Certified Technology Specialist (CTS) Job Task Analysis Final Report* that provided the basis for the CTS exam scope and questions is available from the InfoComm website (www.infocomm.org/cts).

Domain/Task	Percent of Exam	Number of Items
Domain A: Creating AV Solutions	61%	61
Task 1: Conduct site survey	6%	6
Task 2: Gather customer information	6%	6
Task 3: Evaluate site environment (acoustics, lighting, seating, finishing, etc.)	6%	6
Task 4: Maintain awareness of changes to site environment (acoustics, lighting, seating, finishing, network security, etc.)	6%	6
Task 5: Define the functional AV scope	8%	8
Task 6: Design AV solutions	12%	12
Task 7: Sell AV solutions	5%	5
Task 8: Conduct vendor-selection process	4%	4
Task 9: Provide AV solutions	8%	8
Domain B: Operating AV Solutions	16%	16
Task 1: Operate AV solutions	6%	6
Task 2: Conduct maintenance activities	5%	5
Task 3: Manage AV solutions/operations	5%	5
Domain C: Conducting AV Management Activities	11%	11
Task 1: Project manage AV projects	5%	5
Task 2: Perform AV finance and job-costing activities	6%	6
Domain D: Servicing AV Solutions	12%	12
Task 1: Troubleshoot AV solutions	7%	7
Task 2: Repair AV solutions	5%	5
Total	100%	100

Table 2-1 CTS Exam Domains and Tasks

The CTS Exam Over Time

Like AV technology, the CTS exam evolves over time. In the case of the CTS exam, evolution is a must. ANSI requires that InfoComm review the JTA that is the basis of the CTS exam every five years in order to determine whether it still accurately reflects the job of a certified AV professional. Through the process, SMEs introduce changes to the exam that make it better and ensure that CTS-certified AV professionals can perform the tasks necessary to keep pace with changes in technology, best practices, industry standards, and more.

The most recent revalidation of the CTS exam began in 2011, starting with four JTA focus groups, held both in the U.S. and internationally. The focus groups were supplemented by a survey of current CTS holders around the world. In September 2012, the results of that research were presented to the CTS Scheme Committee. This committee reports to the Certification Committee and reviews psychometric reports on CTS exam performance, decides scoring matters based on recommendations of test-company experts, and works with the CTS Technical Committee when new items are required for the CTS exam. That same month, the Scheme Committee presented its recommendations to the Certification Committee, which voted to approve the recommended changes to the CTS exam.

With the recommended changes in hand, the CTS Technical Committee is responsible for creating new exam questions, if necessary. New questions are written by SMEs, evaluated by psychometricians, and tested for validity.

However, recommended changes also include reweighting certain domains or tasks. This is necessary when the JTA determines that particular domains or tasks have become more important to the job of a CTS, or that they have become less important. For example, reflecting ongoing change in the professional AV industry toward companies deriving more of their revenue from services, Domain D (Servicing AV Solutions) will now account for 12% of the CTS exam, up from 8%.

Also, as a result of the most recent JTA revalidation, it was determined that the process of conducting a site survey, a task under Domain A (Creating AV Solutions), was worthy of an additional question on the exam, based in part on its importance to a successful AV project. Other tasks in Domain A deemed worthy of additional exam questions include the following:

- Maintain awareness of changes to site environment
- Define the functional AV scope
- Define an AV solution

In a related change, SMEs and CTS credential holders determined that evaluating AV personnel competencies was not central to the job of a CTS, and the task was removed from the CTS exam, giving the Scheme Committee the opportunity to add questions in more important areas.

In addition, three business-related tasks under Domain C (Conducting AV Management Activities) of the previous CTS exam—conduct purchasing activities, conduct estimating activities, and conduct job-costing activities—were combined into one task: perform

AV finance and job-costing activities. The number of questions devoted to this task was reduced to six from the previous nine.

As the CTS exam evolves, InfoComm strives to be as transparent about changes as ANSI requirements allow. You can find the results of the most recent JTA on the Info-Comm website.

Studying for the Exam

There is no single way to prepare for the CTS exam, including studying this book. Instead, there are a number of different ways to determine whether you are ready for the exam, starting with a free practice test composed of questions that are similar to the questions presented on the CTS exam. The practice exam can be found on the accompanying CD, with additional resources available at www.infocomm.org/certification.

 NOTE Because of the way the CTS program and CTS exam are designed in order to comply with ANSI standards, no CTS practice exam is allowed to include actual exam questions, and practice questions may not be informed by the exam itself. Any practice question you find here or elsewhere is written to be *similar* to an actual CTS exam question.

Prior to studying for and taking the exam, it's wise to perform a self-assessment of your AV industry knowledge in order to identify your strengths and weaknesses. The practice exam is a good place to start. You may also want to review the actual AV JTA study that provided the basis for the CTS exam questions.

In reviewing the practice exam and the JTA, focus on the content areas (tasks) addressed under each domain, as well as the number of exam questions devoted to each area. The greater the number of possible questions for a task, the more emphasis you might place on studying for that content area/task. For example, if only 2 out of 100 questions address a specific task, it might not make sense to spend 50 percent of your study time focused on that task.

Reviewing the practice exam and the JTA can also help you assess your strengths and weaknesses, and more finely hone your study plan. For example, you may feel comfortable with your knowledge level in certain tasks, such as conducting a site survey (Task 1 in Domain A), based on previous experience or study. Understanding that, you may be wise to focus on areas where your knowledge is not as strong.

In addition, InfoComm offers what is known as the InfoComm-Recognized AV Technologist Certificate, which is an entry-level designation designed as a bridge to CTS certification. The program is not ANSI accredited and does not carry most of the benefits of the CTS, but it can be a good measure of your AV knowledge base. The program requires passing a 100-question exam. The AV Technologist exam is based on InfoComm's GEN102 *Essentials of AV Technology* online course, which is also one of many ways to prepare for the CTS exam. Visit the InfoComm website for more information about this certification.

When it comes to actual studying, hands-on learners may prefer attending a class-room course or training with an experienced mentor in a manner that provides opportunities to work directly with the gear or discuss concepts. Others may prefer a reference book, like this one, that they can highlight and annotate.

InfoComm offers classroom-based CTS preparation courses to meet the needs of hands-on learners. It also provides a CTS virtual classroom course suitable for those without the time and/or resources to travel to a class, and who enjoy the opportunity to repeat lessons until they feel comfortable with the topic.

Visit the InfoComm website to review the full range of current resources that are available to assist you in preparing for the CTS exam.

Sample CTS Exam Questions

The CTS exam is composed of 110 multiple-choice questions that address each of the domains and tasks listed in Table 2-1. Ten of the questions are pilot questions, used by InfoComm to evaluate and select new questions for the CTS exam. You will not know which are the pilot questions, and these questions will not be scored.

CTS exam questions focus primarily on issues that an AV professional may encounter when working on a specific job or task, rather than on general AV technology knowledge. In other words, the majority of questions are designed to prompt the candidate to apply AV knowledge to a specific job/task.

Let's look at a pair of sample questions that illustrate this difference between AV technology questions and applied AV knowledge questions that might appear on the CTS exam. Here is a sample AV technology question:

Aspect ratio is the _____.

    a. Relationship of the width to the height of a displayed image

    b. Viewer's attitude when viewing an image

    c. Relationship of the viewer distance to the screen size

    d. Relationship of the number of viewers to the screen diagonal measurement

 NOTE Including sample questions in this book does not mean the specific questions are or are not on the actual CTS exam.

Notice that this sample aspect ratio question verifies that the candidate knows a specific fact. However, a job/task-oriented question would address the same issue from a slightly different perspective. Here is an example of a job/task-oriented question that addresses aspect ratio:

When determining the width of a standard HD display, given the desired screen height, which of the following aspect ratios would you use in your calculations to determine the screen width?

    A. 1.33:1 (4:3)

    B. 3:2

    C. 1.78:1 (16:9)

    D. 2.39:1

The following are some other examples of the types of questions that may be found on the general CTS exam. For each practice question, the domain and task from which the question is drawn are identified within brackets, preceding the question, in the form [*domain/task*].

1. [Creating AV Solutions/Design AV solutions] What method of wiring is depicted in the illustration below?

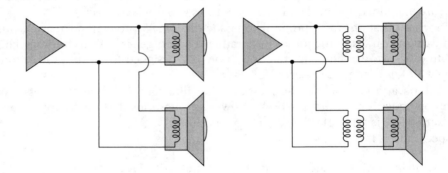

    A. Series circuit

    B. Parallel circuit

    C. Stereo pair circuit

    D. Combination parallel/series circuit

2. [Creating AV Solutions/Design AV solutions] Direct coupled loudspeakers are usually listed by size, frequency response, power handling capacity, and _____.

    A. Impedance

    B. Constant voltage

    C. Resonant frequency

    D. Thiele/Small parameters

3. [Creating AV Solutions/Sell AV solutions] When developing an AV solution for a client, whose input is *most* crucial for determining system requirements?

    A. The end user's

    B. The system designer's

    C. The project manager's

    D. The audience's

4. [Operating AV Solutions/Manage AV solutions/operations] Bluetooth technology can be useful in AV systems because it ____.

    A. Can communicate over long distances

    B. Connects similar devices via LAN

    C. Uses MAC addressing to connect the network

    D. Allows additional connections for control of source devices

5. [Servicing AV Solutions/Troubleshoot AV solutions] A client has reported that a laptop input is not working. The interface is connected to a 5 BNC wall plate. No connectors appear damaged. It has been verified that the interface is receiving a signal and the projector is on the correct input. What is the next logical step in troubleshooting the issue?

    A. Replace the interface

    B. Re-terminate the wall plate

    C. Reverse the H and V sync cables

    D. Unplug the cables, and using a TDR device, test all the cables

6. [Creating AV Solutions/Design AV solutions] What is the formula for calculating current using Ohm's law?

    A. $I = V / R$

    B. $I = R / V$

    C. $I = V * R$

    D. $I = V + R$

7. [Creating AV Solutions/Evaluate site environment] What is one of the major considerations in creating an AV solution for an important historical site?

    A. Price

    B. Site location

    C. Weather conditions

    D. Preservation of the site

8. [Creating AV Solutions/Define the functional AV scope] Which drawings does one usually provide for customer approval during the bid process?

    A. Final drawings

    B. Concept drawings

    C. Marked-up drawings

    D. As-built drawings

9. [Creating AV Solutions/Sell AV solutions] Good customer service is crucial for ____.

   A. Networking

   B. Hiring new employees

   C. Obtaining new business

   D. Building good customer relationships

10. [Creating AV Solutions/Provide AV solutions] What type of connector is shown below?

   A. BNC

   B. XLR

   C. PS/2

   D. Mini-DIN

11. [Conducting AV Management Activities/Perform AV finance and job-costing activities] When negotiating a contract for a job that will take a long time to complete, one should consider ____.

   A. Progress, stage, or interim invoicing

   B. Progress, stage, or internal invoicing

   C. Progress, strategy, or interim invoicing

   D. Progress, strategy, or internal invoicing

The following are the correct answers for the preceding sample questions.

1. B

2. A

3. A

4. D

5. C

6. A

7. D

8. B

9. D

10. B

11. A

Question-Evaluation Strategies

As noted earlier, the CTS exam is made up of multiple-choice questions. For exam-takers, there are several strategies to help identify the correct answers. These tips can save you time when trying to distinguish the correct answer from the *distracters* (incorrect answers), as they're known in exam development. By adopting the following strategies, you may be able to either reinforce your knowledge of the correct answer or eliminate one or more wrong answers.

- Read the question without looking at the answers. You may also want to cover the answers before reading the question to help you focus on exactly what the question is asking. By reading only the question, you can formulate an answer in your head before getting distracted by the wrong answers.

- Read the entire question. It's common to begin making inferences before you've finished reading the full question. By reading the entire question first, you may keep from missing important information and answering the question incorrectly.

- Determine the root meaning of each question by identifying key terms. First, cover the answers. Next, read the entire question and look for key terms that could indicate what the question is asking. Finally, reread the question, emphasizing the key terms.

- Determine the root meaning of each question by removing distracters. A candidate who has mastered a topic should be able to eliminate unnecessary words and focus on what the question is asking.

- Categorize the answers. Look at the answers and determine if there are similarities among them. Do some answers have the same meaning or a similar relationship? Also look for dissimilarities. Is there one answer in particular that stands out from the others, and if so, why?

- Read all of the answer options before selecting your choice. Needless to say, you're more likely to pick the wrong answer—even if you think you know the right answer—if you have not reviewed all the answers.

- Eliminate answers that are obviously incorrect.

- Even if you are not sure about the correct answer, select the one you think is the correct answer. There is no added penalty for selecting an incorrect answer versus not answering the question at all.

What to Expect on Exam Day

On the scheduled day of the CTS exam, you should report to the exam center as instructed in your appointment confirmation letter. Plan to arrive at least 30 minutes prior to the scheduled start time. It is not necessary (although it is preferred) to bring your e-mail or letter of confirmation with you. However, you must have proper identification (as described shortly). The name and address on the ID must match the information on file with InfoComm and the vendor responsible for presenting the exam.

If you live more than an hour from the exam center, you might consider staying at a nearby hotel the night before so you can get a good night's rest and make sure you arrive on time. It may also be a good idea to actually visit the testing center prior to the exam to ensure you know exactly where to go and how to get there. On the day of the exam, make sure to allow extra time for unforeseen events, such as traffic delays. These measures can help reduce unnecessary stress on exam day. If you arrive after your assigned exam time, you will be considered a "no-show" and will not be admitted. To take the exam, you will need to reapply by contacting InfoComm and paying a reinstatement fee.

Identification Requirements

CTS exam candidates must check in using two forms of valid identification, one of which must be a government-issued photo ID with signature (driver's license, other government-issued photo ID, or passport). The name on the photo ID must match exactly the name submitted on the exam application.

If you do not have proper identification (government-issued photo with signature ID), or the name on your ID does not match your application information, you will not be allowed to take the exam, and you will need to have your exam authorization reinstated by contacting InfoComm. Doing so will incur an additional fee.

 NOTE The candidate demographic information used in exam application/ eligibility documents, the certification database, the certification, and elsewhere is recorded in English. For certain Asian countries—specifically, China, Hong Kong, and Taiwan—if candidates do not have their English name printed on their passport, the candidates must use standard Pinyin to translate their name into English to meet the required identification policies.

For ID purposes at the testing center, the candidate must fulfill *one* of these options:

- **Valid passport** For the exam application and certification process, candidates should use the name as displayed on a valid passport. For ID purposes at the testing center, the candidate must present a valid passport that matches the exam application and eligibility notice.

- **Government-issued ID and one other ID** The standard Pinyin English translated name must be used for the exam application and certification process. For testing center identification purposes, the candidate must provide both a valid government-issued national ID for the photo verification and either a valid credit card or military ID with a signature that matches the name on the national ID for the signature portion of the verification of ID. Testing center staff will verify that the standard Pinyin English translated names on the two presented IDs match the exam application/eligibility documents.

At the testing center, candidates will be required to provide an electronic signature and have their digital photo taken. This information is retained in a secure database for no more than five years from the last exam date. It is not linked to a candidate's personal identification information, such as address or credit card information.

Items Restricted from the Exam Room

You are not allowed to bring anything into the exam room. Secure lockers are provided to store personal items while taking the exam. The following are *not* permitted in the exam room or testing center:

- Slide rules, papers, dictionaries, or other reference materials
- Phones and signaling devices such as pagers
- Alarms
- Recording/playback devices of any kind
- Calculators
- Photographic or image-copying devices
- Electronic devices of any kind
- Jewelry or watches (time will be displayed on computer screen and wall clocks in each testing center)
- Caps or hats (except for religious reasons)

The Exam

The exam will be presented via computer. The computer displays each question, along with four possible answers (A, B, C, and D). One of the answers represents the single correct response, and credit is granted only if you select that response.

Candidates have 150 minutes to answer 110 questions. Remember, 10 of the questions are unscored pilot questions used by InfoComm to evaluate and select new questions for the CTS exam, but you won't know which they are. There is a brief on-screen computer-based tutorial prior to starting the exam, and a brief online survey at the end of the exam. The tutorial and survey do not count against your 150 minutes.

A tutorial and practice exam are available online from Pearson VUE's InfoComm International Testing page, at www.pearsonvue.com/infocomm. Candidates can access these any time prior to taking the exam.

During the Exam

The following points pertain to how things are handled during the exam:

- Candidates should listen carefully to the instructions given by the exam supervisor and read all directions thoroughly.

- Questions concerning the content of the exam will not be answered during the exam.

- The exam center supervisor will keep the official time and ensure that the proper amount of time is provided for the exam.

- Restroom breaks are permitted, but they are included as a part of the 150 minutes allotted for the actual exam.

- Candidates will be reminded when logging in to the testing center computer screen, and prior to being allowed to take the exam, that they have agreed to follow the CTS Code of Ethics and Conduct and nondisclosure agreements presented earlier in the application process.

- Candidates will have access to a computer-based calculator and a wipe-off note board provided by the testing center.

- Candidates will have the capability to provide comments for any question, as well as mark questions and return to them for review.

- There will be an on-screen reminder when only five minutes remain to complete the exam.

- No exam materials, notes, documents, or memoranda of any kind may be taken from the exam room.

For best results, pace yourself by periodically checking your progress. This will allow you to make adjustments to the speed at which you answer the questions, if necessary. Remember that the more questions you answer, the better your chance of achieving a passing score. If you are unsure of a response, eliminate as many options as possible and choose from the answers that remain. You will also be allowed to mark questions for review prior to the end of the exam.

Be sure to record an answer for each question, even if you are not sure the answer is correct. Again, you can note which questions you wish to review and return to them later. There is no penalty for guessing.

Dismissal or Removal from the Exam

During the exam, the exam supervisor may dismiss a candidate from the exam for any of the following reasons:

- The candidate's admission to the exam is unauthorized.
- A candidate creates a disturbance, or gives or receives help.
- A candidate attempts to remove exam materials or notes from the testing room.
- A candidate attempts to take the exam for someone else.
- A candidate possesses any item excluded from the exam center, as specified in the "Items Restricted from the Exam Room" section earlier in this chapter.
- A candidate exhibits behavior consistent with attempting to memorize or copy exam items.

Any individual who removes or attempts to remove exam materials, or is observed cheating in any manner while taking the exam, will be subject to disciplinary and/or legal action. Sanctions could result in removing the credential or denying the candidate's application for any InfoComm credential.

Any unauthorized individual found in possession of exam materials will be subject to disciplinary procedures in addition to possible legal action. If the individual has CTS certification, sanctions could result in the removal of certification.

Candidates in violation of InfoComm testing policies are subject to forfeiture of the exam fee, as well as disciplinary and/or legal action.

Hazardous Weather or Local Emergencies

In the event of hazardous weather, or any other unforeseen emergencies occurring on the day of an exam, the exam presentation vendor will determine whether circumstances require cancellation. Every attempt will be made to administer all exams as scheduled.

When an exam center must be closed, the vendor will contact all affected candidates and ask them to reschedule. Under those circumstances, candidates will be contacted through every means available: e-mail and all phone numbers. This is an important reason for candidates to provide and maintain up-to-date contact information with InfoComm and the exam vendor.

Special Accommodations for Exams

InfoComm complies with the Americans with Disabilities Act (or country equivalent) and is interested in ensuring that no individual is deprived of the opportunity to take the exam solely by reason of a disability as defined under the American with Disabilities Act (or equivalent). Two forms must be submitted to receive special accommodations:

- Request for InfoComm (CTS, CTS-D, CTS-I) Exam Special Accommodations
- InfoComm (CTS, CTS-D, CTS-I) Exam, Healthcare Documentation of Disability Related Needs

Applicants must complete both forms and submit them with their application information to the InfoComm Certification Office no later than 45 days prior to the desired exam date.

Requests for special testing accommodations require documentation of a formally diagnosed and qualified disability by a qualified professional who has provided evaluation or treatment for the candidate.

These forms, along with more information about the process, can be found on the InfoComm CTS website.

Exam Scoring

The final passing score for each examination form is established by a panel of SMEs using a criterion-referenced process. This process defines the minimally acceptable level of competence, and takes into consideration the difficulty of the questions used on each examination form.

Candidates who do not pass the exam receive their score and the percentages of questions they answered correctly in each domain. InfoComm provides these percentages in order to help candidates identify their strengths and weaknesses, which may assist them in studying for a retest. It is not possible to arrive at your total exam score by averaging these percentages because there are different numbers of exam items in each domain on the exam.

Retesting

Candidates who do not pass the CTS exam may retake it two more times, waiting a minimum of 30 days between exams. Once approved for a retest, you have up to 120 days to retake the exam.

After two retests, if you still have not passed the exam, you must wait 90 days before restarting the application process. This period allows the applicant time to adequately prepare and prevents overexposure to the exam.

Currently certified CTS individuals may not retake the CTS exam, except as specified by InfoComm's CTS renewal policy.

Candidates must meet all eligibility requirements in effect at the time of any subsequent application. You can find the CTS Exam Retest Application form and current retest fees at the InfoComm website.

Chapter Review

This chapter briefly reviewed some of the characteristics of the InfoComm CTS exam, covering the following:

- The scope of the CTS exam, including how the exam was created and the topic areas addressed within the exam
- How and why the CTS exam changes over time, including changes made to the most recent CTS exam

- How to study for the exam in the most effective manner
- Sample questions that give you a better idea of the general types of questions that are presented on the exam
- Strategies for evaluating questions while taking the exam
- What to expect on the day of the exam, including how the exam is conducted, what you need to bring to the exam, and other relevant issues

PART II

Essentials of AV Technology

Part II of the *CTS Exam Guide* is based largely on InfoComm's *Essentials of AV Technology*, an online course and accompanying publication that offers a comprehensive overview of the science and technology of audio, visual, and AV systems integration. It represents years of work by many SMEs in the AV industry, who grappled with complex concepts in an effort to refine them to their most important essence without losing their accuracy. It is the cornerstone for AV knowledge that applies to all of InfoComm's subsequent works.

The goal of *Essentials of AV Technology* is to document the fundamentals that AV professionals use on a daily basis and to illustrate the mechanics of how these AV fundamentals work. InfoComm's training philosophy has always been that if you understand the fundamentals and their mechanics, you will be able to apply this knowledge to new technologies and the applications that follow. With this knowledge base, you are able to solve problems and meet your clients' needs.

But *Essentials of AV Technology* does not exist in a vacuum. You cannot simply study the *Essentials of AV Technology* course material and expect to succeed on the CTS exam. As noted earlier, CTS certification is not merely a measure of technical prowess. Understanding these essentials is a key component of your CTS exam preparation, but it is only the beginning.

Domain Check

As a general overview of AV technology, the material in this part of the book pertains to many of the domains and tasks on the CTS exam. Much of the technical material in this part informs Domain A (Creating AV Solutions), particularly Task 6 (Design AV solutions) and Task 9 (Provide AV solutions), as those exam tasks have the most to do with actual AV systems and account for 20% of the exam's questions. This material may also aid CTS candidates in preparation for Domain D (Servicing AV Solutions). This part of the book focuses on the technical knowledge required of CTS holders, not soft skills.

What Is an AV System?

In its simplest form, AV is about helping people communicate an idea effectively. Whether it is a flat-panel display in a hotel lobby or a wedding video on YouTube, AV tools and technology are used to help people relate to and understand one another. Adding AV to an idea helps people pay attention, learn, laugh, enjoy, make decisions, and remember.

How you define AV communication varies based on your interest and experience. For some people, AV goes back to black-and-white television; for others, AV is a video chat on a smartphone or the latest video game in 3D.

AV technology is used to communicate ideas everywhere: to doctors in operating rooms, teachers in classrooms, students in dormitories, corporations in boardrooms, lawyers in courtrooms, marketers in retail stores, and rock stars on the stage. It would be a dull world indeed without AV communication.

The human being is an analog creature living in a digital age. If these terms mean little to you, after reading the following chapters, you will understand their significance and impact. You will also learn about the science and technology of AV, where and how it is used, and why working in the multibillion-dollar AV industry is one of the best jobs around.

An AV system is two or more pieces of AV equipment designed to work together to meet a communication need. These systems can be connected with cable or wirelessly. The equipment used in the system may be passive (not powered) or active (powered).

The markets in which professional AV systems are used include schools and universities, government, the military, businesses, health care, legal, retail, museums, houses of worship, sports arenas, entertainment, transportation, and many other areas—in short, everywhere. The actual applications of AV technology are often similar across markets, including presentations, conferencing (web, audio, video, and data), education, advertising, retail signage, dynamic displays, command-and-control systems, concerts, and public information systems. Sharing information using audio and video has become essential.

AV System Goals

Although an AV system is made up of specific equipment, the equipment itself is less important than the system's ultimate purpose. The purpose of an AV system is to meet a communication objective; it is the task that a user wants to accomplish—through AV—to communicate an idea. Here are some examples of an AV system's purpose:

- To train a group of 100 people about sales techniques or new products
- To debrief a small task force
- To facilitate changes to company policies globally
- To monitor aircraft locations

An AV system allows people to communicate and share information. It can create or reproduce a complete experience using sound, images, and environmental control. It should fit in with its environment and not be the focus.

The quality of an AV system design dictates how well it meets the objective. Good integration requires careful thought and planning. Sometimes users buy individual pieces of equipment at different times and without forethought. Without a plan in mind, the power of a good system may be lost.

Analog and Digital Signals

In this chapter, you will learn about
- The difference between analog and digital waveforms
- Digital signal processing, sampling, bit depths, and bit rates
- Signal compression and digital formats
- Noise and signal transmission

The technology that supports much of the AV industry is changing from analog to digital. Why is this important? In the AV industry, you will work with some analog equipment, more digital equipment, and some equipment that combines analog and digital. You need to understand both technologies.

One way to understand analog signals is to create a mental image. Think of a dimmer light switch, which offers many possible positions. Analog is usually represented as a continuous, varying wave. In fact, the term *analog* is used because the analog wave is analogous to fluctuations of the voice.

However, digital information requires a different mental image. Think of a standard light switch, which has only two positions: on and off. In the world of signals, a digital signal is either on or off. These two states are numerically represented with a one (on) or a zero (off).

Figure 3-1 shows the difference between analog and digital waves.

Figure 3-1
Analog waves are continuous, and digital waves are on or off.

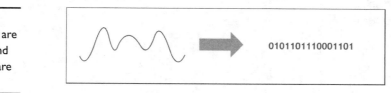

Analog and Digital Waveforms

Figure 3-2 shows two characteristics of an analog waveform. The first characteristic is the smooth, constantly flowing line. It looks like an ocean wave, and it is similar to the dimmer light switch analogy. The second characteristic is the constantly changing amplitude over time. It alternates smoothly from a high level to a low level and back up again.

We know that digital information is represented by two numbers: one and zero. In a digital waveform, the wave looks like a series of squares with high- and low-voltage states. Described another way, a digital waveform is a sudden signal with voltage (on) followed by no voltage at all (off).

Digital Signal Basics

As noted at the beginning of the chapter, AV technology has been quickly moving from mostly analog to predominantly digital. AV professionals need to know about digital signal processing, sampling, bit depths, and bit rates. They also need an understanding of signal compression and digital formats.

Digital Signal Processing and Sampling

You may have analog source material that you need to make digital. For example, you might need a digital copy for streaming content over the Internet, for archiving purposes, or for sending through an AV system for use elsewhere.

Sampling is one important step of analog-to-digital conversion. The goal is to make a reasonable digital copy of the original analog signal.

A digital sample of the analog signal is made at certain time intervals. How often a sample is taken is called the *sampling rate*. Determining an acceptable sampling rate is crucial to obtaining a digital representation that resembles the original signal.

Figure 3-2
Digital samples of
an analog signal

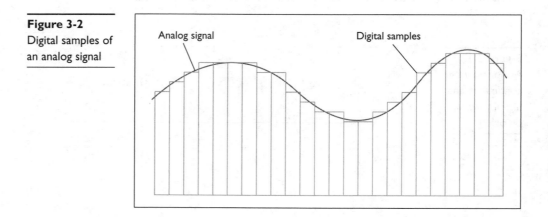

For example, the sampling rate of an audio CD is 44.1 kilohertz (kHz). That means 44,100 samples are taken every second. These samples are recorded and stored for later use in replication or playback. The technique is called *pulse-code modulation* (PCM).

When you take a sample at twice the frequency of the signal, you are using the Nyquist rate. The resultant digital representation of the analog signal more closely resembles the original analog signal. You can also begin to see how higher sampling rates produce a much better representation.

The two images in Figure 3-3 show the differences between an analog signal and a digital sample of the analog signal. The analog signal is on the right. Notice how there is a smooth transition from white at the top to black at the bottom. The image on the left represents the digital sample of the analog image. Notice that the image appears to be made of four different blocks, or four samples.

As another example, assume you want to sample an analog signal for a CD-quality recording. A typical sampling rate for CD-quality audio is 44,100 samples per second (44.1 kHz). Since the audio frequency range extends to 20 kHz, you would need a sampling rate of at least 40 kHz to apply the Nyquist rate.

Bit Depth of a Digital Signal

The sampling rate determines the digital signal's accuracy (how closely it captures the frequency of the original signal). The precision (how close the digital signal is to the analog signal's amplitude values) is determined by the sample's *bit depth*.

Bit depth is defined as the number of states you have in which to describe the sampled voltage level. If you have 1 bit, whose value can be 0 or 1, you can describe the signal as being only on or off (two states). If you have 2 bits, you have four possible states (00, 01, 10, or 11) in which to describe a signal. A bit depth of 3 bits will have eight possible states.

As the number of bits increases, the number of possible states increases exponentially. In fact, the number of possible states increases by 2 to the power of the number of bits. A bit depth of 16 will have 65,536 (2^{16}) possible states to describe the signal. Standard audio CD resolution is 16 bits.

Figure 3-3
A digital representation of an analog image

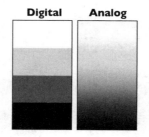

Digital   Analog

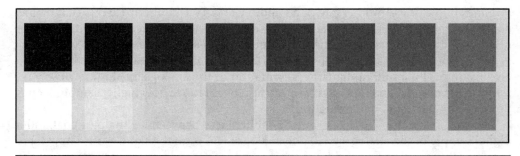

Figure 3-4 A 4-bit grayscale showing all 16 potential states

Bit depth in digital video signals is a measurement of how many shades can be displayed. Each possible color is a state. The greater the bit depth, the more realistic the image will look. A black-and-white display is called 1-bit because it allows only two states (black and white), which is 2 to the power of 1. A 16-color display is called 4-bit because this is 2 to the power of 4. The next level is 256 colors, which is 2 to the power of 8, or 8-bit, and so forth.

Figure 3-4 shows a 4-bit grayscale displaying all 16 potential shades. Table 3-1 lists the bit depths and the number of potential states.

Bit Rate

Bit rate is a measurement of the quantity of information over time in a digital signal stream. It is quantified using bits per second (bit/s or bps), where 1,000 bits equals 1 kilobit and its rate measured in kilobits per second (Kbps), and 1,000,000 bits equals 1 megabit and its rate measured in megabits per second (Mbps). Generally speaking, the higher the bit rate, the better the quality.

Table 3-2 lists the bit rates of several common formats. As you can see, the bit rate for high-definition television (HDTV) is much higher than the bit rate for an audio CD. This means that if you want to send an HDTV signal from one location to another, you will need a lot more bandwidth.

Depth	Multiplier	Number of States
1-bit	2^1	2
4-bit	2^4	16
8-bit	2^8	256
16-bit	2^{16}	65,536
24-bit	2^{24}	16,777,216
32-bit	2^{32}	4,294,967,296

Table 3-1 Bit Depths and the Number of Potential States

Table 3-2	Format	Bit Rate
Common Formats and Their Required Bit Rates	MP3	128–160 Kbps
	Audio CD	1411.2 Kbps
	Video CD	1 Mbps
	DVD	5 Mbps
	HDTV	20 Mbps

Signal Compression

Compression allows us to reduce very large original files to practical, more manageable sizes. To compress something is to make it smaller in some way.

The process of digitally compressing content is used extensively in computer applications, such as streaming audio or video content over the Internet. Compression is also useful for text files. By discarding data that is not required, we get compressed files. Text files are full of spaces between words that are represented by groups of data. These groups of data can be compressed very tightly without affecting the characters, resulting in a much smaller file size.

Compression technology reduces the size of digital files and makes them easier to transmit and store. However, the more you compress a signal, the more you affect its quality. If you remove too much data, the effects can be seen in image files and heard in audio files.

One frame of uncompressed digital video at a resolution of 720-by-480 requires about 1MB of storage. At 30 frames per second, that's about 1.9GB per minute of video! Imagine how much space a 10-minute video would occupy. Standard digital video (DV) cameras usually compress at a ratio of 5:1. Some video formats offer much higher compression rates.

Two important steps in the compression process are encoding the digital file to make it smaller and decoding the file so it plays back properly. These steps are accomplished using one of many codecs. A *codec* is a piece of equipment or software that employs a computer algorithm, or a set of procedures, to encode or decode file information. There are different types of codecs that are specialized for different purposes, including audio codecs and video codecs. A codec encodes a data stream for transmission, storage, or encryption, or it decodes a data stream for playback or editing.

NOTE Although technically speaking, a codec encodes and decodes a variety of AV information, codec is also a generic term for a videoconferencing unit.

Digital Formats

A digital file contains two elements:

- **Container** The container (such as WMV, for Windows Media Video) is the structure of the file where the data is stored. The container defines how the data is arranged to increase performance and which codecs are used.

- **Codec** The codec provides the method for encoding (compressing) and decoding (decompressing) the file. Many video and audio codecs are in use today, and new ones are constantly being created. In most cases, the codec must be installed in the operating system to play the file.

Formats can be confusing because the term *codec* is used interchangeably to describe the container and the codecs used within the container. In addition, some codec names describe both a codec and a container. An AVI container, for instance, could contain data encoded with an MPEG codec.

Programs are available to convert a file from one format to another. For example, a converter can be handy when adding video to a PowerPoint presentation.

Noise and Signal Transmission

Analog signals are challenging to transport from one place to another. Consider the largest audio system there is: the telephone system. To carry the audio signal corresponding to a human voice, the signal may need to be reamplified (with repeater amplifiers) many times in order to overcome the losses found in cable and other devices. Every time you reamplify a signal, you also amplify noise, as illustrated in Figure 3-5. After many generations of reamplification, the noise overcomes the signal. For a medium to rely on a continuous signal, degradation is inevitable.

Digital signals are not affected by noise in the same way as analog signals. As noise is introduced to a digital signal, discerning circuitry can determine if the signal is intended to be high or low, and then retransmit a solid signal without the imposed noise, as illustrated in Figure 3-6.

Figure 3-5
Distance and noise degrade analog signal quality.

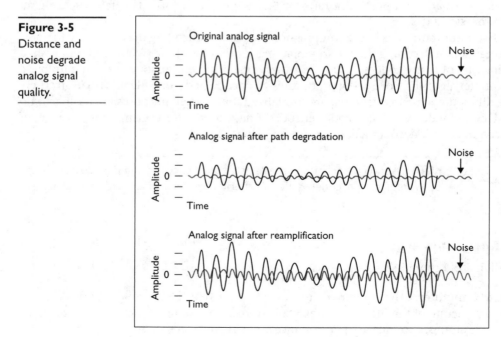

Figure 3-6
Effects of
degradation on
a digital signal

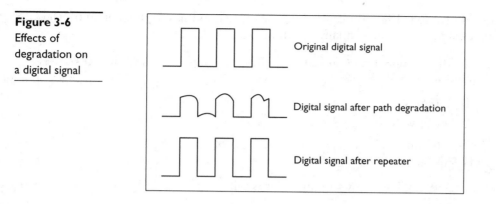

Original digital signal

Digital signal after path degradation

Digital signal after repeater

Analog and Digital Signal Considerations

One problem with analog signals is that they can be subjected to noise generated by the recording device. This can result in a poor-quality recording. Also, if anything happens to the analog recording media (for example, tapes warp over time), the true signal on the media will still exist on the tape, but it will not exactly resemble what was originally recorded. The lifetime and safety of the recording media can be a concern. When making copies of an analog recording, the noise introduced during replication will increase with each copy. After a large number of copies, the file quality may become unacceptable.

Duplicating a digital signal is essentially replicating a list of numbers. The numbers will always be the same, and the data is more likely to be free of noise. A major advantage is that when copies are made of a digital recording, the quality of the recording remains intact. In addition, you can replay a digital recording many times, and the quality will remain the same.

Chapter Review

When it comes to professional AV, CTS candidates must be prepared to work in what was once an all-analog world as it rapidly transitions to digital. Analog and digital waveforms are fundamentally different, and those differences help dictate how AV pros design the systems that create, transport, manage, store, and play back analog and digital media.

Review Questions

The following review questions are not CTS exam questions, nor are they CTS practice exam questions. Material covered in Part II of this book provides foundational knowledge of the technology behind AV systems, but it does not map directly to the domains/tasks covered on the CTS exam. These questions may resemble questions that could appear on the CTS exam, but may also cover material the exam does not. They are included here to help reinforce what you've learned in this chapter. For an official CTS practice exam, see the accompanying CD.

1. In a digital signal, the on state is represented by _____, and the off state is represented by _____.

   A. Two; one

   B. One; two

   C. Zero; one

   D. One; zero

2. A signal that has many varying states is called a(n) _____ signal.

   A. Analog

   B. Fluctuating

   C. Dimmer

   D. Digital

3. Bit depth is defined as the number of _____ you have in which to describe the value.

   A. Signals

   B. Speeds

   C. States

   D. Rates

4. Standard DV cameras usually compress at a ratio of _____.

   A. 10:1

   B. 5:1

   C. 3:2

   D. 2:1

5. What is a codec?

   A. A structure of data containment

   B. A formatting system

   C. A program that holds data

   D. A device or computer program that encodes and decodes file information

6. As noise is introduced to a(n) _____ signal, discerning circuitry can determine if the signal is intended to be high or low, and then retransmit a solid signal without the imposed noise.

   A. Digital

   B. Dirty

   C. Analog

   D. Clean

7. Noise overcomes the signal after many generations of reamplification of a(n) _____ signal.

   A. Digital

   B. Analog

   C. Low

   D. High

Answers

1. **D.** In a digital signal, the on state is represented by one, and the off state is represented by zero.

2. **A.** A signal that has many varying states is called an analog signal.

3. **C.** Bit depth is defined as the number of states you have in which to describe the value.

4. **B.** Standard DV cameras usually compress at a ratio of 5:1.

5. **D.** A codec is a device or computer program that encodes and decodes file information.

6. **A.** As noise is introduced to a digital signal, discerning circuitry can determine if the signal is intended to be high or low, and then retransmit a solid signal without the imposed noise.

7. **B.** Noise overcomes the signal after many generations of reamplification of an analog signal.

Audio Systems

In this chapter, you will learn about
- The basics of sound propagation
- Sound wave frequency and wavelength
- Harmonics, the decibel, and the sound environment
- How these basics apply to the electrical pathway used to amplify sound
- The electrical audio-signal chain from start to finish—microphones to loudspeakers
- The various signal levels, cables used, and types of circuits that are preferred for professional audio

This chapter covers the basics of sound propagation, sound wave frequency and wavelength, harmonics, the decibel, and the sound environment. It explains how these basics apply to the electrical pathway used to amplify sound. We will review the electrical audio-signal chain from start to finish—microphones to loudspeakers—and discuss the various signal levels, the cables used, and the types of circuits that are preferred for professional audio.

Sound Waves

Sound that you receive can be generated in different ways. Vibration is a common method.

Two good examples of vibration-induced sound are the playing of a stringed instrument and a loudspeaker cone that moves when the loudspeaker is in use. Air molecules, normally at rest, are displaced (moved) by the vibrations. As the string or loudspeaker cone moves, it compresses, or pushes against, the air molecules next to it. As the string or loudspeaker cone reverses direction, it pulls at the molecules that surround it. Because air is an elastic medium, the displaced molecules transmit this back-and-forth motion to the molecules surrounding them. The result is that the displacement, or disturbance, moves the sound as a waveform out and away from the source that generated it.

The pushing together of molecules is called *compression*, and the pulling apart is called *rarefaction*. These are areas of high and low pressure in the elastic medium (the air). For an AV system, you are usually concerned with sound moving in air, but there may be occasions when you are interested in its movement through other elastic mediums.

A visual example of the principle of sound waves is a rock thrown into another elastic medium: water. The rock creates a disturbance (displacement of water molecules) in the medium. Water molecules are pushed together in compression and pulled apart in rarefaction, creating the peaks and valleys, or ripples, we see on the surface of the water. We also see that these waves of energy move away from the source in concentric circles as the energy is transferred to nearby molecules.

In addition to vibration, sound in air can also be generated by sudden increases in velocity (speed) or turbulence. This helps explain why you sometimes hear sound coming from the ducts in an air-ventilation system.

Wavelength

Sound waves have a physical length, as well as a particular loudness or intensity. The length is called *wavelength*. Wavelength is the physical distance between two points exactly one cycle apart in a waveform. Wavelength measures the distance between two points that occur at the same place.

To understand wavelength, pick any point on the wave shown in Figure 4-1. Now move along the wave and find the next occurrence of that exact spot. That is a wavelength.

The middle horizontal line, known as the *zero-reference line* or *reference level*, represents the molecules at their rest position. Parts of the sine wave above the reference level represent molecules being compressed (compression). Maximum compression occurs at the top of the waveform. Parts of the wave below the reference level represent molecules being rarified (rarefaction). Maximum rarefaction occurs at the bottom of the waveform.

When the molecules have moved from the rest position, through compression, back to their rest state, then through rarefaction, and back to the same distance from one another as they were at the beginning, that is one complete cycle.

Frequency

The number of times this cycle occurs per second—from rest position through maximum compression, back to rest position, maximum rarefaction, and finally back to rest

Figure 4-1
This wave is one simple, single smooth wave, or sine wave.

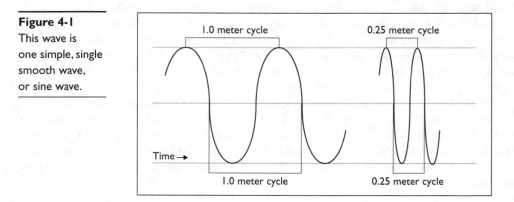

Figure 4-2
Frequency is the
number of cycles
per second.

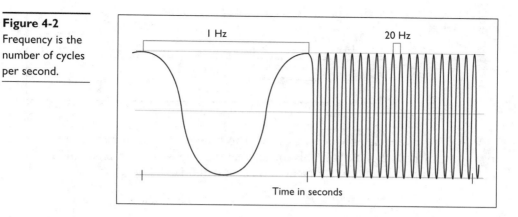

position—is called *frequency*. If the complete cycle occurs 100 times per second, the sine wave has a frequency of 100 cycles per second (cps). Figure 4-2 illustrates frequency.

The standard term used to describe frequency (number of cycles per second) is *hertz* (Hz). Therefore, we usually talk about a sound wave that has 100 cps as having a frequency of 100 Hz.

Although the sounds you hear every day are complex waveforms, each particular sound can be broken down into individual sine waves of all the frequencies that make up that sound. The length of a sound wave depends on the speed of sound in the medium through which the wave is propagating, as well as the frequency of the sound. The wavelengths of the frequencies most humans can hear (20 Hz to 20 kHz) range from 56.5 feet to less than 3/4 inch (17.2 meters to 17.2 mm).

Frequency and wavelength are inversely proportional. This means that as one gets larger, the other gets smaller. The lowest frequencies have the longest wavelengths, and the highest frequencies have the shortest wavelengths.

Octaves and Bands

To study and measure sound, the frequencies we can hear (20 Hz to 20,000 Hz) are divided into groups. The most common example of this division is in music.

On the treble staff, each space and line indicates a musical letter A through G, as shown in Figure 4-3. The A note above middle C has a fundamental frequency of 440 Hz. Each A has a similar sound, but a different frequency. If you start at A and count up or down the staff eight notes, you will find another A. This interval is called an *octave*.

Figure 4-3
Octaves on a
musical scale

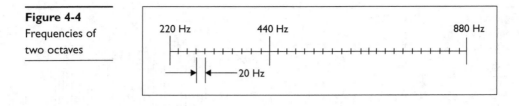

Figure 4-4
Frequencies of
two octaves

Why are frequencies divided this way? The human ear's response to frequency is logarithmic (think 2, 4, 8, 16). This means that the way we hear is exponential. A tone at 220 Hz sounds similar to a tone at 440 Hz and 880 Hz.

$$220 \times 2 = 440 \qquad\qquad 440 \times 2 = 880$$

Even though there is more frequency room between 440 and 880 Hz than 220 and 440 Hz, our ears hear the same interval between each sound. The width of an octave is exponential. Figure 4-4 illustrates the frequencies of two octaves.

Frequencies also are divided into bands. In general, human hearing covers frequencies from 20 Hz to 20,000 Hz. This spectrum of hearing is divided into ten bands, and each band is one octave. As with octaves, the bands do not have equal numbers of frequencies. The frequencies are divided logarithmically. Each band is identified by its center frequency.

The spectrum of hearing can be divided into even smaller bands. In the AV industry, you may encounter 1 octave, 1/3 octave, or 1/10 octave bands.

Harmonics

In the previous discussion, you learned that the sounds you hear are actually complex waveforms and that each complex waveform can be broken down into individual sine waves.

A complex waveform is made up of a fundamental frequency plus whole-number multiples of that fundamental frequency. The whole-number multiples are called *harmonics*.

For example, let's look at a fundamental frequency of 4,000 Hz. Whole-number multiples of 4,000 are 8,000, 12,000, 16,000, and so on. A complex waveform with a fundamental frequency of 4,000 Hz would also have varying levels of harmonic energy at 8,000 Hz, 12,000 Hz, 16,000 Hz, 20,000 Hz, and so forth.

The fundamental frequency is also the first harmonic, so we would say that 4,000 Hz is the first harmonic in our example. The second harmonic is 8,000 Hz, and the fourth harmonic is 16,000 Hz. These are the even-order harmonics (second, fourth, and so on). The third harmonic is 12,000 Hz, and the fifth harmonic is 20,000 Hz. These are the odd-order harmonics (first, third, fifth, and so on).

The varying proportional energy levels of the fundamental frequency, combined with the harmonic frequencies, help us distinguish one sound from another. Consider the fundamental frequency of the musical note of A4 (the A above middle C), which is 440 Hz. Although two different types of stringed instruments might both play the note A4, the amount of energy at the fundamental frequency, combined with the amount of energy at the various harmonic frequencies, helps you distinguish the note played on a guitar from the same note played on a piano.

Logarithms

A *logarithm* (log) of a number is how many times the number 10 must be multiplied by itself to get a certain value. For example, the log of 10,000 is 4 (10^4), and the log of 0.0001 is −4 (10^{-4}).

Logarithmic scales make the ratio values easier to express. For example, the ratio between the threshold of hearing (when sounds become audible) and the threshold of pain is 1 to 1,000,000. No one wants to count that many zeros.

Think of a standard ruler. Each unit on the ruler represents a unit of one, wherever it's located on the ruler (whether it's an inch, millimeter, or some other unit). There is a one-to-one relationship between the units shown and the units represented. This is a *linear scale*.

What if the value of each unit on that ruler represented something other than a single unit of one? What if each time you moved to the right, each unit represented ten times as much as before? Or each time you moved one unit to the left, it represented one-tenth as much as before? In this case, comparing adjacent units on a scale would represent a ratio of 1:10 or 10:1. This is a *logarithmic scale*.

Look at the graph in Figure 4-5. The top row represents a linear scale, and the bottom row represents a logarithmic scale.

We can write $10 \times 10 \times 10 = 1,000$ or $10^3 = 1,000$ (10 multiplied by itself three times). We have just used a logarithm with a base of 10 and an exponent of 3 as a shortcut for an equation that uses multiplication. Therefore, we could substitute 10^1, 10^2, and 10^3 at the bottom of the graph for the numbers 10, 100, and 1,000.

Why is all this important? Because humans perceive differences in sound levels logarithmically, not linearly. Therefore, a base-10 logarithmic scale is used to measure, record, and discuss sound level differences.

If you used a linear scale to describe the perceived difference in sound pressure level from the threshold of hearing to the threshold of pain, you would need to use numbers from 1 to well over 3 million.

 NOTE Not only is our perception of sound intensity logarithmic, but so are our other senses. The psychophysical law (or Weber-Fechner law) describes this generalization in psychology. It states that the intensity of a sensation is proportional to the logarithm of the intensity of the stimulus causing it. So the relationship between our sensations—such as sight, hearing, and touch—and the stimulus is logarithmic.

Figure 4-5
Comparing a
linear scale and
logarithmic scale

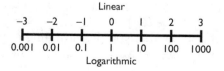

Decibels

Because the AV industry frequently uses decibel measurements, it is important for you to understand what decibels are and how they apply to AV systems.

The basic unit, the Bel, was named after telecommunications pioneer Alexander Graham Bell. As a unit of measurement, the Bel was too large for practical purposes, so the decibel (one-tenth of a Bel), or dB, was used instead. It is a logarithmic scale used to describe ratios with a very large range of values.

Decibel Measurements

The decibel is a unit of measurement used to describe a base-10 or base-20 logarithmic relationship of a power ratio between two numbers. Decibels are also used for quantifying differences in voltage, distance, and sound pressure as they relate to power. When you are quantifying differences, the numbers being compared to one another must be of the same type. For example, you could compare one voltage to another voltage. You cannot compare voltage to wattage.

You could also compare a number or measurement to a known reference level. The decibel system measures an amount of increase or decrease above or below the chosen reference level. Sound pressure levels are measured this way, when you compare a sound pressure level measurement to the threshold of hearing reference of 0 dB SPL. And remember, because it's a logarithm, whatever you're measuring—whether the increase is from 1 to 100 or 100 to 10,000—the increase in both cases (in base 10) is still 20 dB.

We can state the difference in decibels for two powers or voltages or distances using equations. Ten times the logarithm of the ratio (of the two numbers we're comparing) describes the difference in decibels if we're comparing two power values. For example, this equation:

$$dB = 10 \times \log (P_1 / P_2)$$

would give us the difference in decibels if we compared one power in watts (P_1) against another power in watts (P_2).

Twenty times the logarithm of the ratio describes the difference in decibels if we are comparing two voltages or distances. For example, this equation:

$$dB = 20 \times \log (V_1 / V_2)$$

would give us the difference in decibels if we compared one voltage (V_1) against another voltage (V_2).

This equation:

$$dB = 20 \times \log (D_1 / D_2)$$

would give us the difference in decibels if we compared the sound pressure level at one distance from the source (D_1) as compared to the sound pressure level at some other distance from the source (D_2).

Now that you know that you perceive differences in sound levels logarithmically, and that you use the decibel for a logarithmic scale, let's explore further how the decibel is used.

Using the Decibel

Not only can you compare one number against another number of the same type, you can also compare a number against a known reference level. For example, using the 20 log equation, you can compare the difference in decibels between two sound pressure levels. As you learned earlier, you can also compare a sound pressure level measurement in decibels against a known sound pressure reference level in decibels.

For a measured SPL of 85 dB to be meaningful, it must have a reference, because decibels are comparisons, not units of absolute quantity. Without a reference, it would be like saying, "It is ten degrees hotter today than yesterday." Without knowing yesterday's temperature, you cannot know today's temperature because you lack that reference. You need to know what 85 dB means relative to something.

In the case of sound pressure, the reference for 0 dB is 0.00002 Pascals (or 0.0002 dyne/cm^2), and it is considered to be the threshold of hearing. So our 85 dB SPL measurement is 85 dB of sound pressure referenced to the threshold of hearing. We can write it as 85 dB SPL, where dB SPL means dB referenced to the threshold of hearing.

Reference levels are also used for other measurements. We commonly use dBu and dBV to express voltage levels, and dBm for electrical power levels. Some examples are 0 dBu = 0.775 volts, 0 dBV = 1.0 volt, and 0 dBm = 0.001 watts. While –20 dBu is a voltage less than 0 dBu, it does not mean a negative voltage, but rather a voltage less than 0.775 volts.

In relation to human hearing, the following are some accepted generalities:

- A 1 dB change is the smallest perceptible change. Unless they're listening very carefully, most people will not discern a 1 dB change.

- A "just noticeable" change, either louder or softer, requires a 3 dB change (such as 85 dB SPL to 88 dB SPL).

- A 10 dB change is required for people to subjectively perceive either a change as twice as loud as before or one-half as loud as before. (For example, a change from 85 dB SPL to 95 dB SPL is perceived to be twice as loud as before.)

The Inverse Square Law and Sound

As mentioned earlier, sound waves move out and away from the source of generation. Again, envision a rock is thrown into water. The waves of energy move away from the source of molecule disturbance in concentric circles. This occurs as the energy is transferred to nearby molecules. With the rock and water illustration, you see the energy spreading out only on the horizontal plane of the water surface. However, when a sound is generated, the energy is spread spherically in all directions.

Regardless of whether the energy propagates horizontally or spherically, you can follow its spread by using the *inverse square law*. The inverse square law states that sound energy is inversely proportional to the square of the distance from the source. This occurs because every time the distance doubles from the source of the energy, the energy spreads out and covers four times the area it did before.

For example, if you are 5 meters from an energy source, and then double your distance from the source to 10 meters, the energy must now cover an area four times larger than it did at the 5-meter distance. In other words, for every doubling of the distance, there is a fourfold increase in the surface area of the sphere. Subsequently, the energy unit per unit of area is one-quarter of what it was previously.

Simply put, sound pressure is reduced by 6 dB every time the distance from the source is doubled. Conversely, the sound pressure increases by 6 dB when the distance from the source is reduced by one-half. This is often referred to as the *6 dB per doubling rule*.

 NOTE The inverse square law doesn't just apply to sound. It also applies to other things, such as light, gravity, and the electric field.

In practice, a true reduction by 6 dB for every doubling of distance occurs only in a *near-field* or *free-field* environment. These terms describe the space around a sound source that does not include energy reflected back by boundaries, such as walls, ceilings, and floors.

Acoustics

Acoustics is a branch of science that focuses on the qualities and characteristics of sound waves. As you might expect, the study of acoustics covers many topics. It is more than simply hanging some fabrics on a wall to solve an "acoustical problem."

The study of acoustics includes the following:

- How sound is generated
- How sound energy moves through air and other media (such as concrete, steel, and water)
- How dimensions and shapes affect the way sound behaves in an environment
- How sound energy can be prevented from leaving or entering a space through partitions or vibrations
- What happens to sound energy when it encounters a boundary (materials)
- Your perception of a sound as processed by your ear and brain

Earlier, you learned that sound can be generated by vibrations (for example, vibrations from a stringed instrument). Many things besides strings and loudspeakers can vibrate and generate sound. Sound can be generated by the mechanical motion of a loudspeaker, the structural elements of a building, and many other sources.

Sound energy moves out and away from a source in all directions. Unless sound is generated in a completely free space, the energy will, at some point, encounter a boundary or surface. If you're outdoors, the boundaries may be only the ground or nearby buildings. But indoors, there can be many boundaries and surfaces. Along with the walls, ceilings, and floors, furniture and people also affect what happens to sound energy in an environment.

So what happens to the energy produced by a sound-reinforcement system or other source of generated sound? The law of conservation of energy tells us that total energy neither increases nor decreases in any process. Energy can be transformed from one form to another—and transferred from one body to another—but the total amount of energy remains constant. This means that when the sound energy encounters a surface or room boundary, one of three things occurs (it will actually be a combination of all three):

- The sound is reflected. As the sound energy moves away from the source, some of it will be reflected off various surfaces back into the room. The reflections can be specular (direct) or diffused (scattered). Either way, the energy remains in the space.

- The sound is absorbed, either in the air (not much of an issue except in extremely large spaces) or by the materials in the space (sound energy converted into heat).

- The sound is transmitted. In other words, the energy actually passes from one space to another through a partition or other barrier.

Reflected Sound Energy

There are two types of sound reflections:

- **Direct reflection**   Also known as a *specular* or *hard reflection*, this type of reflection is mirror-like; most of the energy is reflected back in a single direction. As with light, the direction of the angle of reflected sound energy is determined by the incoming angle, as the angle of incidence is equal to the angle of reflection.

- **Diffuse reflection**   This type of reflection scatters the energy back equally in all directions. This is similar to the properties of a diffusion-type projection screen.

Whether the energy is reflected in a specular or diffuse fashion depends on how smooth the surface is, relative to the wavelength. Because we're dealing with varying wavelengths, the transition from specular to diffuse is not immediate, and neither is how much energy is reflected, absorbed, or transmitted. Some wavelengths will act one way or another, while others will act somewhere in between. Their behavior is determined by the size, material, and mass of the boundaries in a space.

In every situation involving boundaries or surfaces, you will always have direct sound—sound that arrives at the listener position in a direct, straight line—and reflected sound—sound that takes any indirect path from the source to the listener.

Reflected energy arrives later in time than direct sound. This should be obvious, because the shortest path between two points is a straight (direct) line. Taking any other path requires traveling a longer distance and requires more time. Thus, reflected sound will always arrive later than direct sound. Reflections can be defined as *early* or *late*. A late reflection is called an *echo*.

Reverberation

When a room has many hard, reflective surfaces, the energy level of each sound reflection can remain quite high. As energy reflects off more surfaces around a room, the listener begins to receive reflected energy from all directions. When the energy level remains high and the reflections become dense in relation to one another, it is called *reverberation*. Reverberation simply describes numerous, persistent reflections.

True reverberation is a phenomenon seen in larger rooms with many hard, reflective surfaces. While a typical conference room may have troublesome reflections, it will not exhibit true reverberation.

Absorption

The materials used in the construction and finish of a venue (walls, ceiling, floor, furniture, curtains, windows, seating, etc.), as well as people themselves, play a part in the amount of sound energy absorption available.

One type of absorber you may be familiar with is the porous absorber, or "acoustical fuzz." As the displaced air molecules pass through a porous absorber, the friction between the molecules and the material of the absorber slow down the molecules. Typical porous absorbers include carpets, acoustic tiles, acoustical foams, curtains, upholstered furniture, and people (clothing). Figure 4-6 shows an example of using acoustic tiles.

Though acoustically absorptive material may be applied in a room, effective absorption will occur over a limited frequency range. Energy above or below the effective frequency band of the absorption will be either reflected or transmitted.

Transmission

Sound energy that is not reflected or absorbed will be transmitted into another space through partitions (walls, windows, floors, and ceilings) or structure-borne vibrations. The ability of a partition to transmit or reflect sound energy is limited by factors such as the weight of the partition (mass), stiffness of the partition, air gaps, design of the partition, and materials used in construction of the partition.

Figure 4-6
An acoustic
treatment
using tiles

Ambient Noise

Ambient noise is generally defined as anything other than the desired signal. While an electronic sound system inherently generates noise through its electronic components, rooms are also characterized by noise.

Anything heard in a room other than the desired signal from a sound-reinforcement system is considered *noise*. This can include noise from equipment fans; office machines; heating, ventilation, and air conditioning (HVAC) systems; and people in the room. Noise can intrude from outside the room as well, through partitions or windows. Outside sources may include vehicular traffic, adjoining corridors, or structure-borne vibrations.

Ideally, because excessive noise levels interfere with the message being communicated, an acoustician, AV consultant, or designer should specify background noise level limits appropriate to the type of room and its purpose. In practice, the criteria and limits for background noise levels for a gymnasium will be much different than those for a conference room. With such limits determined, parts of the project—the HVAC system, partitions, and necessary acoustical treatments, for example—may be designed and applied so that the background noise level criteria are not exceeded.

A room's acoustical properties (its reflections and types and amount of transmission allowed) and background noise levels are significant contributors to a sound system's overall effectiveness.

Sound Capture

You have read about the basics of sound propagation, sound wave frequency and wavelength, harmonics, the decibel, and the sound environment. How do these basics apply to the AV industry? They help you understand the electrical pathway used to amplify sound.

In its most basic form, the complete audio-signal path takes acoustic energy and converts it into electrical energy so it can be routed, processed, further amplified, and converted back into acoustic energy, as follows:

1. The sound source creates sound wave vibrations in the air.

2. A microphone picks up the vibrations.

3. The microphone converts the vibrations into an electrical signal.

4. The electrical signal is processed.

5. The signal ends up in an output device (such as an earphone or a loudspeaker).

6. The output device converts the electrical signal back into sound waves.

This conversion of energy from one form to another is called *transduction*. A microphone is a transducer—it converts acoustic energy into electrical energy. A loudspeaker is also a transducer—it converts the electrical energy into acoustic energy. This means you can have transducers on either end of an electrical audio path.

Microphones

Understanding a microphone's construction and its intended usage, as well as its directional, sensitivity, frequency response, and impedance characteristics, will help you select the appropriate microphone for each situation.

Microphones come in a variety of types, sizes, and construction. Carbon microphones were once widely used in telephones. There are also ceramic, fiber-optic, laser, piezoelectric, ribbon, moving coil, condenser, and electret microphones, each with its own application and characteristics.

Dynamic Microphones

In a dynamic microphone (mic), you will find a coil of wire (a *conductor*) attached to a diaphragm and placed in a permanent magnetic field. Sound pressure waves cause the diaphragm to move back and forth, thus moving the coil of wire attached to it.

As the diaphragm-and-coil assembly moves, it cuts across the magnetic lines of flux of the magnetic field, inducing a voltage into the coil of wire. The voltage induced into the coil is proportional to the sound pressure and produces an electrical audio signal. The strength of this signal is very small and is called a *mic-level signal*.

Dynamic microphones are used in many situations because they are economical and durable, and they will handle high sound pressure levels. Dynamic microphones are very versatile because they do not require a power source.

Condenser Microphones

In the study of electricity, you will find that if you have two oppositely charged (*polarized*) conductors separated by an insulator, an electric field exists between the two conductors. The amount of potential charge (*voltage*) that is stored between the conductors will change depending on the distance between the conductors, the surface area of the conductors, and the dielectric strength of the insulating material between the two conductors. An electronic component that uses this principle is called a *capacitor*.

A condenser microphone (see Figure 4-7 for an example) contains a conductive diaphragm and a conductive backplate. Air is used as the insulator to separate the

Figure 4-7
A condenser
microphone on
a podium

diaphragm and backplate. Voltage from a power supply, known as *phantom power*, is used to polarize, or apply, the positive and negative charges to create the electric field between the diaphragm and backplate.

Sound pressure waves cause the diaphragm to move back and forth, subsequently changing the distance (spacing) between the diaphragm and backplate. As the distance changes, the amount of charge, or *capacitance*, stored between the diaphragm and backplate changes. This change in capacitance produces an electrical audio signal.

The strength of the signal from a condenser microphone is not as strong as the mic-level signal from a typical dynamic microphone. To increase the signal, a condenser microphone includes a preamplifier, powered by the same phantom power supply used to charge the plates in the microphone. This preamplifier amplifies the signal in the condenser microphone to a mic-level signal, but is not to be confused with a microphone preamplifier found in a mixing console.

The diaphragm used in a condenser microphone is small, requiring less mass than other microphone types. Generally speaking, because of this, the condenser microphone tends to be more sensitive than other microphone types and responds better to higher frequencies with a wider overall frequency response.

Electret Microphones

An *electret microphone* is a type of condenser microphone. The electret microphone gets its name from the prepolarized material, or the *electret*, applied to the microphone's diaphragm or backplate.

The electret provides a permanent, fixed charge for one side of the capacitor configuration. This permanent charge eliminates the need for the higher voltage required for powering the typical condenser microphone. This allows the electret microphone to be powered using small batteries, as well as normal phantom power. Electrets are small, lending themselves to a variety of uses and quality levels.

Defining Phantom Power

Phantom power is the remote power required to power a condenser microphone. It typically ranges from 12 to 48 volts DC (VDC). Positive voltage is supplied equally to the two signal conductors of a balanced circuit. This is accomplished through standard microphone cable, as standard microphone cable contains two signal conductors plus a shield. Because the voltage is applied equally on both signal conductors, applying phantom power will not cause damage to dynamic microphones.

Phantom power is most often available from an audio mixer. It may be switched on or off at each individual microphone input, or from a single button on the audio mixer that makes phantom power available on all the microphone inputs at once. If phantom power is not available from the audio mixer, separate phantom power supplies may be used.

Microphone Physical Design and Placement

Whether dynamic, condenser, electret, or otherwise, microphones come in an assortment of configurations to meet a variety of uses. The following are some common microphone configurations:

- **Handheld**   This type, shown in Figure 4-8, is used mainly for speech or singing. Because it is constantly moved about, a handheld microphone includes internal shock mounting to reduce handling noise.

- **Surface mount or boundary**   This type of microphone is designed to be mounted directly against a hard boundary or surface, such as a conference table, wall, or sometimes a ceiling. The acoustically reflective properties of the mounting surface affect the microphone's performance. Mounting a microphone on the ceiling typically yields the poorest performance, because the sound source is much farther away from the intended source (for example, conference participants) and much closer to other noise sources, such as ceiling-mounted projectors and HVAC diffusers.

- **Gooseneck**   Used most often on lecterns and sometimes conference tables, this type of microphone is attached to a flexible or bendable stem. The stem comes in varying lengths. Shock mounts are available to isolate the microphone from table or lectern vibrations.

- **Shotgun**   Named for its physical shape and long and narrow polar pattern, this type of microphone is most often used in film, television, and field-production work. You can attach a shotgun microphone to a long boom pole (fishpole), to a studio boom used by a boom operator, or to the top of a camera.

Figure 4-8

Man using a wireless handheld microphone

- **Lavalier and headmic**   These microphones are worn by a user, often in television and theater productions. A lavalier (also called a *lav* or *lapel mic*) is most often attached directly to clothing, such as a necktie or lapel. In the case of a headmic, the microphone is attached to a small, thin boom and fitted around the ear. As size, appearance, and color are critical, lavaliers and headmics are most often electret microphones.

Microphone Polar Patterns

One of the characteristics to look for when selecting a microphone is its *polar pattern*. The polar pattern describes the microphone's directional capabilities—in other words, the microphone's ability to pick up the desired sound in a certain direction while rejecting unwanted sounds from other directions.

Polar patterns are defined by the directions from which the microphone is optimally sensitive. These polar patterns help you determine which microphone type you should use for a given purpose. There will be occasions when you want a microphone to pick up sound from all directions (like an interview), and there will be occasions when you do not want to pick up sounds from sources surrounding the microphone (like people talking or someone rustling papers). The polar pattern is also known as the *pickup pattern* or a microphone's *directionality*.

As a microphone rejects sounds from undesired directions, it also helps to reduce potential feedback through the sound system. The following polar patterns are available.

- **Omnidirectional**   Sound pickup is uniform in all directions.

- **Cardioid (unidirectional)**   Pickup is from the front of the microphone only (one direction) in a cardioid pattern. It rejects sounds coming from the side, but the most rejection is at the rear of the microphone. The term *cardioid* refers to the heart-shaped polar patterns.

- **Hypercardioid**   A variant of the cardioid, this type is more directional than the regular cardioid because it rejects more sound from the side. The trade-off is that some sound will be picked up directly at the rear of the microphone.

- **Supercardioid**   This type provides better directionality than the hypercardioid, as its rejection from the side is better. It also has more rear pickup than the hypercardioid.

- **Bidirectional**   Pickup is equal in opposite directions, with little or no pickup from the sides. This is sometimes also referred to as a *figure-eight pattern*, due to the shape of its polar patterns.

Figure 4-9 shows various types of microphone polar patterns.

Microphone Sensitivity

One performance criterion that characterizes a microphone is its *sensitivity* specification. This defines its electrical output signal level given a reference sound input level. Put another way, sensitivity defines how efficiently a microphone transduces (converts) acoustic energy into electrical energy.

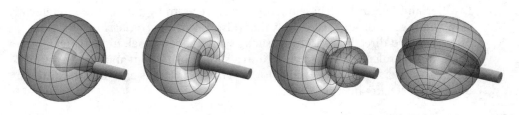

Figure 4-9   Microphone polar patterns: omnidirectional, cardioid, supercardioid, and bidirectional

If you expose two different types of microphones to an identical sound input level, a more sensitive microphone provides a higher electrical output than a less sensitive microphone. Generally speaking, condenser microphones have a greater sensitivity than dynamic microphones.

Does this mean that a lower sensitivity microphone is of lesser quality? Not at all. Microphones are designed and chosen for specific uses. For example, a professional singer typically uses a microphone up close, which can produce a very high sound pressure level. In contrast, a presenter speaking behind a lectern and a foot or two away from the microphone produces a much lower sound pressure level. For the singer, a dynamic microphone may be the best choice, as it will typically handle the higher sound pressure levels without distortion, while still providing more than adequate electrical output. The presenter, using a microphone farther away than a singer's, would certainly benefit from a more sensitive microphone.

Here is an example of a sensitivity specification:

–54.5 dBV/Pa

1 Pa = 94 dB SPL

Pa refers to the Pascal, and it is a unit of pressure. A Pascal is the equivalent of 94 dB SPL. In this example, if we were to put 94 dB SPL into the microphone, we would realize a –54.5 dBV electrical output signal.

Although most manufacturers use 94 dB SPL as the reference input level, you may also find 74 dB SPL (0.1 Pa) used as this reference level. Using a different input reference level would obviously produce a different output level.

Microphone Frequency Response

Another important measure of a microphone's performance is its *frequency response*. This defines the microphone's electrical output level over the audible frequency spectrum, which in turn helps determine how an individual microphone sounds.

A microphone's frequency response gives the range of frequencies, from lowest to highest, that the microphone can transduce. It is often represented as a plot on a two-dimensional graph of electrical output versus frequency, as shown in Figure 4-10. A graphical representation of the microphone's directional and frequency response characteristics is called a *polar plot*.

Figure 4-10

A frequency response chart for a microphone

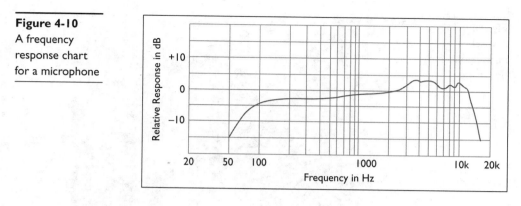

With directional microphones, the overall frequency response will be best on-axis (directly into the front of the microphone). As you move off-axis with a directional microphone, not only will the sound be reduced, but the frequency response will also change.

Microphone Impedance

For a microphone to be of any use, you must plug it into something. How do you know if the microphone is compatible with the device you're plugging it into?

Another microphone specification you must consider is its *output impedance*. Impedance is the opposition to the flow of electrons in an alternating current (AC) circuit. (Your audio signals are AC circuits.)

Back in the early days of the telephone and vacuum tubes, it was necessary to match output impedance with input impedance for maximum power transfer. Modern audio systems use a maximum voltage transfer instead. To accomplish this, a device's output impedance should be one-tenth or less than the input impedance the device is being plugged into. For example, a professional microphone's output impedance specification should be 200 ohms or less. An input to be used with a professional microphone will have an input impedance of 2,000 ohms or more.

Microphones can fall into two categories, based on output impedance:

- Low-impedance, which is 200 ohms or less (some as high as 600 ohms)
- High-impedance, which is more than 25,000 ohms

Professional microphones are low-impedance microphones. Low-impedance microphones are less susceptible to noise and allow for much longer cable runs than high-impedance microphones.

Wireless Microphones

Sometimes called *radio mics*, wireless microphones use RF transmission in place of a microphone cable. Some wireless systems use infrared (IR) transmission.

For a handheld microphone, a standard microphone casing is often integrated onto the top of a transmitter, and the microphone casing and transmitter are finished as one

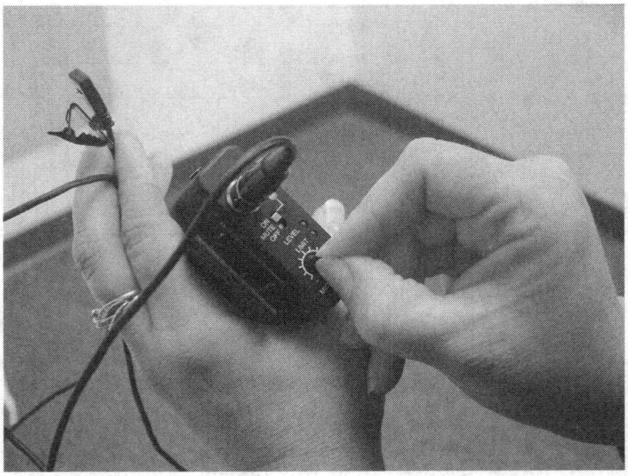

Figure 4-11
A wireless
microphone and
bodypack

unit. At other times, a small plug-on style transmitter is attached to the bottom of a regular handheld microphone.

For hands-free applications, a lavalier or headmic microphone is plugged into a bodypack-style transmitter, as shown in Figure 4-11. The bodypack transmitter is then clipped onto a belt or placed in a pocket or pouch. Either way, at the other end of the RF or IR transmission is a receiver tuned to the transmitter's specific frequency.

Modern RF wireless microphones allow you to change frequencies in order to avoid interference from outside sources, as well as interference from other wireless microphones that may be in use. This is called *frequency coordination* and will be specific to your geographical area. Your wireless microphone manufacturer can provide help in coordinating compatible frequencies for your area.

Microphone Cables and Connectors

Microphones and wireless microphone receivers are connected to audio mixers with a cable and a connector. Professional microphone cables use shielded twisted-pair cable, which contains the following:

- Two small-gauge insulated copper wires (conductors) twisted together
- An aluminum foil or a copper braided shield that covers the twisted conductors
- A protective jacket (rubber or plastic) that covers the twisted pair and shield

Typically, the ends of the shielded twisted-pair cable are finished (*terminated*) with an *XLR connector*. XLR is now a generic term describing the common audio connector. Although available in 3- to 7-pin configurations, the 3-pin XLR is used for almost all microphone cable applications.

When multiple twisted-pair cables are needed, *snake cables* are used. Snake cables combine anywhere from 2 to 58 shielded twisted-pairs under a single protective jacket.

Audio Signal Levels

Now that you've learned how to select the proper microphone and connect it, where do you go from here? First, let's go over some terminology that relates to audio signals:

- **Mic level**   As mentioned earlier, a microphone, regardless of type, produces a signal level that is called the *mic level*. Mic level is a very low-level signal that's only a few millivolts (abbreviated as mV to express one-thousandth of a volt).

- **Microphone preamplifier**   Mic level is a very low-level signal and subject to interference. Therefore, you need to amplify this signal by using a microphone preamplifier. A microphone preamplifier, often called a *preamp* for short, takes the mic-level signal and amplifies it to what is known as *line level*.

- **Line level, professional**   Line level in a professional audio system is about 1 volt. Line level is where all signal routing and processing is performed.

- **Line level, consumer**   Line level in a consumer device is less than line level in a professional device. Consumer line level is 0.316 volt (316 mV). Consumer line level can often be identified by its use of an RCA (phono) connector.

- **Loudspeaker level**   Once you have routed and processed the signal, it is sent to the power amplifier for final signal amplification up to loudspeaker level. The loudspeaker takes that amplified electrical signal and transduces the electrical energy into acoustical energy.

Signal Level Compatibility

As you begin to connect your system, you need to make sure the components of your system are compatible with each other. For example, while you could plug a microphone directly into the input of a power amplifier, you would not get much sound level. The mic-level signal isn't strong enough by itself. You also would not want to connect the output of a power amplifier to a device expecting either a mic or line level—you would almost certainly damage components.

What about plugging your microphone directly into the back of a powered loudspeaker? Some companies manufacture "powered loudspeakers," which are all-in-one devices meant to simplify setup and provide for easy portability. In the case of a powered loudspeaker, all of the signal requirements listed earlier are built in to the loudspeaker. If the powered loudspeaker has microphone inputs, it has a microphone preamplifier, and any internal processing will be done at the line level. It will also have the power amplifier built in to power the loudspeaker.

Signal Level Adjustments

When working with signal levels, you may need to make slight changes in order to provide a more adequate signal or to avoid the signal distortion that results from too much signal.

The following terms are used in association with signal level adjustment:

- Adjustments you make to signal levels are called *gain adjustments*.
- *Gain control* refers to the general ability to make adjustments to the signal levels.

- If you increase the signal level, it is called *gain*, which refers to the amount of amplification applied to the signal.

- If you decrease the signal level, it is called *attenuation*.

- If you apply neither gain nor attenuation, it is called *unity gain*. Unity gain means that the signal is passing through the gain control without any changes to the signal level.

Audio Devices

Now that we've reviewed the basics of capturing sound, we'll look at the devices used to process the audio signal. These include a wide range of equipment, from equalizers to power amplifiers.

Audio Mixers

In its most basic form, an audio system has a sound source at one end and a destination for that sound at the other. In almost all situations, there is more than one source.

Audio technicians deal with multiple and varied sources of sound. The sources could be several vocalists with instruments at a concert; playback devices such as CD, DVD, MP3 players, or media streamers; multiple panelists at a conference; or several actors in a theater performance. All of these signals come together in the *audio mixer*.

All audio mixers serve the same purpose: to combine, control, and route audio signals from a number of inputs to a number of outputs. Usually, the number of inputs will be larger than the number of outputs.

Audio mixers are often identified by the number of available inputs and outputs. For example, an 8-by-2 mixer would have eight inputs and two outputs. Each incoming mic- or line-level signal goes into its own channel. Many mixers provide individual channel equalization adjustments, as well as multiple signal-routing capabilities, called *main* or *auxiliary busses*.

A larger audio mixer is often called a *mixing console*, a *console*, or a *mixing desk*.

Regardless of the size and complexity, any mixer that accepts mic-level inputs will have microphone preamps. Once the mic level is amplified to line level by the preamp, it can be sent to the rest of the mixer.

Between the inputs and outputs, the typical audio mixer provides multiple gain stages for making adjustments. These adjustments allow the mixing console operator to balance or blend the audio sources together for the most realistic sound appropriate for the listening audience.

Some audio mixers will turn microphone channels on and off automatically, like an on/off switch. These are called gated *automatic mixers*. Others will turn up microphone channels that are in use and turn down microphone channels not in use, like a volume knob. These are called *gain-sharing automatic mixers*. The channels set in an automatic mixer for automatic mixing are to be used for speech-only situations. Other sound sources, such as music, would not be set for automatic mixing. Music and other uses still require live personnel for operation.

NOTE *Automatic mixers* should not be confused with *automated* mixers, which are automated by computer and store presets, control settings, and various mixing moves.

Audio Processors

Numerous types of processors can refine audio signals. The type you need is determined by intended use and listening environment. Some common processors include limiters, compressors, expanders, gates, and filters.

Compressors, limiters, and expanders are dynamic processors. They either decrease or increase the overall dynamic range of the signal. The term *dynamic range* refers to the difference between the loudest and quietest levels of a signal. A signal level that varies greatly between the loudest and quietest parts is said to have a wide dynamic range. Compressors and limiters operate in the same way but have different uses. Compressors have the following characteristics:

- They reduce the level of all signals above an adjustable threshold. In other words, they keep loud signals from being too loud.

- The amount of reduction above the threshold is determined by an adjustable ratio.

- The reduction reduces the variation between highest and lowest signal levels, resulting in a compressed (smaller) dynamic range.

- They can be used to prevent signal distortion.

Extreme compression is called *limiting*. Limiters are described as follows:

- They limit the level of all signals above an adjustable threshold. In other words, they prevent high-amplitude signals from getting through.

- Limiting is used to prevent damage to components such as loudspeakers.

- They are triggered by peaks or spikes in the audio signal (like a dropped microphone), and react quickly to cut them off before they exceed a certain point.

- The amount of limiting above the threshold is determined by a more aggressive ratio than a compressor reduction ratio.

- The reduction limits the variation between highest and lowest signal levels, resulting in a limited dynamic range.

Figure 4-12 shows the effects on a signal after using a limiter.

Expanders, which are more properly called *downward expanders*, are described as follows:

- They reduce the level of all signals below an adjustable threshold.

- The amount of reduction below the threshold is determined by an adjustable ratio.

Figure 4-12
The effects on
a signal after
using a limiter

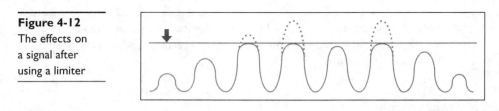

- The signal-level reduction increases the variation between highest and lowest signal levels, resulting in an increased dynamic range.
- They are used for reducing unwanted background noise.

Gates have the following characteristics:

- They can be thought of as an extreme downward expander.
- They mute the level of all signals below an adjustable threshold.
- Signal levels must exceed the threshold setting before they are allowed to pass.
- They can be used to automatically turn off unused microphones.

Figure 4-13 shows the effects of applying a gate.
  Filters are described as follows:

- They filter, remove, or pass certain frequencies from a signal.
- A notch filter "notches out" a specific frequency.
- Low-pass filters pass the low-frequency content of a signal.
- High-pass filters pass the high-frequency content.

Figure 4-14 shows low-pass and high-pass filters.

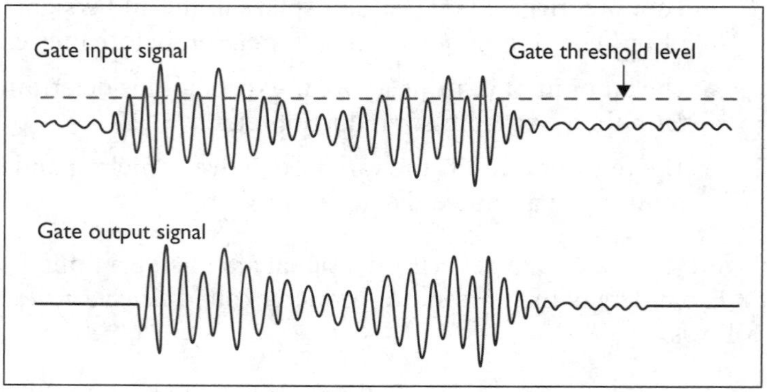

Figure 4-13
Impact of applying
a gate on input
signal amplitudes

Gate input signal   Gate threshold level

Gate output signal

Figure 4-14
Low-pass filters allow lower frequencies to pass through.

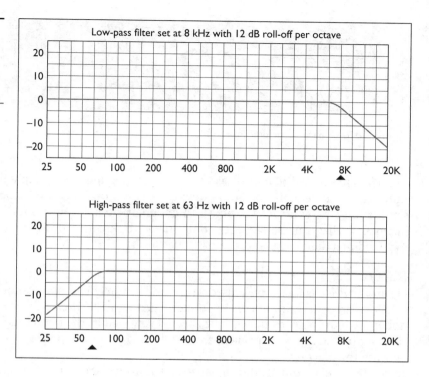

Equalizers

Equalizers (EQs) are frequency controls that allow you to boost (add gain) or cut (attenuate) a specific range of frequencies. The simplest equalizers are the bass and treble tone controls found on your home stereo or surround-sound receiver. The equalizer found on the input channel of a basic audio mixer may provide simple high-, mid-, and low-frequency controls.

Going beyond the home stereo and basic input channel equalizers, two types of sound-system equalizers are common:

- **Graphic equalizer**   A common graphic equalizer is the 1/3-octave equalizer, which provides 30 or 31 slider adjustments corresponding to specific fixed frequencies with fixed bandwidths. The frequencies are centered at every one-third of an octave. The numerous adjustment points allow for shaping the overall frequency response of the system so that the sound system sounds more natural. The graphic equalizer is so named because the adjustments provide a rough visual, or graphic, representation of the frequency response adjustments.

- **Parametric equalizer**   A parametric equalizer, shown in Figure 4-15, offers greater flexibility than a graphic equalizer. Not only does the parametric equalizer provide boost and cut capabilities, as does the graphic equalizer, but it also allows center frequency and bandwidth adjustments. A *semi-parametric equalizer* can be found on the input of many audio mixers. While it still allows adjustment of the center frequency, it does not allow bandwidth adjustment.

Figure 4-15 A parametric equalizer

There are many different types of equalizers, ranging from simple tone controls to fully parametric equalizers. Some mixing consoles offer a combination of fixed and semi-parametric controls. Graphic and parametric equalizers are for system-wide adjustments. They can be separate components, built in to powered audio mixers, or included in a digital signal processor (DSP) unit.

Delays

Electronic *delay* is commonly used in sound-reinforcement applications. For example, consider an auditorium with an under-balcony area. The audience seated directly underneath the balcony may not be covered very well by the main loudspeakers. In this case, supplemental loudspeakers are installed to cover the portion of the auditorium beneath the balcony.

While the electronic audio signal arrives at both the main and under-balcony loudspeakers simultaneously, the sound coming from these two separate loudspeaker locations will arrive at the audience underneath the balcony at different times and sound like an echo. This is because sound travels at about 1,130 feet per second (344 meters per second), which is much slower than the speed of the electronic audio signal.

In this example, an electronic delay would be used on the audio signal going to the under-balcony loudspeakers. The amount of delay would be set so that both the sound from the main loudspeakers and the under-balcony loudspeakers arrive at the audience at the same time.

Power Amplifiers

The amplifier is the last device used before the signal reaches the loudspeakers. Power amplifiers boost, or amplify, electronic audio signals sufficiently to move the loudspeakers. They do this by increasing the gain (the voltage and power) of the signal from line level to loudspeaker level. Line level is around 1 volt. Loudspeaker level depends on the type of system, as follows:

- 4 volts or more for typical sound systems
- Up to 70 or 100 volts in a distributed sound system
- More than 100 volts in extremely large venues

Most amplifiers have only a power switch and input sensitivity controls. Some now include digital signal processing and network monitoring and control.

Potentially, the more powerful the amplifier is, the greater the amplification of the signal and the louder the sound from the loudspeaker.

Power amplifiers are connected to loudspeakers with larger gauge wire than used at the mic or line level. The size of wire will depend on the distance between the power amplifier and the loudspeaker, as well as the current required. Loudspeaker cabling will be unshielded and may or may not be twisted.

A common connector used for loudspeaker cabling is the Speakon connector. Speakon connectors are used exclusively for professional loudspeaker connections. They are commonly employed for AV staging events because they are rugged, durable, and simple to use. They also lock into place.

Loudspeakers

For the purpose of sound reinforcement, loudspeakers are the end of the electrical signal path. The acoustic energy that was transduced into electrical energy by the microphone is transduced back into acoustic energy by the loudspeaker. Most loudspeakers share a common element: They have drivers mounted in an enclosure. The suitability of a loudspeaker depends on its intended use, which may include the following:

- **Communication**   Simple public address, intercom, or radio communications.
- **Sound reinforcement**   Lectures, musical performances, theaters, and more.
- **Sound reproduction**   Playback of prerecorded music, motion picture soundtracks, and more.

Loudspeakers may be designed for portability or permanent installation.

Crossovers

The audio spectrum has wavelengths and frequencies that vary dramatically. No single driver can reproduce the entire frequency range accurately or efficiently. This is handled in professional audio by giving the loudspeaker multiple drivers, as shown in Figure 4-16. A loudspeaker enclosure containing more than one frequency range of drivers is known by the different frequency ranges it covers.

So that each driver is sent only those frequencies that it will transduce efficiently, an electrical frequency-dividing network circuit called a *crossover* is used. A passive crossover

Figure 4-16

Loudspeakers by frequency range

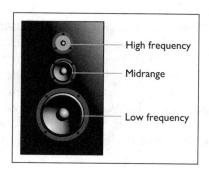

(one that doesn't require powering) is used to take the electrical signal coming into the loudspeaker enclosure and split it into the different frequency ranges.

The following are examples of the different drivers and frequency ranges:

- Tweeters, for high frequencies
- Horns, for mid to high frequencies
- Cone or midrange for midrange frequencies
- Woofers, for low frequencies
- Subwoofers, for lower frequencies

Loudspeaker Sensitivity

Like microphones, loudspeakers are rated based on their ability to convert one energy form into another. With loudspeakers, this is about converting electrical energy into acoustic energy.

As with microphones, this rating is called a *sensitivity specification*. This defines the loudspeaker's acoustic output signal level, given a reference input level. Put another way, sensitivity defines how efficiently a loudspeaker transduces (converts) electrical energy into acoustic energy. Given the same reference electrical input level into two different loudspeakers, a more sensitive loudspeaker provides a higher acoustical energy output than a less sensitive loudspeaker.

Loudspeakers vary quite a bit when it comes to efficiency. Does this mean that lower-sensitivity loudspeakers are always of lesser quality? Not at all. Like microphones, loudspeakers are designed and chosen to meet specific uses. Uses for loudspeakers range from emergency notification, paging, speech reinforcement, music reinforcement, recording studios, sports arenas, touring concerts, houses of worship, and the home-listening environment. There is probably nothing else in an AV system that comes in as many different configurations and prices—and for as many different uses—as loudspeakers.

The following is an example of a loudspeaker sensitivity specification:

88 dB/1 W @ 1 m

This means that with 1 watt input, 88 dB SPL will be measured 1 meter away from the loudspeaker.

Loudspeaker Frequency Response and Polar Patterns

With loudspeakers, as with directional microphones, the overall frequency response will be best on-axis (directly in front of the loudspeaker). As you move off-axis, not only will the sound be reduced, but the frequency response will also change.

A loudspeaker with specifications showing a nominal 90-degree by 40-degree dispersion (coverage) pattern holds that pattern over only a limited frequency range. Lower frequencies will spread out in a more omnidirectional fashion due to their much longer wavelengths. Very large devices are required to control the dispersion pattern at lower frequencies.

Figure 4-17

A loudspeaker polar plot

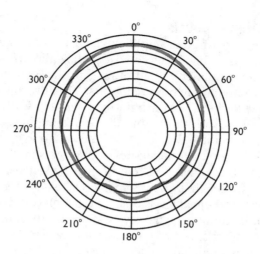

A graphical representation of the loudspeaker's directional versus frequency response characteristics is called a *polar plot*. Figure 4-17 shows an example of a loudspeaker polar plot.

Loudspeaker Impedance

Loudspeakers have a nominal impedance rating. Most are rated at 4, 8, or 16 ohms. Some are rated at 6 ohms. Because impedance is frequency dependent, the impedance will not be the same over the loudspeaker's entire frequency range.

When you connect loudspeakers together, you need to know the total impedance load that you are connecting to the output of the power amplifier. Knowing the total impedance load will help you do the following:

- Get optimum volume
- Avoid wasting power
- Avoid overloading and damaging your power amplifier
- Prevent damage to your loudspeakers
- Reduce distortion and noise
- Avoid uneven sound distribution

Audio Signal-Level Monitoring

Now that you have seen the audio-signal path all the way through and know the signal levels involved (mic, line, and loudspeaker), you need to be able to monitor, and possibly adjust, the signal levels at various points to make sure the signal levels aren't too low or too high.

Checking Signal Levels

Before turning on a power amplifier, check to make sure you are getting signals on all the audio mixer channels to be used. Make the necessary level adjustments to the microphone preamplifier and all other gain stages in the mixer and the other audio equipment leading to the power amplifier. Usually, you will check signal levels in the audio mixer with the aid of the built-in meters and verify them using headphones.

After setting the gain stages, turn the power amplifier input adjustments all the way down and turn on the power amplifier. Slowly bring up the power amplifier input adjustments until you reach the desired sound pressure level at the loudspeakers. Verify, by listening to the system and monitoring the signals, that no signals are distorted. These practices will help prevent signal levels that are too low, resulting in poor signal-to-noise levels, and signal levels that are too high (or hiss), resulting in distortion.

With analog, running signal levels at around 0 dBu for line-level signals is often preferred, and there may be some occasions when the normal signal level exceeds 0 dBu. However, with digital, the level must never exceed 0 dBFS (FS for the full scale of the digital signal). Exceeding 0 dBFS with a digital signal causes immediate distortion.

Frequently, equipment will provide at least a single LED light as some sort of indicator of the signal-level condition. Some LEDs may even change color as the signal level approaches or goes into distortion. LED labels may include the following:

- Signal present
- Overload
- Clip

 NOTE Read the equipment's manual. It will help you to make sure you operate the equipment properly and understand what a specific indicator tells you about the condition of the signal. Quite often, you will find that a red, flashing LED indicates signal distortion, but you need to read the manual to be sure.

Balanced and Unbalanced Circuits

Signal-processing circuitry and cables are continually exposed to noise and interference. In fact, all electrical circuits—and the cabling used to connect them—generate energy fields that can interact with other electrical circuits and cabling, including AV circuitry and cabling. This interference and noise degrades the quality of the audio signal and may introduce hum and buzz into an audio system.

One way to reduce the noise in a circuit or cable is to use a balanced electrical design. To understand the significance of this, you must first understand the difference between *balanced* and *unbalanced circuits* and the cabling used to connect them. First, we'll look at the fundamental difference between balanced and unbalanced, and then we'll cover how you know if a circuit is one or the other.

Basically, cable is subject to electrical noise that is added to the intended signal. In a balanced design, the equipment outputs a balanced signal to the cable, which is sent

Figure 4-18

A balanced
circuit

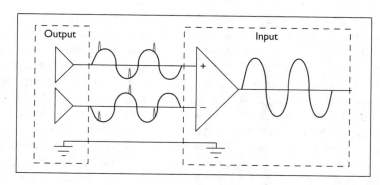

to equipment with a balanced input, as shown in Figure 4-18. The design of balanced circuits offers a defense mechanism against noise. This defense mechanism removes the noise, or most of it, leaving only the intended signal. As a general rule, you should use balanced components whenever you can. The cabling used with balanced circuitry requires two signal conductors. In audio, the two signal conductors are surrounded by a shield.

In an unbalanced design, the equipment outputs an unbalanced signal to the cable that is sent to an unbalanced input, as shown in Figure 4-19. As with a balanced circuit, the cable picks up noise from surrounding sources. However, with an unbalanced circuit design, there are no noise-defense mechanisms. If noise gets onto the signal conductor, it is there to stay. Cabling for unbalanced circuitry uses a single signal conductor. The single conductor is surrounded by a cable shield that also acts as the return electrical path for the circuit.

With either balanced or unbalanced design, the longer the cable run, the more noise the cabling is subjected to. Therefore, unbalanced lines are extremely limited in the distance they can cover and their ability to transfer a usable signal.

A quick way to determine whether a piece of equipment is balanced or unbalanced is to look at its outputs or inputs. Unbalanced audio equipment and cables typically use an RCA (phono) connector. All RCA connectors are unbalanced. However, you need to know the real distinction.

Unbalanced circuits are also called *single-ended circuits*. They require two conductors in the cable and the connector. The first conductor, an insulated wire, transports the signal. The second conductor is a shield around the wire used as the electrical circuit's return path and provides the ground reference for the circuit.

Figure 4-19

An unbalanced
circuit

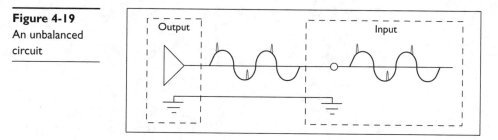

As with most things that offer higher quality, balanced components are more expensive, due to design and manufacturing. Like an unbalanced circuit, a balanced circuit requires two conductors in the cable and connector. However, in this case, both conductors are used to transport the signal. The first signal conductor carries a signal, and the second conductor carries an inverse, or mirror image, of the first signal conductor. In this case, the impedances on the two signal conductors, as well as the input and output circuitry connected to them, are the same with respect to one another. Since the impedances are the same for the two signal conductors, they are said to be equal, or balanced.

Although a balanced circuit requires only two conductors, when it comes to audio, we also use a foil or braided shield wrapped around the two signal conductors that serves as the circuit's ground reference. An audio balanced circuit has a connector that requires three pins: two for the signal conductors and a third for the shield connection.

In audio, balanced circuits always use shielded twisted-pair cable, as shown in Figure 4-20. When noise is induced on the conductors, twisting the conductors around one another ensures that both conductors are subjected to the noise.

The noise is added to the signals on both signal conductors at the same time. When the signal reaches the balanced input, the signal on the second conductor is flipped 180 degrees (inverted) once more, and added to the signal from the first conductor. Now both signals line up and the amplitude increases, as shown in Figure 4-21. This increases the signal strength of the intended signal. The noise, however, is now 180 degrees out of phase (one is negative when the other is positive), and this cancels out the noise. So, a balanced circuit provides greater signal strength for longer distances with less noise.

Feedback

Feedback is the squealing or howling generated between microphones and loudspeakers. Feedback occurs if a microphone is either too close to the front of a loudspeaker or the gain (volume) has been turned up too high somewhere in the sound system.

A sound-reinforcement system is an amplification system. If the microphone "hears" itself through the sound system, it goes through the signal path where the sound is amplified and comes out of the loudspeaker at an even louder level than before. That's why

Figure 4-20
A shielded
twisted-pair
cable

Figure 4-21
Amplitude
increases

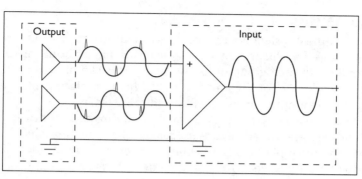

feedback typically gets loud quickly—it is a regenerative amplification loop. The signal continues to receive additional amplification each time it goes through the system.

One way to avoid feedback is through proper microphone and loudspeaker placement. Some best practices for controlling feedback include the following:

- Keep the microphone as close to the sound source as possible.
- Keep the loudspeakers in front of, and as far from, the microphones as is practical.
- Select directional microphones with polar patterns that fit the usage requirements.
- Select loudspeakers with coverage patterns that cover only the audience area.

Examples of Audio System Applications

We've reviewed the audio signal chain from start to finish—from microphones to loudspeakers. You've also learned about the various signal levels, cable used, and types of circuits (balanced) preferred by the audio professional. Now it's time to identify a few common audio system applications.

Sound Reinforcement

If you can't hear something at an adequate level acoustically (unamplified), then microphones, audio mixers, signal processors, power amplifiers, and loudspeakers are used to electronically amplify that sound source so you can hear it and distribute it to a larger or more distant audience.

In general, sound reinforcement comes in two variations: music and speech reinforcement. Music-reinforcement systems must cover a good bit of the audible spectrum, therefore they tend to be full-bandwidth systems, capable of reproducing a wider frequency range with higher sound pressure levels. Speech-reinforcement systems are used in situations where people can't adequately hear an unamplified presenter. Because the human voice has limited bandwidth, the speech-reinforcement system does not need to be designed for full bandwidth.

Mix-Minus

A "mix-minus" system is a type of speech-reinforcement system. When meeting participants, as well as the presenter, must be heard, microphones are distributed and mixed carefully with each group of loudspeakers. This can present challenges in making the system work without becoming unstable, because placing a live microphone near the loudspeaker amplifying its signal would cause feedback.

The approach is to create a separate sound subsystem for each loudspeaker, or group of loudspeakers (called *loudspeaker zones*). The term *mix-minus* means each subsystem mixes the microphone signals, minus (or except) those microphones closest to that group of loudspeakers, so as not to render the system unstable due to feedback.

Intercom and Paging Systems

Intercoms are sound systems used for of intercommunication between people or rooms. Usually, but not always, the people/rooms are located within the same building or complex of buildings. These systems typically comprise a central unit with local (personal) stations. They may also include multiple satellite or remote stations. The purpose is to provide two-way communication in the same way that a telephone provides two-way communication.

Paging systems are for one-way communication only. They are often used for information that needs to be communicated to a large audience. Paging systems can be simple, such as a system used for occasional announcements at a department store, or they can be complex, such as an emergency notification system used during life-threatening situations. With a paging system, the emphasis is on intelligibility.

Audioconferencing and Audio for Videoconferencing

Conferencing—and audioconferencing in particular—is communication between groups of people. Depending on the size of the group, an audioconferencing "pod" may be placed on a table. Larger groups may require multiple individual microphones installed in a table as part of a larger integrated AV system. An audioconferencing system may include either line or acoustic echo-cancellation technology. Larger systems may also include speech-reinforcement and playback capabilities.

Audio for videoconferencing is simply an audioconferencing system to support a videoconference.

Sound Masking

In almost every other type of audio system, it is important to minimize background noise. With a *sound-masking system*, also known as a *speech-privacy system*, background noise is purposely introduced to an environment in order to reduce intelligibility, thus aiding speech privacy. The sound-masking system also helps to reduce distractions caused by other noises, such as those caused by mechanical systems.

Chapter Review

In this chapter, you read about the basics of sound propagation, sound wave frequency and wavelength, harmonics, decibels, and the sound environment. With these as a foundation, you learned how to apply the basics of sound to the electrical pathway used to amplify speech, music, and other audio.

The electrical audio-signal chain, from start to finish, comprises everything from microphones, to mixer and processors, to loudspeakers. Each device in an audio system has varying features and characteristics, which the AV professional chooses among based on client need.

Review Questions

The following review questions are not CTS exam questions, nor are they CTS practice exam questions. Material covered in Part II provides foundational knowledge of the technology behind AV systems, but it does not map directly to the domains/tasks covered on the CTS exam. These questions may resemble questions that could appear on the CTS exam, but may also cover material the exam does not. They are included here to help reinforce what you've learned in this chapter. For an official CTS practice exam, see the accompanying CD.

1. How sound moves through the air is called_____.

   A. Generation

   B. Compression

   C. Acoustical

   D. Propagation

2. Wavelength is the _____.

   A. Number of times a wavelength cycle occurs per second

   B. Intensity or loudness of a sound in a particular medium

   C. Physical distance between two points of a waveform that are exactly one cycle apart

   D. Cycle when molecules move from rest through compression to rest to rarefaction

3. Which of the following does 0 dB SPL describe?

   A. The threshold of human hearing

   B. Ambient noise level

   C. The threshold of pain

   D. Normal listening level

4. A "just noticeable" change in sound pressure level, either louder or softer, requires a _____ change.

   A. +/–10 dB

   B. +/–6 dB

   C. +/–1 dB

   D. +/–3 dB

5. Acoustics covers how sound is _____, and its _____ and _____.

   A. Received; effects; structure

   B. Produced; propagation; control

   C. Controlled; delivery; translation

   D. Produced; amplification; reception

6. Numerous, persistent reflections of sound are called _____.

   A. Directional sound

   B. Echo

   C. Surface reflection

   D. Reverberation

7. Ambient noise is sound that _____ the desired message or signal.

   A. Interferes with

   B. Completely blocks

   C. Enhances

   D. Is louder than

8. The audio signal ends up in a(n) _____ before being converted back into acoustical energy.

   A. Output device

   B. Electrical signal

   C. Processor

   D. Microphone

9. The strength of the electrical audio signal from a microphone is called a(n) _____ -level signal.

   A. Dynamic

   B. Condenser

   C. Electret

   D. Mic

10. Phantom power is the _____ required to power a condenser microphone.

    A. Polarized conductor

    B. Electrical field

    C. Remote power

    D. Internal capacitor

11. Typically, what type of connector finishes the shielded twisted-pair cable?

    A. XRL

    B. RXL

    C. LRX

    D. XLR

12. The simultaneous use of multiple wireless microphone systems requires _____.

    A. Frequency coordination

    B. Multiple receivers all tuned to the same frequency

    C. Using lavalier microphones

    D. Using IR wireless microphones

13. If a technician changes the level of a signal, it is called _____.

    A. Unity gain

    B. Gain adjustment

    C. Attenuation

    D. Signal expansion

14. The amplifier comes right before _____ in the audio system chain.

    A. The equalizer

    B. Everything

    C. The audio processor

    D. The loudspeakers

15. What is a crossover?

    A. A loudspeaker containing multiple drivers

    B. A loudspeaker enclosure with more than one frequency range

    C. An electrical frequency dividing network circuit

    D. A single driver reproducing the entire frequency range

Answers

1. **D.** How sound moves through the air is called propagation.
2. **C.** Wavelength is the physical distance between two points of a waveform that are exactly one cycle apart.
3. **A.** 0 dB SPL describes the threshold of human hearing.
4. **D.** A "just noticeable" change in sound pressure level, either louder or softer, requires a +/–3 dB change.
5. **B.** Acoustics covers how sound is produced, and its propagation and control.
6. **D.** Numerous, persistent reflections of sound are called reverberation.
7. **A.** Ambient noise is sound that interferes with the desired message or signal.
8. **A.** The audio signal ends up in an output device before being converted back into acoustical energy.
9. **D.** The strength of the electrical audio signal from a microphone is called a mic-level signal.
10. **C.** Phantom power is the remote power required to power a condenser microphone.
11. **D.** An XLR connector typically finishes the shielded twisted-pair cable.
12. **A.** The simultaneous use of multiple wireless microphone systems requires frequency coordination.
13. **B.** If a technician changes the level of a signal, it is called gain adjustment.
14. **D.** The amplifier comes right before the loudspeakers in the audio system chain.
15. **C.** A crossover is an electrical frequency dividing network circuit.

Video Systems

In this chapter, you will learn about
- The nature of light as it applies to video systems
- Capturing, encoding, and decoding video signals
- The basics of computer signals
- Optimally projecting images

If the foundation of audio systems is sound waves, video systems are based on light. There are two primary theories about the nature of light. One theory says that light, like sound, is made of waves of energy. Another theory says that light is made of small particles, called *photons*. For the purposes of this chapter, and for most AV applications, it's helpful to think of light as a wave.

Light

There are many different kinds of light waves, each categorized by their wavelength. Figure 5-1 shows the entire range of the electromagnetic spectrum categorized by wavelength.

Wavelength is the physical distance, in meters, between two points exactly one cycle apart in a waveform. With light, that distance is typically measured in nanometers. A nanometer is equal to one billionth of a meter (10^{-9} meters).

At the right edge of the electromagnetic spectrum are *gamma rays*. They have the shortest wavelengths of 10^{-6} nanometers. At the left edge are television and radio signals with long wavelengths of 10^8 nanometers.

Frequency is the number of cycles in a given time period. Most often, frequency is measured in hertz (Hz). One hertz is equal to one cycle per second. If you are viewing light with a frequency of 10^{14} Hz, your eyes receive just more than *400 trillion* light waves every second. Gamma rays have a frequency of 10^{18} Hz. Radio waves have a frequency of 10^4 Hz.

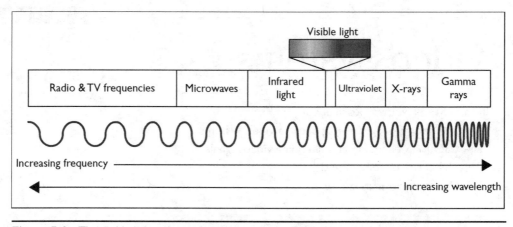

Figure 5-1 The visible light spectrum within the entire electromagnetic spectrum

The frequency of a light wave determines its color. Green is at the center of the visible-light color spectrum. Blue light waves have a higher frequency and are hardest to see. Red light waves have a lower frequency.

Frequency and wavelength are inversely related. As wavelength increases, the frequency decreases, as illustrated in Figure 5-2. Think of it this way: If there are a lot of little waves hitting the beach quickly, they could be said to have a short wavelength and a high frequency. If there are only a few big waves hitting the beach slowly, they have a long wavelength and lower frequency.

Amplitude is the magnitude of a signal. As represented on a sine wave, it is the intensity of a wave. Amplitude of a light wave is what your eyes perceive as brightness. The greater the amplitude of a light wave, the farther the wave is displaced from the midline, and the brighter the light will be. Figures 5-3 and 5-4 show amplitudes of a sine wave and similar amplitudes at different frequencies.

Figure 5-2
Frequency
increases as
wavelength
decreases.

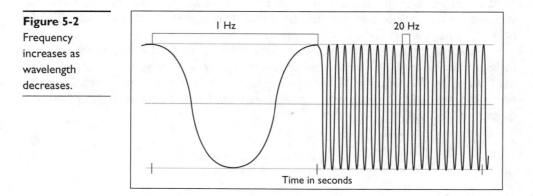

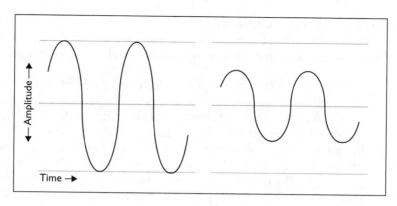

Figure 5-3
Higher and lower
amplitudes of a
sine wave

Units of Light Measure

You may think of light in terms of visual systems, but light affects every aspect of the user's experience in the environment. It is used for everything from reading and writing to walking through a building. You must therefore be able to accurately measure and quantify the types of light in your environment.

Light is measured using two types of meters:

- *Incident meters* measure light coming directly from a source such as a light bulb, projector, or monitor.

- *Reflected meters*, or *spot meters*, measure the light that bounces off an object like a projection screen or work surface.

You can also use these types of meters to measure the light emitting from a monitor, rear-projection screen, or light-emitting diode (LED) sign. A designer can use spot measurements and total surface area to calculate the proper lamp size required for a specific task.

The units of measurement for light vary by geographic region. You need to be able to recognize several measurement terms for both direct light and reflected light.

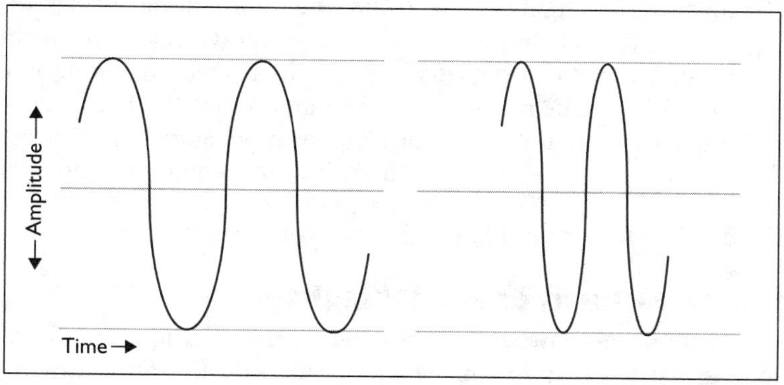

Figure 5-4
Similar ampli-
tudes at different
frequencies

Direct Light Measurement

Using an incident light meter, you can measure the brightness of an emitting light source. Three units of measure are commonly used in the AV industry:

- **Lumen**   A lumen is a measure of the quantity of light emitted from a constant light source across a 1-square-meter area. Lumens are the most common measurement term used for describing light output from a projector or light bulb. Manufacturers state brightness measurements in lumens in their product literature. A higher lumen rating means a brighter displayed image. However, organizations measure lumens by different methods. The most common standard for the lumen measurement worldwide is from the American National Standards Institute (ANSI). The International Electrotechnical Commission (IEC) has a similar standard, IEC 61947-1.

 NOTE   Officially, the two ANSI lumen standards, ANSI/NAPM IT7.228-1997 and ANSI/PIMA IT7.227-1998, were retired on July 25, 2003. However, most manufacturers still use them today.

- **Lux**   Lux is a combination of the words *luminance* and *flux*. Lux is usually associated with metric measurements. A lux is equal to 1 lumen/square meter, or 0.093 footcandles. Generally, the lux measurement is taken at a task area. The task area could be a video screen, note-taking location, or reading area.
- **Footcandle**   The footcandle (fc) is a U.S. customary measurement. One footcandle equals approximately 610.78 lux.

Reflected Light Measurements

Two units of measurement are commonly used in the AV industry to describe reflected light:

- **Footlambert**   The footlambert (fl) is a U.S. customary unit of measurement for luminance. It is equal to $1/\pi$ candela per square foot.
- **Nit or candela per square meter**   Originally, light measurements were based on a candle's light output. Today, the International System of Units (SI) unit called a candela has replaced the candlepower standard. Technically, a candela is defined as the luminous intensity, in a given direction, of a source that emits monochromatic radiation of frequency 540×10^{12} Hz and that has a radiant intensity in that direction of 1/683 watt per steradian. Nits are the unit used to measure candela per square meter. A nit is equal to 1 candela/meter^2.

Figure 5-5 shows the U.S. and SI units for measuring light.

Inverse Square Law and Light

The inverse square law states that as you move farther away from a light source, the brightness of the light at the viewing point will diminish. Moving closer to the light source, the light will appear brighter.

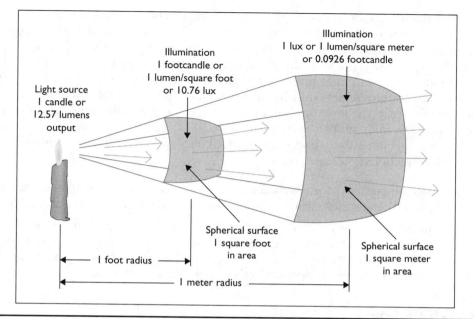

Figure 5-5 U.S. and SI units for measuring light

The inverse square law of light has two parts:

- The *illumination* from a light source varies inversely as the square of its distance from the measuring point. As the distance from the light source is doubled, the perceived illumination decreases by 75 percent. Light radiates in all directions from a point source, such as a candle or a lamp. As it disperses, it has the effect of diminishing with distance. The first part of the law simply states that the farther away you place a light source, the less the illumination is received.

- The *illuminance* of a surface varies inversely as the square of its distance from a light source. Doubling the distance between the light source and the illuminated surface will increase the illuminated area by four times. If you shine a flashlight at a wall while standing very close to it, the area of illumination is small. As you back away from the wall, the area of illumination gets larger. However, because the light energy disperses over a larger area, it is not as bright.

Color Mixing and Temperature

You may have learned by playing with paints that yellow and blue make green, and mixing all your paints together usually creates brown. The colors of light, however, combine differently.

White is the perceived color when red, blue, and green light are added together in equal proportions. Black is the absence of light. There is no such thing as black light.

While many light sources may appear to emit white light, they have different distributions of colors, so each white has a different tint—some are actually reddish white; others are bluish white. This difference in whites is referred to as a light source's *color temperature*.

Color temperature is a scientific measurement for expressing the distribution of the colors radiating from a light source, expressed in the Kelvin (K) scale. (It is a common error to refer to color temperature in degrees Kelvin.) The higher the color temperature, the bluer the light. The lower the color temperature, the redder the light. Table 5-1 lists some common color temperature ranges.

Color temperature is very important to visual display. Here are two examples of how you will see color temperature applied:

- Video cameras notice more changes in light's color temperature than the human eye can. Therefore, electronic adjustments are made for what white should be.

- A monitor or projector may have color-temperature selection features. These allow you to alter the "whiteness" of an image for the lighting conditions in the environment. For example, you can make the white of the displayed image match the white of a piece of paper under the room's lighting conditions.

Ambient Light

Ambient light is any light in a presentation environment other than light generated by the displayed image. Ambient light may strike the screen or the walls in the room and reflect onto the viewing area, competing with the displayed image by reducing contrast and washing out the picture.

The amount of ambient light in a display environment significantly impacts the quality of the displayed image, just as the acoustics of a room affect the sound quality in audio. When viewing a projected image in a dark room, such as a movie theater, light from outside can reflect off the screen and interfere with the viewers' perception of the image. Higher ambient light reduces contrast on the display.

The following are examples of how you can take precautions to minimize ambient light problems:

- Movie theaters often have two sets of doors leading into the theater. The hallways are dimly lit, and aisle lighting is low to the ground. The walls are painted a dark color to absorb any light that strikes them.

	Temperature	Source
Table 5-1	1900K	Candle light
Color Temperature Ranges	2200K	High-pressure sodium lamp
	3200K–3400K	Tungsten lamp
	4200K	Cool-white fluorescent bulb
	5400K	Daylight
	5500K	Noon direct daylight, cloudless sky
	6500K	SMPTE reference white

- In a corporate environment, ambient light can enter a room through windows or skylights. Custom window treatments may be used to shut out unwanted light. Ambient light can also come from other lamps in the room through surface reflections.

You will learn more about dealing with ambient light in Chapter 13.

 NOTE There are times when some ambient light is necessary to see some visual displays (for example, a flip chart), to read, or to take notes.

Image Capture

Having learned about light, the next step is to capture an image and create a video signal. A video signal must be created before you can send it to a display. Any discussion of video processes, equipment, or AV applications presupposes the capture of video signals.

Video is made of a series of still images, like the frames on a filmstrip. A video camera uses a lens as an eye to focus on an image. Just like your eye, a camera needs light to "see," and whatever it sees is turned into a video signal. Figure 5-6 illustrates the video-broadcast process.

However, a video camera does not record an image exactly like a movie camera captures an image on a filmstrip. A traditional movie camera passes film in front of a lens, capturing individual images directly on the strip. A video camera functions more like a translator (transduction), converting light to electronic information. *Transduction* is the conversion of light energy into electrical energy or electrical energy back into light energy.

There are many ways to form a video signal, and one video signal type builds up to the next. This section will walk you through the steps of the video-signal chain, following the translation of light through the video camera lens into the five levels of video signal types. Figure 5-7 illustrates the video-signal chain.

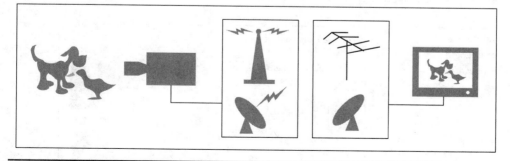

Figure 5-6 The video-broadcast process

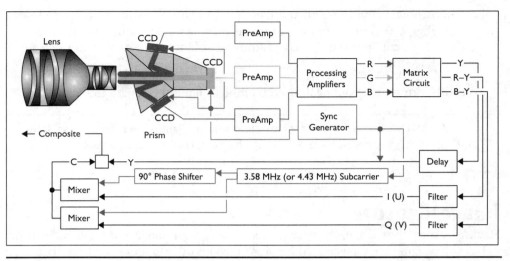

Figure 5-7 The video-signal chain

Camera and Focus

We'll start with how the camera processes and projects the image, defining focal length, aperture, depth of field, and other related terms.

Focal Length

Focal length is the distance, in millimeters, light must travel after it passes through the optical center of a lens. If someone placed a tape measure from your pupil to the retina inside your eye, that person would be measuring the focal length of your eye.

In a camera, focal length determines how much of a scene is visible to the lens, as illustrated in Figure 5-8. This is usually called the *angle of view*. A wide-angle lens has a short focal length and shows a greater area than a telephoto lens. Telephoto lenses have a narrow angle of view and show less of the scene.

A zoom lens varies its focal length, which alters the angle of view. Lens specifications usually provide the angle of view for the lens. Zoom lenses have angle-of-view specifications for both wide and telephoto extremes.

Figure 5-8
Focal length
changes image
size.

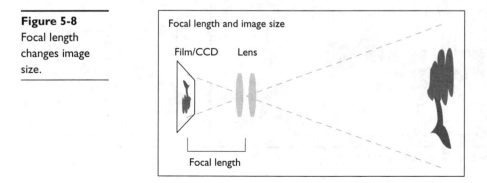

Back Focus

Zoom lenses must stay in focus as the lens is zoomed from wide angle to telephoto. This adjustment, used to ensure focus tracking, is called *back focus*.

To manually set the back focus for a lens, you must repeat a two-step process:

1. Zoom into a back focus chart, as shown in Figure 5-9. The image will be fuzzy in the center. Adjust the lens until the image is sharp and the blurred area smallest.

2. Set the lens to wide angle and adjust again.

Repeat these two adjustments until the image remains focused from wide to telephoto.

The back focus adjustment may be made by an adjustment ring on the lens, close to the camera body, or it may be an adjustment to the image-sensor position in the camera. Adjusting the image sensor is usually done only to single-chip cameras, such as surveillance cameras.

Aperture

The *aperture* is the opening in a lens. The iris controls the opening, regulating the amount of light passing through the lens to the imager (discussed in the next section).

The aperture of a lens is measured in f-stops. An f-stop is based on the relationship between the diameter of the lens opening and the focal length of the lens. Just like the pupil in your eye, when the iris is stopped down, the aperture is very small, letting in very little light. The larger the f-stop number, the smaller the lens opening, and the less light allowed in.

Common f-stops include f/1.4, f/2, f/2.8, f/4, f/5.6, f/8, f/11, f/16, and f/22. An f-stop of f/2 lets in more light than f/2.8, but less than f/1.4. Each f-stop in this sequence lets in half or twice as much light as the one next to it.

Depth of Field

The distance between the nearest object in focus and the farthest object in focus is called *depth of field*. Anything out of the range of a particular lens's depth of field will be out of focus.

Figure 5-9
A back focus
chart

Figure 5-10   Using depth of field to highlight an apple

The flowers in the front and back of the image in Figure 5-10 are out of focus. The area that is in focus is the depth of field. When capturing an image, you can adjust the iris of the lens to ensure you are capturing the desired depth of field.

The depth of field is dependent on the aperture. The smaller the aperture, the larger the depth of field. For example, when you are trying to see something far away, you may squint to bring it into focus. Squinting reduces the size of your pupils, increasing your depth of field and bringing more objects in focus. Dimly lit scenes have a smaller depth of field, because the iris of the lens is opened larger. A pinhole camera lens has a very large depth of field and requires no focus adjustment.

Prism and Imager

Once light has passed through a video camera's lens, it travels to the next part of the camera's "eye": the light beam splitter.

A camera with three charge-coupled devices (discussed in the next section) has a prism, or beam splitter, that refracts the light into red, green, and blue elements, as shown in Figure 5-11. This filtered light is focused and routed onto light-sensitive electronic chips called *imagers*.

These imagers convert the light's optical image into a stream of electrons: the video signal. These signals are electrical representations of the still image. This process is repeated many times per second, creating the illusion of motion.

Figure 5-11
A lens and three charge-coupled devices

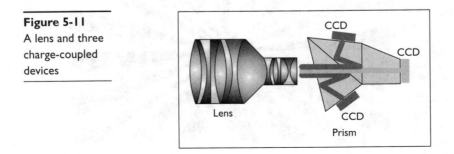

In a single-imaging device camera, there is no prism. The imaging device has color filters applied to the imager on a pixel-by-pixel basis. The pattern of the filter layout differs depending on the intended use of the camera. These filters are applied at the factory.

Imagers

You just learned that the light-sensitive electronic chip behind the video camera's lens is called an imager. There are two common types of imagers: charged-coupled device (CCD) and complementary metal-oxide semiconductor (CMOS). These devices are made up of thousands of sensors, called *pixels*. The pixels convert the light input into an electrical output.

Light sensors vary in sensitivity and resolution, creating a difference in image quality and camera cost. Less expensive cameras have only one chip; most consumer cameras are single-chip cameras. Cameras that are more expensive will use three imaging chips: one each for red, blue, and green light. Broadcast and digital cinema cameras use three imager arrays for the highest image quality.

In normal operation, an imager will output a frame of captured video at the frame rate of the video standard. NTSC will output at 29.94 frames per second, while PAL will output at 25 frames per second. That means the imager will collect light that particular number of times, and then send out the electronic signal. Higher frame rates create a larger signal.

Video Shutter

Too much light can overload an imager. The electronic shutter helps avoid overloading by regulating how much time the incoming light can charge the imager. Shortening the time the imager charges will reduce the voltage level output of the imager, so overloading is greatly reduced.

Shutters are set using frame rates or speeds. It is important to remember that this frame rate affects only the imagers, not the video output frame rate.

NOTE Be careful to choose the appropriate shutter for your application. A scene involving fast motion calls for a higher shutter speed. If the shutter is too slow, the resulting image will look jerky, or stuttered.

Video Signals

The term *video signal* is used to refer to two types of data:

- The output of a video device such as a camera
- The output of a computer to a display device

First, we'll discuss the kinds of video signals from video devices. We'll cover computer signals in the "Digital Signals" section, a little later in this chapter.

Several different video signals form the building blocks of an image. Some signals indicate color. All light can be broken down into the three primary colors—red, green, and blue—the only color signals needed to create every hue in an image. All other colors of light are mixtures of these three. The light focused on a camera's imager is separated into red, green, and blue signals. These three signals are commonly referred to as RGB.

These color signals must be combined with synchronizing information, typically called *sync*. Sync preserves the time relationship between video frames, and correctly positions the image horizontally and vertically. This information is called *horizontal sync* and *vertical sync*, usually abbreviated H and V. So, R, G, B, H, and V are the five basic elements of video signals. They contain all the information needed for a complete video image.

Horizontal and Vertical Signals

Video signals were originally scanned line by line, in both the camera used for recording images and the television used for displaying them. While the technology used to capture and display images has changed dramatically, a form of image scanning still takes place in all cameras.

Horizontal and vertical sync signals define the edges of an image. Without these signals, your picture would roll from side to side and top to bottom.

Inside each camera is a sync generator that produces the horizontal and vertical sync pulses. Horizontal sync pulses are used to time the edges of the image. They determine the point where the pixels end on the right edge of the screen. Vertical sync pulses are used to time the top and bottom of the screen. Vertical sync begins at the bottom-right pixel.

Syncing signals are also used to produce smooth switching between two or more cameras. In an analog camera, a signal called *black burst* is fed to each camera. The sync generators in each camera use this reference to lock the cameras in time. Digital cameras may use a reference clock signal to do the same thing.

Scan Rates

Horizontal and vertical sync signals define the edges of the image. The *horizontal scan rate* describes the number of horizontal lines that a display can draw per second. The *vertical scan rate* is the amount of time it takes for the entire image, or frame, to appear, since all the horizontal lines must be present in order for the image to be complete vertically. Together, the horizontal and vertical scan rates are known as the *scan rate*.

The number of horizontal scan lines in an image depends on the video content. For example, a DVD might be 780 × 480, with a refresh rate of 30 frames per second, while a computer might be 1,440 × 900, with a refresh rate of 60 frames per second. For both of these types of content to be viewed on the same display, the display must be capable of playing back both scan rates.

Table 5-2 shows a select sample of scan rates for analog video, digital video, and computer graphics.

Each standard also defines the scan rate according to a specific horizontal frequency. You will recall that frequency is cycles per second, measured in kilohertz (kHz). A kilohertz is equal to 1,000 cycles per second. The horizontal scan rate, also known as

Analog Video			
Format	**Width**	**Height**	**Frames/Sec**
NTSC	720	486	29.97
PAL	720	576	25
Digital Video			
480i	720	486	29.91
576i	720	576	25
480p	720	480	29.97
575p	720	576	25
720p	1,280	720	25
1080i	1,920	1,080	29.97
1080p	1,920	1,080	25
Computer Graphics			
XGA	1,024	768	60
WXGA	1,280	768	60
SXGA	1,280	960	60
WSXGA	1,440	900	60
WUXGA	1,920	1,200	60

Table 5-2 Video Format Characteristics

the *horizontal frequency*, is the number of horizontal lines a display device draws each second, measured in kilohertz.

Because many horizontal scans are completed in just one vertical scan, the horizontal scan has a higher frequency than the vertical scan. The vertical scan rate describes the number of complete fields a device draws in a second. This may also be called the *vertical sync rate* or *refresh rate*. The vertical scan rate is measured in hertz, or cycles per second. The rate for an interlaced image is one-half that of the vertical scan rate.

Signal Quality and Bandwidth

After video has been captured and converted into signals, you need to transport those signals to a device for display. The image quality requirements will determine how this is done.

High-resolution video images are like large computer files—they take up a lot of space. To preserve image detail, you need high-quality cable, as well as video equipment that can switch, process, and distribute these signals without degrading them.

Signal quality is determined by bandwidth. *Bandwidth* is a measure of the amount of data or signal that can pass through a system during a given time interval. The path that a signal travels is like a highway, and the bandwidth of that path determines how much data can travel along that highway at once. High-resolution video signals contain a lot of information, requiring a highway with more lanes—higher bandwidth. The highest quality analog video signals are called *full-bandwidth signals*. Lower-quality signals are bandwidth-limited signals and require slightly less room.

Technically speaking, bandwidth is the width of the band of frequencies the signal requires. If your start frequency is 0 Hz and the end frequency is 6 MHz, then the bandwidth is the difference between six and zero: 6 MHz.

Bandwidth is usually expressed as a range of frequencies. For example, light travels at varying frequencies, only some of which we can see. The bandwidth of human vision—that is, the lowest frequency of light we can see to the highest—is 430 trillion Hz to 750 trillion Hz.

In electronics, bandwidth is expressed in terms of the frequency of voltage, rather than frequency of light. It is the range of frequencies that can pass through a circuit. Every electronic circuit can read frequencies in a specific range. The difference between the highest and lowest frequencies a circuit can detect and react to is that circuit's bandwidth. Just as with the light frequencies we cannot see, when electronic frequencies fall outside the circuit's bandwidth, the frequencies are lost and not passed on.

High-quality video contains a lot of detail. More detail requires more information. More information requires more bandwidth.

A video signal covers a range of frequencies:

- Lower frequencies in the hertz range include vertical sync.

- Middle frequencies in the kilohertz range include horizontal sync.

- High frequencies in the megahertz range include detailed picture information.

Full-Bandwidth Signals

The combination of three signals plus sync or timing signals makes up all of the information contained in the image converted by the camera's sensors. This is a full-bandwidth RGB signal, or an unencoded signal, as shown in Figure 5-12.

Signals that are this level of quality include RGBHV and RGBS, as shown in Figure 5-13. RGBHV is typically used for computer and signal processors. It includes red, green, blue, horizontal sync, and vertical sync, and is referred to as *full-bandwidth*

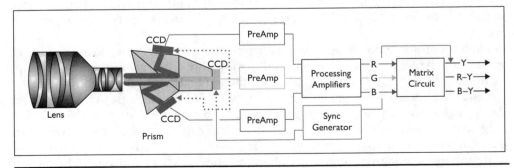

Figure 5-12   Creation of a full-bandwidth video signal

Figure 5-13

Three types of full-bandwidth quality signals

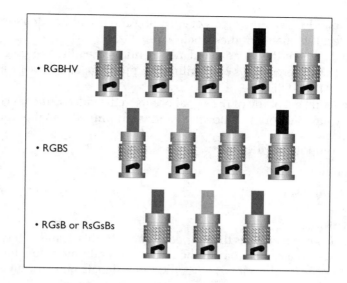

- RGBHV

- RGBS

- RGsB or RsGsBs

RGB with separate sync. RGBHV requires five cables to carry the signal—one cable for each signal. This is the height of signal quality and stability. Everything after this is lower quality.

RGBS is typically a camera output. It includes red, green, blue, and composite sync—a combination of horizontal and vertical sync. RGBS is referred to as *full-bandwidth RGB with composite sync.* It requires four cables to carry the entire signal: one for each signal. This is close to the height of signal quality, but stability may suffer because the two sync signals are added together.

When you use one of these signals, you have not applied any bandwidth limiting, or encoding, to the camera signal, so the overall image quality has not been compromised. However, the signal stability may be affected if you choose not to use separate sync. Signals such as RGsB and RsGsBs combine sync signals with color signals.

Digital variations of these signals have been adopted for next-generation display technologies.

Bandwidth Limiting

Bandwidth limiting is necessary to transmit complex signals over a limited band of frequencies. When the first analog color television systems were proposed in the 1950s, they needed to be backward-compatible with black-and-white television systems. This meant that all three of the color channels—red, green, and blue—needed to be squeezed into the same space as that of an existing black-and-white television channel, which was about 4.5 MHz. Full-bandwidth RGB signals required 10 MHz or more. Adding color signals into the space of a black-and-white television channel was like trying to squeeze a size-10 foot into a size-4 shoe. Bandwidth limiting was needed.

Bandwidth limiting is achieved through filtering and encoding. Values of red, green, and blue color channels are reduced to smaller values. While an encoded signal takes up less bandwidth, the process of encoding and decoding adds some noise and artifacts that affect the quality of the signal.

This is where the video standards mentioned earlier are used. Different countries decided on different ways of handling this problem, and three primary standards arose: NTSC, PAL, and SECAM.

The actual encoding of the signal occurs in the video camera's circuitry. A video camera may have any of the following bandwidth-limited signal outputs:

- Component video
- S-Video
- Composite video

Luminance

In video, *luminance* (Y) is the monochromatic, or black and white, picture information. The imager of a camera passes the red, green, and blue color signals into the camera's matrix circuitry. This circuit takes the three signals and combines them into a single luminance signal. The matrix circuit also produces color-difference signals, which are discussed in the following sections.

The human eye is not equally sensitive to all three primary colors. Studies completed in 1931 by the International Commission on Illumination (CIE) show this fact. Humans are most sensitive to the yellow/green spectrum of light, followed by red and then blue.

The composite color video television standards take advantage of this color-weighted system. Expressed in proportions, human color sensitivity is about 59 percent to green, 30 percent to red, and 11 percent to blue. With this in mind, the Y (luminance) signal for composite video is encoded using the following formula: $Y = 0.30R + 0.59G + 0.11B$.

Component Video

Component video has three signals: luminance (Y), red minus luminance (R–Y), and blue minus luminance (B–Y). It therefore requires three cables to carry the signal. It is used extensively in production environments due to its quality and lack of artifacts.

Y is created using proportions of red, green, and blue. Subtracting the R and B signals from the luminance signal creates what are known as the *color-difference signals*: R–Y and B–Y. These three signals (Y, R–Y, and B–Y) can mathematically reproduce RGB at the display device. This approach reduces the total bandwidth needed to send or store the image.

The Y signal may have a bandwidth of 6 MHz or greater. However, the color-difference signals will have a bandwidth of only about 1.5 MHz. Reducing the size of the color-difference signals saves bandwidth and makes handling the signals much easier.

You may have seen other letters used to indicate color-difference signals, such as Y/Pb/Pr and Y/U/V. These formats use different mathematical formulas to produce the signals.

Composite Video

Composite video, also referred to as *baseband video*, is the most common video signal format and requires one cable. A composite signal begins with the S-Video signal. The chrominance signal (C) is added to the luminance signal (Y) to form a composite signal.

With composite video, the chrominance shares the same space as the luminance. This makes it difficult to separate the two signals at the display. You may need to use sophisticated filtering to avoid artifacts such as cross-luminance—"dot crawling" around the edges of objects—and cross-color, or moiré, patterns.

Current digital video standards do not include either Y/C or composite formats. You will find digital RGB and component video formats, however.

Video Signal Decoding

To be shown on a display, a composite (analog) signal—one that combines all the information necessary to create a color picture on one cable, whether using NTSC, PAL, or SECAM—must be *decoded*, or processed. The process of decoding reverses the encoding process by converting to a different signal type, as shown in Figure 5-14.

Recovering the luminance and color information during the video-decoding process is a real challenge. The resulting images will be only as good as the decoding process allows. Remember that fine detail in a video image requires higher frequencies and sits at the upper end of the video signal. However, the color subcarrier is located at 3.58 or 4.43 MHz, which is also in the range of high-frequency image detail. The challenge is to extract this color information without sacrificing luminance detail.

> **NOTE** A *color subcarrier* is simply the frequency in a video signal that carries color information.

The first step is to decode the composite video signal. The sync will be stripped from the signal and supplied to the sync circuitry. (The sync separator is not shown in the diagram in Figure 5-14.)

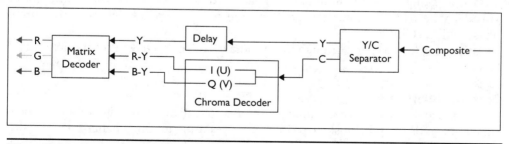

Figure 5-14 Decoding process for a video signal

Y/C Separation and Filtering

Y (luminance) and C (chrominance) can be separated by using a number of different methods. The quality of a Y/C separator has a large impact on the picture quality produced by a display device.

Several techniques are used to separate the C and Y information from a composite video signal. The following are the three filtering methods that are currently used, ranked in order of performance, from lowest to highest quality:

- **Low-pass filter** This simple filtering method was used in early color televisions. It was very effective at separating the Y and C signals, but fine detail was lost, as all picture information above the color subcarrier was discarded.

- **Notch filter** This method is currently used in low-cost display devices to decode composite video. The performance of a notch filter is just what you would expect from the name: It "notches out" a specific frequency or band of frequencies.

- **Comb filter** This method is used in displays for improved composite video decoding. A comb filter can more precisely separate the luminance from the chrominance than a notch or low-pass filter, and it eliminates the majority of the cross-color and luminance artifacts. Chrominance and luminance do not actually share the same space. They are interleaved, right next to each other. A comb filter has the ability to "comb" out the luminance and chrominance signals.

After being processed by one of these three methods, the composite signal has been decoded to an S-Video signal, which is a two-channel signal with Y and C on separate channels.

Chroma Decoder

Composite video must go through a Y/C filter to extract the C, or chroma, information, which is then sent to the chroma decoder. The *chroma decoder*, or *chroma demodulator*, simply takes this phase-encoded color information and recovers the component signals.

Once C has been successfully extracted from the luminance signal, Y, it must be further broken down into the two color-difference signals, R–Y and B–Y. While this process is taking place, the luminance signal is fed to a delay line—the waiting room. This delay keeps the luminance signal and the color-difference signals in the correct phase with each other.

After the video has gone through the chroma decoder, you have a component signal: Y, R–Y, and B–Y.

Matrix Decoder

Digital video starts at the component level. Because digital video is not a composite signal, it avoids the previously described steps, resulting in much better image quality than composite signals.

The video color matrix decoder takes the R–Y and B–Y color-difference signals coming from the chroma decoder and the Y signal, and converts them back to discrete red, green, and blue channels. The color-difference decoding process determines the values of red, green, and blue by adding and subtracting the component signals Y/R–Y/B–Y. The final output is RGB. In a display device, these are the signals that drive the circuitry and devices that produce the pictures that we see, including liquid crystal display (LCD) panels, plasma monitors, and projectors.

At this point, the video signal is completely decoded.

Digital Signals

Computers and digital media devices output red, green, and blue signals that are used for external displays. To maintain quality, the RGB computer signals must remain separate.

Motion video can be reduced in quality or encoded to reduce bandwidth because encoding errors are difficult to spot in quickly changing images. Computers generally display still information; viewers have time to notice any encoding errors while they study the image.

Along with the RGB signals, there are the horizontal and vertical (H and V) sync signals. These are transmitted as separate signals on an analog output. This helps with signal loss over distance and makes recovering the syncing information easier for the display device.

Computers and related devices, such as digital signage players and media servers, can also have digital outputs that carry much more information. These outputs still provide the RGBHV signal, but may also include audio and control signals.

Figures 5-15 and 5-16 show cable and connectors used with computers.

The following are the three major formats of digital connections:

- **Digital Visual Interface (DVI)**   DVI was the first accepted digital standard for computers. DVI comes in several versions: DVI, DVI-D, DVI-I, and Mini-DVI.

- **High-Definition Multimedia Interface (HDMI)**   HDMI builds on the DVI standard by adding audio, device control, content protection, and Ethernet. HDMI is backward-compatible with DVI. HDMI comes in HDMI, Mini-HDMI, and Micro-HDMI versions.

- **DisplayPort**   DisplayPort also builds on the success of DVI and uses a similar yet not completely compatible format to send video and audio. DisplayPort is also available in a Mini-DisplayPort version.

Figure 5-15
An RGBHV
cable with BNC
connectors

Figure 5-16
A digital visual
(DVI) connector

Creating the Digital Image

Signals sent from different computers and digital devices may result in widely varying images. Graphic cards can produce many different resolutions and aspect ratios. The display device should always be considered when selecting the output resolution and aspect ratio. You will learn more about aspect ratios later in this chapter. Not all displays can synchronize to all resolutions, and not all resolutions will look good on all displays.

A display device has a native resolution, which is the resolution that is optimized for the display. For example, if a display has 1,366 pixels horizontally and 768 pixels vertically, it has a native resolution of 1,366 × 768. The native resolution is primarily thought of as the resolution of the display engine or pixel count of the screen or projection device. Selecting the native resolution of your display device for your computer output will generally produce the best picture quality.

The Video Electronics Standards Association (VESA) coordinates computer resolution standards. There are approximately 50 different computer resolution standards, not including variations of each standard. VESA also defines the extended display identification data (EDID) standard. EDID communicates a digital display's capabilities to attached video sources. Managing digital video equipment that utilize EDID can be a challenge for AV professionals.

Computer Output Bandwidth

A characteristic of computer signals is the ability to change the resolution of the computer graphics output to meet your needs. As you increase or decrease the resolution, the amount of bandwidth required to support the signal changes. This is a direct relationship: An increase in resolution means an increase in bandwidth.

The output of the computer is RGB, which has a much larger bandwidth than most video signals. A computer can display many different resolutions. How can you determine the bandwidth of each? There are two basic ways: measure it or calculate it.

To measure the bandwidth, you will need a spectrum analyzer. With this test device, you can monitor the lowest and highest frequencies used. This is your bandwidth.

Spectrum analyzers are expensive. If you do not have access to a spectrum analyzer, you can use the following formula to calculate the bandwidth:

$$[(Tp * Frate)/2] * 3 = Signal\ bandwidth\ in\ Hz$$

where *Tp* is the total number of pixels in one frame and *Frate* is the frame rate.

To calculate signal bandwidth, take the following steps:

1. Determine the total pixels in the picture by multiplying the number of horizontal pixels by the number of vertical pixels.

2. Multiply the total pixels by the frame rate.

3. Divide the result by 2 to indicate half the pixels on (white) and half off (black). This is the highest video resolution possible.

4. Multiply the result by 3 to leave room for the third harmonic. Accounting for the third harmonic ensures that the signal edges remain sharp at the display device.

The result is the bandwidth in hertz of your computer signal output. Red, green, and blue all share the same bandwidth value.

Here is an example for a $1,024 \times 768 \times 60$ screen:

Bandwidth = $[(Tp \times Frate)/2] \times 3$
Bandwidth = $[(1,024 \times 768 \times 60)/2] \times 3$
Bandwidth = $[(47,185,920)/2] \times 3$
Bandwidth = $(23,592,960) \times 3$
Bandwidth = $70,778,880$ Hz
Bandwidth = 70.8 MHz

If you perform several of these calculations using different resolutions, you will discover that bandwidth increases with resolution.

Producing an Image

Up to this point, you have captured an image from either a camera or computer and have encoded and decoded the signal. In order to figure out what kind of display device to use, you need to know what you want to display. Think about the kind of media you are viewing. Is it video, computer data, analog data, or digital?

The content affects how large an image must be for viewers to see the necessary detail. This is especially important when displaying images like spreadsheets. In projection, image size determines screen size and display placement.

Display and Projection Technology

Once you know what you want to display—and the environmental conditions under which you'll display them—you can proceed to display types and technologies. Display types include monitors (sometimes called *direct view*), projectors, and traditional print. The AV industry is primarily concerned with electronic display, but traditional, printed visual displays, like flip charts and whiteboards, are still used in some applications. Most AV systems incorporate a monitor, a projector and screen, or both.

Visual displays send light to our eyes for viewing. Displays can use three methods:

- **Transmissive**   Rear-screen projection uses the transmissive method, as shown in Figure 5-17. The light from a projector is projected onto a rear screen, and the light transmits through the screen. The advantages of transmissive displays include good contrast ratio and the possibility of large displays. However, they require more space for projection than front-screen projection, and hot spotting is possible. Liquid crystal displays (LCDs) that have backlights are also considered transmissive.

Figure 5-17
A transmissive
projection screen

- **Reflective** Front-screen projection uses the reflective method, as shown in Figure 5-18. The light from a projector bounces off a screen, and the reflective light hits our eyes. Reflective applications allow for the possibility of large displays and are easy to install. However, front projection has a lower contrast ratio than other methods, and because the projector is located in the room, it creates potential for fan and projector noise in the audience area.

- **Emissive** LED and plasma displays use the emissive method. The light is created by the display and sent directly outward. Silicon or phosphors are excited and glow, emitting light.

Figure 5-18
An AV space
with a reflective
projection system

Different types of technologies can be employed for either projectors or displays. For example, a plasma display uses emissive technology, while an LCD uses transmissive technology. To the users, both flat-panel displays appear similar.

Lasers are used in very small and very large applications. On the small end, lasers are sometimes used in *pico-projectors*. This is the technology used in pocket-sized devices such as mobile phones. On the large end, lasers are used for very big screens, such as for displaying images on buildings and large theaters. Laser projection is becoming more economical for corporate use.

Flat-Panel Display Technologies

Monitors, flat-panel displays, flat-screens—they are all terms for a common type of one-piece display technology. We are not talking about projectors, but rather the same type of flat-panel technology used for watching television at home. The prominent flat-panel technologies include plasma, LCD, and organic light-emitting display (OLED) technology.

Plasma

Plasma display panels (PDPs), also called *plasma displays*, are fixed-resolution emissive imaging devices. Plasma displays combine phosphors in pixels and are available as both fully integrated televisions or as monitors. They have wide viewing angles, high contrast, excellent color saturation, and low black levels. They offer very good picture quality under normal room-lighting conditions.

Plasma displays contain hundreds of thousands of tiny pixels. Each pixel is made up of three cells, or subpixels, consisting of red, green, and blue phosphors. Each pixel is also filled with neon and xenon gas. When a high voltage is discharged through a pixel, the gas inside ionizes and gives off ultraviolet rays. These rays strike the phosphors, causing them to glow.

Because plasma pixels can operate only in a switched on/off mode, they are considered to be digital display systems, even though the phosphor response to the discharged voltage is very much analog.

Plasma monitors can handle a wide variety of computer and video signals. However, because plasma monitors have a fixed-resolution pixel matrix, they must scale images up or down to match their native resolution.

LCD

LCDs use a grid of pixels to create images on a screen. Each pixel on the screen is filled with a liquid crystal compound. LCD is an example of a fixed-resolution display technology. The pixel resolution is usually based on computer display standards. The pixel resolution is known as the *native resolution*.

To create an image in an LCD, light must first pass through a polarizer. A polarizer is a set of extremely fine parallel lines that act like a net, or filter, blocking all light waves that are not parallel to those lines. The polarized light then travels through a sandwich of transparent switching transistors and liquid crystals. LCDs use two sheets of polarizing material. The liquid crystal solution is sealed between them. When a transistor is turned on, a specific voltage is applied to its pixel. The tiny liquid crystals within each pixel act like light shutters, passing or blocking varying amounts of light.

Each transistor must be shielded from the light that goes through the LCD, so it is covered with a nontransparent chemical layer called the *black matrix*. The black matrix also creates a defined edge around every pixel, causing a visible "screen door" effect on lower resolution displays. The ratio between the remaining transmissive surface of each pixel and the total surface of the LCD is called the *aperture ratio*. The smaller the transistors are compared to the size of the pixel, the higher the aperture ratio, and the more light that will pass through. Higher aperture ratios result in fewer image artifacts, such as the screen door effect.

A color LCD has three filters in each imaging pixel array, one for each primary color: red, green, and blue. The backlight can be compact fluorescent tubes or LEDs. The LEDs can be clustered in groups with red, green, and blue elements, or only white LEDs. LEDs have lower power consumption than compact fluorescent tubes, higher brightness, wider color range, better white balance, and local area dimming for higher contrast.

NOTE LCD technology also applies to projectors, which use light from the projection lamp and refract it into three colors—red, green, and blue—by special dichroic mirrors. When all three colors are recombined in a special prism, the additive color behavior of light results in full-color images that are passed through the projection lens to the screen.

OLED

OLED technology is based on layers of organic, carbon-based, chemical compounds that emit light when an electric current flows through the device. There are separate organic layers for red, green, and blue.

OLEDs are emissive devices, meaning that they create their own light, as opposed to LCDs, which require a separate light source: the backlight. As a result, OLED devices use less power and may be capable of higher brightness and fuller color than LCDs.

OLEDs are imprinted on a very thin silicon substrate. The active matrix silicon-integrated circuits are imprinted directly under the display, controlling the power to each organic point of light diode (pixel), performing certain image control functions at a very high speed. OLEDs' capability to refresh in microseconds rather than milliseconds, as LCDs do, creates highly dynamic motion video.

Positioning a Display Within a Room

Before deciding where to place a display, you should consider the size of the display and how far away viewers are likely to be. The optimum viewing area is determined by the video display size, location, and orientation within the room. The size of the display is proportional to the farthest viewer in the room. Either dimension can be the starting point when laying out a room.

The height of the image determines how far away from the center of the display the farthest viewer can acceptably be. When displaying images that require detail-viewing with clues, such as presentations with text, use a factor of six. In this case, the farthest viewer should be no more than six times the height (H) of the image from the center of the display, as in the following formula:

Farthest viewer distance = 6 × H

For tasks requiring more detail, such as architectural drawings, use a factor of four, increasing the minimum screen height and/or decreasing the farthest viewer distance. For video applications that do not require the study of text or detailed images, a factor of up to eight may be used, which allows for a smaller image size or longer viewing distances. Figure 5-19 shows the three distance multipliers for farthest viewer distance.

In addition to the farthest viewer limitations, there are also limits to how far off-axis viewers can be from the center of the image. The optimum viewing angle is within 45 degrees horizontally to each side of the center axis of the display. The less optimum, but still acceptable, viewing angles include 45 degrees horizontally to the left and right edges of the display. You'll learn more about viewing angles in the "Viewer Placement" section later in this chapter. You'll also learn about sightlines in Chapters 13 and 14.

Aspect Ratio

Human vision is oriented horizontally. Your eyes' angle of view is wider than it is high—landscape rather than portrait. Therefore, people naturally frame a scene horizontally. This perceptual quality is translated into a mathematical equivalent known as *aspect ratio*, which is the ratio of image width to image height, as shown in Figure 5-20.

Use aspect ratio to help determine appropriate screens, image areas, display devices, and lenses. For example, a video monitor has an aspect ratio of 4:3—four units wide and three units high. A widescreen display has an aspect ratio of 16:9. An alternate way to represent an aspect ratio is by dividing the first number by the second number; for example, 4:3 is also stated as 1.33:1.

Different presentation media have different aspect ratios. While widescreen video is 16:9, a computer display might be 16:10. And that same computer screen can be mounted in a portrait orientation, causing the aspect ratio to change to 10:16, as illustrated in Figure 5-21.

Figure 5-19
The three distance multipliers for farthest viewer distance

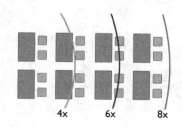

4x   6x   8x

Figure 5-20
The two
dimensions
that determine
aspect ratio

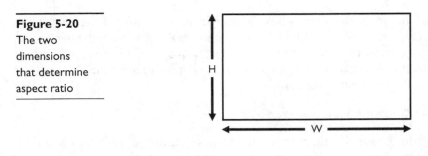

NOTE Some architectural specifications will reverse the aspect ratio from what you might expect. Always double-check the language in specifications.

Table 5-3 shows some common media formats and sizes.

When expressed as a ratio with whole numbers, aspect ratio is expressed as *units*, such as 16 units by 9 units. These units are proportional ratios that you can use to determine screen size in millimeters or inches, as follows:

$W{:}H = 16{:}9$

which can be written like this:

$W/H = 16/9$

Figure 5-21   Portrait orientation (left) and widescreen orientation (right)

Name	Description	Width	Height
XGA	eXtended Graphics Array	1,024	768
WXGA	Widescreen eXtended Graphics Array	1,152	864
WXGA	Widescreen eXtended Graphics Array	1,280	768
WXGA	Widescreen eXtended Graphics Array	1,280	800
SXGA	Super eXtended Graphics Array	1,280	960
SXGA	Super eXtended Graphics Array	1,280	1,024
WXGA	Widescreen eXtended Graphics Array	1,366	768
WSXGA	Widescreen Super eXtended Graphics Array	1,440	900
UXGA	Ultra eXtended Graphics Array	1,600	1,200
WSXGA	Widescreen Super eXtended Graphics Array Plus	1,680	1,050
HD-1080	High-Definition Television	1,920	1,080
WUXGA	Widescreen Ultra eXtended Graphics Array	1,920	1,200

Table 5-3 Media Formats and Sizes

Using this formula, you can substitute a number for either W or H, and then calculate the other value. For example, if you know the width of an image is 125 inches, you can calculate the height using the following formula:

$W/H = 16/9$
$125/H = 16/9$

Now cross-multiply and divide:

$9 \times 125/16 = H$
$70.3 = H$

If you know the width and height of an image, you can find the aspect ratio using this formula:

Aspect ratio = Screen width/screen height

Let's demonstrate this by determining the aspect ratio of the two monitors shown in Figure 5-22.
For the smaller monitor, calculate the aspect ratio as follows:

6.12 inches/3.45 inches   155.45 mm/87.63 mm
1.77   1.77

Because this is the formula for aspect *ratio*, the answer is expressed as follows:

1.77:1

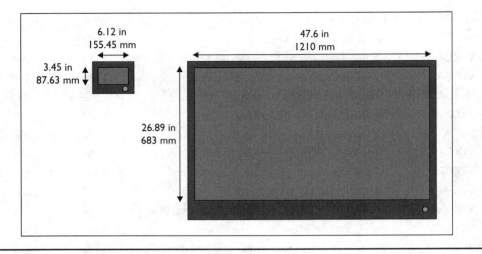

Figure 5-22   Examples of two different monitor sizes

However, in this industry, we like to work with whole numbers. It's hard for people to visualize 1.77:1. By multiplying each side of the ratio by a number between 2 and 9, you can get close to a whole number. Finding a whole number will take several calculations to achieve the best match. For example, multiply 1.77:1 by 9:

1.77 × 9:1 × 9
15.93:9

The value 15.93 rounds to 16. The aspect ratio for this monitor is expressed as 16:9. For the larger monitor in Figure 5-22, calculate the aspect ratio as follows:

47.6 inches/26.89 inches   1210 mm/683 mm
1.77   1.77
1.77:1
1.77 × 9 : 1 × 9
16:9

The Projected Image

Projection is used when large images are necessary and lighting can be controlled. The design of a projection environment is critical to creating a successful projection system. The environment significantly impacts the perceived quality of the displayed image. In projection systems, room-lighting designs are very important to achieve high-quality images.

Projected images use additive color; adding colors makes the image brighter. The projection system's brightness is extremely important. The dark areas of the images are equally important. Black cannot be projected. This means the screen can get no darker than when the projector is turned off.

A projection system includes the projector and its optics system, screen, and quality of setup.

Projector Optics

Projector optics comprise a system of lamps and lenses that carries and focuses an image on a screen for a viewer to see. The lamp is the projector's primary light source. It's a precision device, designed to work as part of a projector's optical system. This seemingly simple device can be very complex.

A projection device operates at optimum performance only when used with the lamp it was designed to use. The lamp works with a reflector to direct the light. The lamp reflector may be internal or external. Figure 5-23 shows three lamp and reflector types.

The amount of light the lamp emits impacts the images' brightness and color temperature. Low color temperatures are those near the reddish side, or warm side of the visible color spectrum. High color temperatures are on the blue side, or cool side of the color spectrum. Illuminating an object with a light source that has a very high color temperature causes a shift toward blue. Generally speaking, low color temperature, warm color lamps are best for flesh tones, while high color temperature, cool color lamps are better for showing graphics and presentations.

The lens that focuses the image onto the screen is referred to as the *primary optic*. Four factors related to primary optics influence the quality of the projected image: refraction, dispersion, spherical aberration, and curvature of field.

Front Projection and Rear Projection

With front projection, the viewer and the projector are positioned on the same side of the screen. A front screen is simply a reflecting device. It reflects both the desired projected light and the unwanted ambient light back into the viewing area, so front-screen contrast is dependent on the control of ambient light.

Front-screen projection generally requires less space than rear projection. It offers simplified equipment placement. In front projection, the projector is in the same space as the viewer. This can result in problematic ambient noise from cooling fans in the projector. Another downside is that people, especially the presenter or instructor, can

Figure 5-23

Three types of reflectors

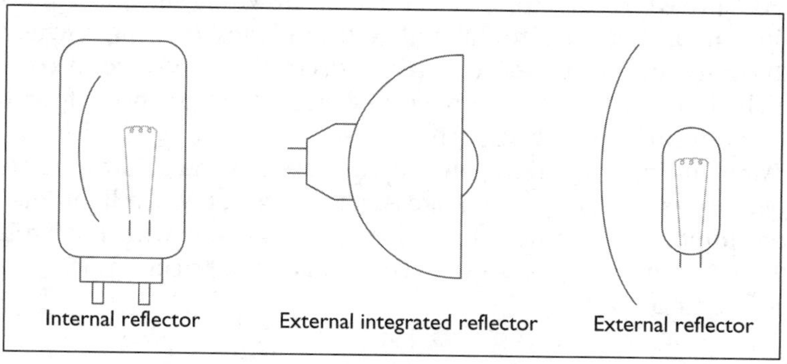

Internal reflector   External integrated reflector   External reflector

block all or part of the image by walking in front of the projected light. Front projection may be used because of its lower cost or because of space limitations behind the screen.

Rear projection is a transmissive system where light passes through a translucent screen toward the viewer. Rear projection needs more space for installation because it requires a projection room or cabinet behind the screen to accommodate the projector. Mirrors are commonly used to save space in rear-projection applications.

The primary advantage of using rear projection is that the equipment is not in the same room with the viewer, eliminating equipment noise and shadows on the screen when the presenter walks in front of the screen. Rear projection also handles ambient light issues better than front-screen projection.

Screens

Projectors require a surface to project upon. This surface is typically a screen. Optimum performance from any projection system is equally dependent on the screen and the projector.

A screen's surface coating greatly influences the quality and brightness of a video projector's image. A well-designed projection system incorporates a screen type that reflects projected light over a wide angle to the audience, while minimizing the reflection of stray light. Stray light causes a loss of contrast and detail, most noticeably in the dark areas of the image.

The commercial AV industry uses many types of screens, each suited to a particular purpose. The two major categories of screens are those used in front-projection applications and those used in rear-projection applications.

Screen Gain

Projection screens are passive devices and cannot amplify or create light rays. They can reflect light rays back at wide or narrow angles, thereby providing gain, or brighter images.

Screen gain is the ability of a screen to redirect light rays in a narrower viewing area, making projected images appear brighter to viewers sitting on-axis to the screen. The higher the gain number of a screen, the narrower the viewing angle over which it provides optimal brightness.

A matte white screen, such as the one shown in Figure 5-24, evenly disperses light uniformly in both horizontal and vertical planes, creating a wide viewing area, up to 180 degrees in both axes. This type of screen provides good color rendition, can be matched to specific color temperatures, and is the best choice for all projection applications with sufficiently bright projectors.

Ambient light rejection generally increases as the gain of the screen increases. This is because as the screen gain increases, the angle at which light hits the screen becomes more important. Ambient light that is not on-axis with the projector and viewer is reflected away from the viewer. This makes the screen appear darker, increasing the contrast ratio.

Figure 5-24
A matte white
screen

Front-Projection Surfaces

A front-projection screen is a passive reflector of light. You can manipulate how it reflects the light to improve brightness. You can also apply the laws of physics by changing the screen's surface, manipulating its position, and changing its contour.

Front-projection screens can be made of a variety of surfaces and are usually chosen based on application.

A matte white screen evenly disperses light 180 degrees horizontally and vertically, creating a large viewing area. A matte screen has a smooth, nongloss surface, similar to a white sheet. These screens provide good color rendition, and they are the best for data and graphic applications.

Angularly reflective screen surfaces provide performance similar to that of a mirror. The light is reflected back at the same angle that it strikes the screen, but on the other side of the screen's axis. If a projector is mounted at a height equal to the top center of the screen, the viewing cone's axis would be directed back at the audience in a downward direction. Most screens with a gain greater than five are of this type. This screen type works well for video.

Rear-Projection Surfaces

In general, rear-projection technology provides better contrast and color saturation in high ambient light environments. Rear-screen projection is more dependent on screen materials and installation design than front-screen projection. A wide range of screen materials is available for rear-projection configurations, each with different gain and directional properties. Rear-projection screens have either a narrow viewing area with high brightness or a wide viewing area with reduced brightness.

Some rear-projection screen applications use diffusion screen material. The screen substructure may be rigid acrylic, glass, or—for portable applications—a vinyl fabric. This material provides a diffused, coated, or frosted surface on which the image is focused.

It provides a wide viewing angle, both horizontally and vertically, but with little or no gain. There may be some hot spotting due to the transparency of the screen fabric, depending on the vertical placement of the projector with relationship to the audience. The light from the projector is transmitted through the screen with relatively little refraction. The ambient light rejection of this material is moderate and based on the viewer-side material's reflectivity or sheen.

Projection and the Inverse Square Law

When you increase projection distance in an effort to create a larger image, the larger image is not as bright. The brightness of the image noticeably drops off, even when you move the projector back only a few feet. The principle behind this, which applies to all radiated energy, is known as the *inverse square law*. The law simply states that the farther away a light source is from an object, the less illumination will fall on the object, as follows:

- The illumination from a light source varies inversely with the square of its distance from the measuring point.
- The illuminance of a surface such as a screen varies inversely with the square of its distance from a light source.

As you move a projector closer to the screen, the surface area illuminated decreases. Also, the brightness of the image surface is not uniform. Typically, the brightness levels at the edge of the screen are lower than the brightness levels at the center of the screen.

As an example of the inverse square law in action, suppose that you set up a front-screen projector for a small group presentation. There's a lot of ambient light, but the projector's brightness for the screen surface is acceptable. A larger group than expected joins the meeting, and it becomes necessary to pull the projector back and fit the image onto a bigger screen. But now the image is not as bright, and the overhead lights must be dimmed.

Perceived Quality

How an image looks on a screen is called *perceived quality*. Perceived quality is usually based on the contrast ratio of the system. The difference between system black and the brightest image, taking the environment into consideration, is the system *contrast ratio*. The greater the contrast ratio, the better the perceived image quality.

NOTE ANSI/INFOCOMM 3M-2011, Projected Image System Contrast Ratio, is a standard determining minimum contrast ratios for rear- and front-projection systems. You can learn more about it at www.infocomm.org/standards.

Perceived quality also depends on image *uniformity*. Human eyes are good at averaging, so viewers can tolerate some variations in image brightness on a screen, but an image must look consistent across the screen. Some of the causes of inconsistent image brightness stem from a projector's ability to produce an image of equal brightness from center to edge. It can also be caused by a screen-related problem.

Your goal is to provide the largest acceptable viewing area by selecting the best projection method, screen type, and placement for your project.

Viewer Placement

Viewers want to see the best possible image. To ensure this, you must consider the following:

- The viewing cone
- The viewing angle
- The viewing distance
- The overall perceived quality of the final displayed image

Viewing Cone

The best viewing area for the audience is referred to as the *viewing cone*. The term *cone* is used because there is width, height, and depth to the best viewing area, and this area comes from the center of the screen. Figure 5-25 illustrates the viewing cone.

Figure 5-25

A section and plan view of a viewing cone

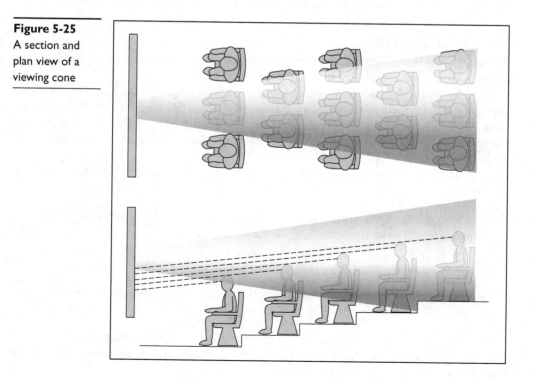

Viewing Angle

Generally, projected images should not be viewed at angles greater than 45 degrees off the projection axis, or outside a 90-degree viewing cone. These criteria vary for each system, and are called the *viewing angle*.

Viewing Distance

The maximum viewing distance is based on the amount of detail in the image. There are a number of guidelines for determining viewing distance, but generally, the best for all types of images is based on text size on the screen. Here are some general rules:

- For the majority of the viewing public, the viewing distance should be no greater than 150 times the character height.

- For video, the closest viewer should be no closer than the width of the screen, and the farthest viewer should be no farther than eight times the height of the screen. This is known as the *1w8h rule*.

- For computer data, the closest viewer should be no closer than the width of the screen, and the farthest viewer should be no farther than six times the height of the screen. This is known as the *1w6h rule*.

You will learn more about addressing viewer placement in Chapters 13 and 14, where we'll discuss how to assess the viewing environment and make suggested changes.

Throw Distance

Throw distance is the distance between the projector and the screen. Improper throw distance causes the image size to be different than required.

Most projectors today come with a zoom lens. The zoom lens allows you to create the same size image from a range of throw distances. The closest acceptable distance from the projector to the screen is called the *minimum projection distance*. The farthest the projector can be from the screen and still create the same image size is the *maximum projection distance*.

You can determine throw distance by referencing a projector's specifications, where multiple formulas are given:

- A formula for minimum distance to the screen from a specified point on the projector, like the tip of the lens or the front faceplate

- A formula for maximum distance to the screen from a specified point on the projector

- Possibly, a formula for the offset, which is the distance vertically that the projector can be from the bottom of the screen

Each projector formula is different. If the projector has multiple lenses, data will be given for each lens. Some formulas will require the use of the screen width; others may require the diagonal screen measurement.

The following is a sample formula. (The boldface information represents specs from a projector manufacturer.)

Max throw in meters = (**5.05** × SW) + **0.2**
Min throw in meters = (**3.2** × SW) + **0.26**
Offset = **0.03** × SW

where *SW* is screen width.

For a screen 1.83 meters high by 2.44 meters wide, calculate the maximum throw as follows:

(5.05 × 2.44) + 0.2 =
12.32 + 0.2 =
12.52 meters

And calculate the minimum throw like this:

(3.2 × 2.44) + 0.26 =
7.8 + 0.26 =
8.068 meters

Calculate the offset as follows:

0.03 × 2.44 =
0.07 meter

To make things easier, some manufacturers provide software that calculates throw distance for you. Most projector manuals will include a simplified chart showing minimum and maximum projection distances for selected screen sizes.

Keystone Error and Correction

In order to provide an accurate, focused image on a screen, your projector must be set up in the proper location. If the projector is not located correctly, the image on the screen can become misshapen and lose its focus from corner to corner. This is due to increased distance to a portion of the screen, allowing the projected light to spread out and cover a larger area. For example, this problem would occur if you set up a tripod screen, put the projector on the floor, tilted the projector's legs up high, and aimed it at the screen. This type of image distortion is called a *keystone error*.

The following are some adjustments you can make to correct keystone errors:

- Tilt the screen toward the projector.
- Adjust the height of the projector, if the visual environment allows.
- Use electronic keystone adjustment, which reduces the number of activated pixels in an attempt to square the picture.

Figure 5-26
Different adjustment methods for keystone correction

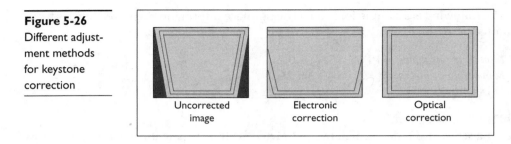

| Uncorrected image | Electronic correction | Optical correction |

- Use optical correction by lens shift. Some projectors have motorized lens shift available in both vertical and horizontal directions. Vertical lens-shift keystone correction is the most common. This method of keystone correction is preferable to electronic keystone correction because it does not remove any pixels.

Figure 5-26 illustrates these adjustment methods.

Digital Display Alignment

A projector lens typically does not have an autofocus feature, so it needs to be manually focused for a sharp image. You adjust the focus on some lenses by twisting the focus ring. Others are motorized and can be adjusted by pressing a button on the remote or the projector, as shown in Figure 5-27. The same is true for adjusting zoom—you may need to move the zoom ring or manipulate a motorized zoom using buttons.

You may need to make adjustments to the centering, clock, and phase.

Centering

Centering a display does not mean centering a projected image on a screen. It actually refers to centering the signal onto the imaging device—the LCD or digital light processing (DLP) inside the projector.

Figure 5-27
Using a remote control to align a display

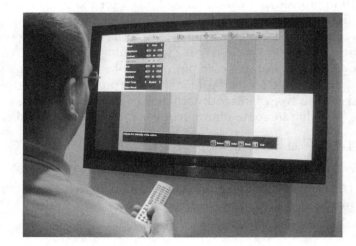

You can tell if an image is centered by looking at the screen. If any image detail seems to fall off the edge of the projected image, it means that a few lines of pixels are actually not lined up correctly on the pixel matrix. In effect, they fell off the matrix. You need to shift the image so that all the pixels align correctly on the matrix. Using the centering function, the image is manipulated left, right, up, or down to shift the image one row of pixels at a time.

Professionals use a test pattern with a single-pixel-width outside border. This allows you to easily see the edges of the picture.

Clock Adjustments

A clock inside the projector defines the digital timing. The clock is the master timing for putting a picture on a digital display. The clock's frequency needs to be synchronized with the incoming signal to display the image properly.

When connecting a source to a digital display, the image may be too large or too small for the device's fixed matrix. By adjusting the clock, usually using a menu function, you can either stretch or shrink the image to properly fit the device's matrix of pixels.

Phase Adjustments

After centering the display and adjusting the clock, it is time to fine-tune the image. If less than a pixel of visual information is falling off the edge of the projected image, it can be adjusted with the phase function. The full range of a phase adjustment is generally 1 pixel; larger adjustments are made with the centering function.

An error in an image phase occurs when the image mapping starts at the wrong time, misaligning the pixels. If the phase is not adjusted correctly, fine detail in the image will be noisy and fuzzy.

 NOTE The clock, phase, and centering controls are used together. After adjusting the phase, you may need to readjust the clock or centering.

Automatic Adjustments

Some projectors may include an automatic adjustment button. This will attempt to adjust the clock, phase, and centering for the incoming signal. Some fine-tuning may still be needed.

Projection System Brightness

The brighter a projection system, the better it stands up under ambient light. Today, thanks to technical advancements, more projection displays distribute light rays equally over the screen's surface. However, not long ago, projection systems created images that were less uniform, with dark spots in some areas of the screen. Manufacturers use brightness ratings to compare their products with other products.

Display brightness specifications for projection systems are usually stated in lumens. Lumens are a U.S. customary measurement; the metric equivalent is lux. Most commercial light meters use the lux measurement. Neither measurement takes into account the screen type or the type of image to be projected, so lumens and lux readings are meaningful only when you know how they are measured.

The average of full-field brightness is a good way to state a display's brightness. To determine full-white-field brightness, a full-white field is projected, and the lumens are measured at the brightest spot. This number gives you a realistic idea of the amount of light actually produced by the display. However, it is not the most accurate measurement.

The nine-zone brightness measurement method used in the original ANSI standard is still the most accurate way to measure display brightness, because it averages in the darkest parts of the projected test pattern image. For this measurement, you project a full-white pattern onto the screen and focus at maximum aperture. For projectors with zoom lenses, set them to their midpoint before measuring. Divide the screen into nine equal areas, as shown in Figure 5-28.

Take an incidental light-meter reading in the center of each area. Add the readings together. Divide the total by 9 to get the average footcandle reading. Then use the following formula to get the ANSI lumen rating:

$$L = SH \times SW \times AF$$

where:

    L = Lumens
    SH = Screen height in feet
    SW = Screen width in feet
    AF = Average footcandles

Brightness and Contrast Adjustments

Once the monitor or projector image is centered and the viewing environment is set, final adjustments can be made to perfect image brightness and contrast. Setting

Figure 5-28
Nine-zone tool
for measuring
lumens

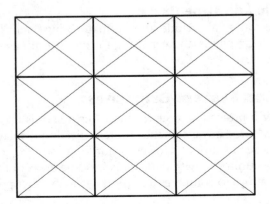

brightness and contrast properly is important so that all detail in an image is represented accurately.

The brightness control is sometimes called the *black-level control* because it determines the black level of the image.

- If the brightness level is too high, the intended black areas of the image will appear gray.
- If the brightness level is too low, then areas with subtle differences in levels of black will all be represented as one shade.

The contrast control sets the range of difference between the lightest and darkest areas of the picture.

- If the contrast control is too low, the white details will appear gray.
- If the contrast control is set too high, then areas with subtle differences in levels of white will all be represented as one shade.

Many technicians use test patterns that incorporate a grayscale divided into bars or blocks. A grayscale test pattern permits you to verify that the complete range of intensities will be available on the screen. Each area of the test pattern should be distinguishable from other areas. Grayscale is used because equal amounts of red, green, and blue light form its levels. Adjusting light levels at grayscale ensures that all three colors are represented equally.

A grayscale test pattern displays the broadest range of intensities between black and white on the screen. Figure 5-29 shows an example of a 4-bit grayscale chart.

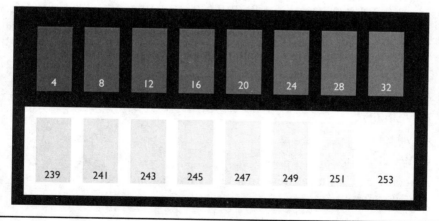

Figure 5-29 A 4-bit grayscale chart

Here is one method to set brightness and contrast accurately:

1. Start with the projector's brightness and contrast controls, often called the picture controls, set to the midrange.

2. While looking at the white levels of the image, adjust the contrast control until the light-gray blocks are visible, stopping before areas show detail loss and become all white, indicating the contrast control is set too high.

3. Set the brightness control until the dark-gray blocks are distinguishable, stopping before the true black value becomes gray.

Contrast and brightness adjustments typically interact with each other, so you may adjust each more than once. This is not a scientific adjustment, but you want to make sure that black is truly black and white is truly white.

Chapter Review

Video is a major component of modern AV systems. To incorporate video correctly, it helps to understand the fundamentals, from how people see things to the basics of a video signal. With this foundation, you can learn how best to present video content in an AV system.

Review Questions

The following review questions are not CTS exam questions, nor are they CTS practice exam questions. Material covered in Part II of this book provides foundational knowledge of the technology behind AV systems, but it does not map directly to the domains/ tasks covered on the CTS exam. These questions may resemble questions that could appear on the CTS exam, but may also cover material the exam does not. They are included here to help reinforce what you've learned in this chapter. For an official CTS practice exam, see the accompanying CD.

1. Light waves are categorized by their _____.

   A. Spectrum

   B. Visibility

   C. Vectors

   D. Wavelength

2. Generally, a _____ measurement is taken at a task area like a video screen.

   A. Lumen

   B. LED

   C. Footcandle

   D. Lux

3. Perceived illumination decreases by _____ when the distance from a light source is doubled.

    A. 95 percent

    B. 75 percent

    C. 50 percent

    D. 25 percent

4. The amount of ambient light in a displayed environment _____.

    A. Negatively affects the quality of the displayed image

    B. Does not affect the quality of the displayed image

    C. Improves the quality of the displayed image

    D. Complements the quality of the displayed image

5. In the component video signal, the _____ signal is combined with the _____ information.

    A. Sync; R–Y

    B. Sync; Y

    C. Chroma; Y

    D. Chroma; B–Y

6. In composite video, which of the following "shares" the luminance space?

    A. S-Video signal

    B. Digital signal

    C. Subcarrier channel

    D. Chrominance

7. Four factors related to primary optics that influence the quality of the projected image are _____.

    A. Reflection, curvature, spherical aberration, and dispersion of field

    B. Reflection, dispersion, spherical aberration, and curvature of field

    C. Refraction, presentation, spherical aberration, and curvature of field

    D. Refraction, dispersion, spherical aberration, and curvature of field

8. Three major formats of digital connections are _____.

    A. DVI, HDVI, and DisplayPort

    B. DMI, HDVI, and DisplayPort

    C. DVI, HDMI, and DisplayPort

    D. DMI, HDMI, and DisplayPort

9. To measure the bandwidth of an image, you will need a _____.

   A. Spectrum analyzer

   B. Frequency analyzer

   C. Output monitor

   D. Bandwidth monitor

10. Rear-screen display applications are considered _____.

    A. Remissive

    B. Emissive

    C. Transmissive

    D. Reflective

11. LCDs first pass light through a ____, which blocks certain light waves.

    A. Transistor mask

    B. Polarizer

    C. Pixel grid

    D. Resistor network

12. When selecting a display type, what should be your first step?

    A. Decide if you should use an analog or digital display.

    B. Figure out what type of mount you will need.

    C. Determine the distance of the farthest viewer.

    D. Determine what you want to display.

13. The _____ the gain number of a screen, the _____ the image.

    A. Higher; brighter

    B. Lower; sharper

    C. Higher; softer

    D. Lower; brighter

14. What quality would the image have if the phase setting of a display needs adjusting?

    A. Blurry only at the image center

    B. Oversaturation of red and green

    C. Edge of image in sharp focus

    D. Fuzzy details

15. The formula for an ANSI lumen rating is _____.

    A. $L = SH \times SW \times AF$

    B. $L = SH \times SW \times AH$

    C. $L = SH \times SA \times SF$

    D. $L = SH \times SW \times AS$

Answers

1. **D.** Light waves are categorized by their wavelength.

2. **D.** Generally, a lux measurement is taken at a task area like a video screen.

3. **B.** Perceived illumination decreases by 75 percent when the distance from a light source is doubled.

4. **A.** The amount of ambient light in a displayed environment negatively affects the quality of the displayed image.

5. **B.** In the component video signal, the sync signal is combined with the Y information.

6. **D.** In composite video, chrominance "shares" the luminance space.

7. **D.** The four factors related to primary optics that influence the quality of the projected image are refraction, dispersion, spherical aberration, and curvature of field.

8. **C.** Three major formats of digital connections are DVI, HDMI, and DisplayPort.

9. **A.** To measure the bandwidth of an image, you will need a spectrum analyzer.

10. **C.** Rear-screen display applications are considered transmissive.

11. **B.** LCDs first pass light through a polarizer, which blocks certain light waves.

12. **D.** When selecting a display type, your first step should be to determine what you want to display.

13. **A.** The higher the gain number of a screen, the brighter the image.

14. **D.** If the phase setting of a display needs adjusting, the details of the image will be fuzzy.

15. **A.** The formula for an ANSI lumen rating is $L = SH \times SW \times AF$.

CHAPTER 6

Networks

In this chapter, you will learn about
- The types and topologies of networks
- Ethernet and Internet Protocol
- The OSI Reference Model
- IP addressing, subnetting, and the Domain Name System
- Network hardware and security

One of the most important enhancements to the CTS exam is its coverage of IT networking knowledge. This is because of the convergence between AV technologies and IT. Today, more AV systems use IT networks, therefore CTS-certified professionals need to be well versed in IT and AV, whether to design and install networked AV systems or to coordinate with IT staff.

In layman's terms, the exam covers things that a CTS holder should be able to *do*, as well as things the holder should *know*. In the process of the 2012 JTA, it was determined that CTS-certified professionals must know significantly more about networking and network security. As a result, throughout this edition of the *CTS Exam Guide*, you will see fresh references to IT-related issues. Specifically, Chapters 16 and 19 include information about networks and networked AV systems that InfoComm's Certification Committee volunteers have determined a CTS holder should know.

This chapter covers fundamentals of networking and network security issues. If your career path to date has included formal training or work designing and installing networks, this chapter may be for review. If not, the information here will help you earn your CTS certification and enjoy a successful career in pro AV.

Types of Networks

In general terms, a *network* is a group or system of things that are interconnected. Examples of networks are all around us. Networks can be informally connected, as with your friends and colleagues, or formally connected, as with networks of television stations or computers.

In the AV and IT communications industries, a network is a group of devices connected in a manner that allows communication among them. Networks are categorized by the area they cover, or their scale.

To understand and talk to others about networks, you need to be able to use the basic vocabulary of the IT networking professional. We'll address those terms as we discuss network types in this section.

The following are types of area networks:

- **Local area network (LAN)**   A LAN connects devices within a confined geographical area, such as a building or living complex. A LAN is typically used to connect network devices over a short distance and generally owned or controlled by the end user.

- **Wide area network (WAN)**   A WAN covers a wide geographic area, such as a state or country. The Internet is the largest WAN—it covers the earth. LANs are connected to WANs through routers, which are discussed later in this chapter.

- **Wireless local area network (WLAN)**   A WLAN is a wireless LAN.

- **Metropolitan area network (MAN)**   A MAN is a communications network that covers a geographic area, such as a suburb or city.

- **Storage area network (SAN)**   A SAN is a high-speed, special-purpose network (or subnetwork) that interconnects different kinds of data-storage devices.

- **Virtual local area network (VLAN)**   A VLAN is created when network devices on separate LAN segments are joined together to form a logical group, thereby spanning the logical LANs to which they are connected.

- **Personal area network (PAN)**   A PAN is a limited-range wireless network that serves a single person or small workgroup.

So, a network can be as small as two connected computers or as large as the Internet, which spans the world. The LAN, WAN, and WLAN are the most common types. A LAN can be part of a WAN, or there can be multiple LANs on a WAN.

Network Topology

The way that the physical connections are made to accomplish communication among devices in area networks is called *network topology*. Physical network topology describes the general shape of the network when it is connected—it's the configuration of the parts.

The network topology determines how each individual device on a network will be arranged and connected. Logical network topology describes the way the information flows through the network.

Here, we'll discuss three common network topologies: star, bus, and ring.

Star Topology

In a star topology, all of the devices are connected to a central point that may be a hub, switch, or router, as illustrated in Figure 6-1. Hubs, switches, and routers (discussed later in this chapter) have multiple connection capabilities. The star topology is common in home networks.

Figure 6-1

A star network
topology

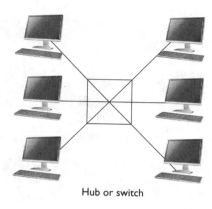

Hub or switch

If a printer is added to the hub, all of the other computers on the star network will be able to access it through the hub.

Bus Topology

In a bus network, all devices are connected to each other through a single cable, as illustrated in Figure 6-2. This type of network may be used for different types of control systems, such as lighting controls. As with other network topologies, each device is identified by a unique number. This number is set either in software or by a mechanical switch.

Ring Topology

Similar to a bus network, a ring network connects devices one after another in sequence. The biggest difference between these technologies is that with a ring topology, the ends of the network are then connected together to form a continuous loop, as illustrated in Figure 6-3. Information can flow around the loop or ring, and if there is a break in the line, the information will still flow, because redundant paths can be created.

Telecommunication utilities use ring networks for phone services. WAN Internet services use a mesh ring topology to create redundancy.

As you can see, communication between devices can be complicated. Making these networks work as intended would be especially difficult if standards were not developed. One of these standards is Ethernet, as discussed in the next section.

Figure 6-2

A bus network
topology

Figure 6-3
A ring network
topology

Ethernet

Ethernet is the most commonly used method of transferring data on a LAN. It is an Institute of Electrical and Electronics Engineers (IEEE) standard that governs how computers exchange information over the network medium.

The Ethernet standard specifies the physical transmission media (cable) and how signal information is handled (frames or packets) on a network. The standard is called IEEE 802, and it is revised periodically to include improved technology and applications. Such revisions are indicated as 802.*x*.*x*, where the *x*s represent the revision numbers.

Variations of Ethernet operate in all three LAN topologies and use a variety of network media (cabling). The topology and the cable medium determine the speed and distance of data transfer.

Ethernet is a packet-based system. Large files or streams of data are broken into smaller chunks called *packets*. These packets are 1,500 bits or smaller. Due to the nature of Ethernet, it is possible for packets to take different paths and arrive at the destination at different times or out of order. The receiving device must keep track of missing packets and either request the missing information again or have the ability to ignore missing packets. Ethernet is designed as a "best-effort" delivery system.

Ethernet Connections

As with other communication devices, Ethernet cabling must be connected to the device by attaching some type of connector. A common Ethernet connection is made with the eight position, eight conductor (8P8C) modular connector, as shown in Figure 6-4, which is attached, or terminated, to the cabling. An 8P8C connector is commonly referred to as an RJ-45 connector. The 8P8C is relatively inexpensive and easy to install in the field.

There are two wiring formats within the IEEE 802 standard: T568-A and T568-B. The AV installation technician needs to ask the IT manager which format is used within the facility.

Figure 6-4

An 8P8C modular jack (RJ-45) termination

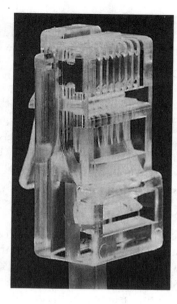

If a cable is terminated with a T568-A on one end and T568-B on the other, the cable is known as a *crossover cable*. A crossover cable allows two devices, such as two computers, to connect and share information without the use of a switcher or router (which normally does the crossover electronically).

Fiber-Optic Connections

The use of fiber-optic cabling for Ethernet is growing in popularity. Fiber-optic cables work by sending information coded in a beam of light through a glass or plastic pipe.

A fiber-optic cable is made up of strands of optical fiber that are as thin as human hair. Consider this: The speed of light is approximately 300,000 kilometers per second (or 186,000 miles). Fiber can reach LAN at speeds of 10 and 40 gigabits per second (Gbps). Fiber-optic speeds can be much greater than LANs in larger communication systems, but these are not typically found in the office environment.

There are two types of fiber-optic cable:

- **Single-mode**   Single-mode means that the transmitted light travels a single light path, You can typically identify single-mode fiber-optic cable by its yellow outer protective jacket.

- **Multimode**   Multimode means it travels multiple light paths. Multimode fiber-optic cable is identified by an orange jacket.

The signals that travel along single-mode fiber can reach farther distances than signals on multimode due to the construction of the cable.

Several types of terminations are used for fiber-optic cable. Two of the most commonly used in office systems are SC and ST. The SC connector inserts and clicks into place. The ST connector uses an insert-and-twist connection, much like a BNC connector.

Wireless Connections

In addition to transmitting signals over a cabled networked system, you can transmit signals wirelessly, with special devices on each end of the wireless signal to translate it back into wired Ethernet. The wireless connection, known as Wi-Fi, is defined by the IEEE 802.11 standard and has been revised several times to keep up with the growing demand for wireless communication. Currently, IEEE developers are working on 802.11 standards that would provide even greater speed (throughput)—theoretically, up to 7 Gbps.

The network device that handles the wireless connection is called an *access point*. For two devices to establish a wireless connection, the radio frequency (RF) signals for transmission must have a minimum level of signal strength. To maintain proper signal strength, the devices should be as close together as reasonably possible. Even though some wireless systems claim distance limitation of 50 meters (about 150 feet), the real distance limitations have more to do with the transmitting antenna, the receiving antenna, the obstructions between the two, and the power output of the transmitters on both ends.

In summary, the speed of a Wi-Fi connection depends on the RF signal strength and the revision of 802.11 with which you connect. As signal strength weakens, the speed of the connection slows. Another factor affecting connection speed is the number of users accessing the wireless devices.

Table 6-1 lists the specifications of the 802.11 standard revisions that have been incorporated into networked products.

 NOTE In 2012, the IEEE published a standard known as 802.11-2012, which aggregates all previous amendments to the 802.11 wireless standard. As of this writing, products using this new standard have not been introduced.

Table 6-1 802.11 Standard Revisions	Revision	Release Date	Frequency Band	Typical Throughput	Maximum Throughput
	802.11a	October 1999	5 GHz	27 Mbit/s	54 Mbit/s
	802.11b	October 1999	2.4 GHz	~5 Mbit/s	11 Mbit/s
	802.11g	June 2003	2.4 GHz	~22 Mbit/s	54 Mbit/s
	802.11n	September 2009	5 GHz and/or 2.4 GHz	~144 Mbit/s	600 Mbit/s

The OSI Reference Model

No matter what kind of connection you are using for your network, the framework that enables communication between devices must be standard, and you must have a common language to describe the process. The Open Systems Interconnection (OSI) model was developed by the ISO to standardize communication between devices all over the world, including those on the Internet.

The OSI model separates communication connectivity into seven different layers, each with a specific duty, as shown in Figure 6-5. This allows for a variety of connection types, as well as the development of specific hardware and software to optimize the network. The layers are processed in a specific, sequential manner.

As data is sent out, each layer adds some information to keep track of the file or data stream as it passes through a network. This system allows packets to be tracked and recombined at the receiving end to recover the file or stream.

When an application sends data, the data moves down the layers until it finally reaches the network medium at the Physical layer (layer 1). As the data is processed by protocols at each layer, it is divided into smaller units that may be easily transported and reassembled at the other end.

Information starts at the uppermost layers (layers 7, 6, and 5) and continues to the lower layers. The large files, or streams, in the upper layers are broken down into smaller chunks of information, called *segments*, in the Transport layer (layer 4). At the Network layer (layer 3), destination and control information is added to the segment, forming a packet. These packets are sent across the network. The destination and control information provides for error checking. At the Data Link layer (layer 2), packets are encapsulated into frames that contain information about the physical destination, such as the MAC address. The final data frame (in this case for Ethernet) is put on the network medium by the Physical layer (layer 1).

Figure 6-5

The seven layers
of the OSI model

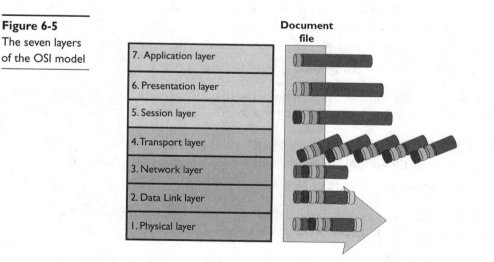

Document
file

7. Application layer

6. Presentation layer

5. Session layer

4. Transport layer

3. Network layer

2. Data Link layer

1. Physical layer

The following is a brief summary of what happens in each layer, plus how each layer relates to AV/IT systems. AV professionals will want to pay particular attention to layers 7 and 1.

Layer 1, the Physical layer, covers cabling and other connection mediums, such as patchbays and more. The Physical layer does the following:

- Defines the relationship between the device and a physical means of sending data over network devices (such as a cable)
- Defines optical, electrical, and mechanical characteristics

Layer 2, the Data Link layer, includes the Ethernet standard and unique hardware addresses. The Data Link layer does the following:

- Defines procedures for operating the communication links
- Encapsulates data into Ethernet frames
- Detects and corrects packet-transmission errors

Internet Protocol (IP) comes into play at layer 3, the Network layer. The Network layer does the following:

- Determines how data is transferred between network devices
- Routes packets according to unique network device addresses
- May provide flow and congestion control to prevent network resource depletion

Layer 4, the Transport layer, governs the transfer of data. The Transport layer does the following:

- Provides reliable and sequential packet delivery through error-recovery and flow-control mechanisms
- Provides connection-oriented or connectionless packet delivery
- Provides flow and congestion control to prevent network resource depletion

Layer 5, the Session layer, is the first of the upper layers and concerns the data itself, ensuring data passes properly through the network. The Session layer does the following:

- Manages user sessions and dialogues
- Controls the establishment and termination of connections between users
- Reports upper-layer errors

In a networked AV system, layer 6, the Presentation layer, unpackages data for use by the Application layer. It also does the following:

- Masks data format differences between dissimilar systems so they can communicate

- Specifies an architecture-independent data-transfer format
- Encodes and decodes data, encrypts and decrypts data, and compresses and decompresses data

Layer 7, the Application layer, presents data to the application software for use. The Application layer also does the following:

- Defines an interface to user processes for communication and data transfer in a network
- Provides standardized services, such as file and job transfer, and operations

Network Interface Cards and MAC Addresses

To connect a device to a network, an interface of some kind is required. This interface is called a *network interface card* (NIC). At one time, most devices had a separate card or adapter. While separate cards are still common, in some cases they have been replaced by integrating the NIC into the device's main circuitry. Even wireless connections are considered NICs because they interface between RF transmissions and traditional wired cabling.

A *media access control* (MAC) *address* is the actual hardware address, or number, of a NIC device. Each device has a globally unique MAC address to identify its connection on a network. It is part of the IEEE 802 standard.

The MAC address uses a 48-bit (2^{48}) number that consists of six groups of two hexadecimal numbers, separated by a hyphen or colon. Here's an example:

01-23-45-67-89-ab or, 01:23:45:67:89:ab

The numbers come from a table of numbers assigned to manufacturers of Ethernet-capable devices. The first part of the number indicates the manufacturer, and the second part is a serial number for the product or circuit component. Due to the nature of a MAC address, there can be only one device with that number on any given network.

The MAC address is one of the lowest levels of communication on the network. It is one of the first pieces of the network communication structure. Because each address is unique, switches and routers can store MAC address locations in order to efficiently send data traffic throughout the network to a specific device.

Internet Protocol Addressing

How do you get a network to acknowledge a networked device? Your device has a means of connecting (the NIC)—via cable, fiber, or wirelessly—and a unique number to identify it (MAC), so how do you connect it to the network? This is the role of the *IP address*. The IP address defines the exact device and its location on a network.

When describing a network, there are distinctions between names, addresses, and routes. A name indicates what we seek. An address indicates where it is. A route indicates how to get there. IP deals primarily with addresses. It is on layer 3 of the OSI model.

Two versions of IP are used today: IP version 4 (IPv4) and IP version 6 (IPv6). Since 1982, IPv4 has been the standard IP addressing scheme, and it may still be for some time. However, IPv6 offers improvements over IPv4, including increasing the total number of devices that can have unique IP addresses, better routing capabilities, and greater efficiency in transferring data. Modern operating systems include support for IPv6, and certain markets, such as the U.S. government, have programs in place to encourage (or mandate) the switch from IPv4 to IPv6.

Regardless of the IP version you use, there are two types of IP addresses: static and dynamic. A dynamic IP address is a temporary address automatically assigned to a network device upon connection to the network. A static address is one that is assigned manually and stays with the device.

We'll cover the addresses for IPv4 and IPv6, subnetting, and static and dynamic addresses in this section.

IPv4

Every networked device needs an IP address in order to transfer data over an IP-based network. The IPv4 format requires a number made up of four 8-bit "chunks," called *octets*, such as the following:

192.168.1.25

This number is actually a 32-digit binary number that looks like this:

11000000.10101000.00000001.00011001

The IPv4 address (number) given to a network device has a structure that includes a network prefix and a network host number. The left part of the IP address is the network prefix, and the right part is the network host.

IPv6

IPv6 has a similar structure to IPv4. However, instead of using a binary numbering system consisting of four groups of octets, IPv6 uses eight groups of four hexadecimal numbers. This change in number structure raises the number of unique network addresses:

- IPv4 has 2^{32}, or 4,294,967,296 potential IP addresses.
- IPv6 has 2^{128}, or 340,282,366,920,938,000,000,000,000,000,000,000,000 potential addresses.

As you can see, IPv6 allows for many, many more addresses, reducing the probability that we would run out of potential IP addresses.

In practice, an IPv6 address looks something like this:

FEDC:BA98:7654:3210:FEDC:BA98:7654:3210

This format employs hexadecimal numbering, with letters representing the digits 10 (A) through 15 (F). To help ensure that two devices won't receive the same host or node address, IPv6 can use the MAC address as part of its numbering scheme. Because a MAC address uniquely identifies an Ethernet connection, using this number should also uniquely identify the IP address. This provides an easy way for systems to keep track of addresses, though the method isn't recommended in all cases.

Subnet Masks

A subnet allows IP networks to be logically subdivided, increasing performance and enhancing network security. Subnetting is a common practice, but it can actually cause performance problems under heavy traffic loads unless IT applies *subnet masks* to the IP addresses on its network. In short, in addition to an IP address, a networked AV device also needs a subnet mask, or it will not be recognized properly on a network.

As you'll recall, an IP address has two components: the network address and the host address. A subnet mask separates the IP address into the network and host addresses. It "masks" or hides the network part of the address and leaves only the host number to identify the device. A subnet mask accompanies an IP address, and the two values work together to identify a device on a network. A subnet mask, with its four-octet form, appears similar to an IP address when written out. The difference comes to light when you decode the octets into binary.

A subnet mask of 255.255.255.0 is common for so-called Class C IP addresses and indicates the subnet can include 254 devices. A subnet mask of 255.0.0.0 indicates a network that can handle more than 16 million devices.

Static IP Addressing

Some networked devices require an IP address that will not change, so that users or other devices can easily and always find it on the network. Examples are a videoconferencing system and certain IP-controlled AV equipment. Such IP addresses are called *static addresses*. In some cases, you may have no choice but to assign a static IP address to a device.

Commonly, when configuring a networked device for operation, you will find a check box to "obtain IP address automatically" (or similar wording). When you check this box and the device connects to a network, the network sees the new device. If there is a Dynamic Host Configuration Protocol (DHCP) server on the network, it will automatically assign an address to the device. This is a dynamic IP address, and it can be different every time the device connects (as described in the next section).

If there is no DHCP server on the network, you will need to manually set a static address. Whatever the situation, in order to set a static address, you will need the following information:

- IP address (required)
- Subnet mask (required)
- Device name
- DNS server
- Gateway

Dynamic IP Addressing

When a device connects to the network and the device has the "obtain IP address automatically" option activated, the DHCP service or server will read the MAC address of the device and assign it an IP address. The pool of available IP addresses is based on the subnet size and the number of addresses that already have been allocated.

A DHCP server will allow a device to hold the IP address for only so long; the amount of time is called the *lease time*. After the lease time has expired, the lease will usually be renewed automatically if the device is still connected to the network; otherwise another device connecting to the network can use that same address.

The advantage of using DHCP is that it is rather easy to manage. It takes care of making sure no two devices get the same address, relieving potential conflicts. It allows for more people to connect to the network, as the pool of addresses is continually updated and allocated.

The disadvantage of using DHCP is that you never know what your IP address will be from connection to connection. If you need to reach a certain device by IP address, you must have a high level of confidence that the number will be there all the time, and DHCP will not give you that confidence.

A hybrid approach to DHCP is to reserve a block of addresses for static addresses and dynamic addresses. The pool of addresses for DHCP is reduced by the number of addresses reserved for static devices. To make this happen, an IT manager will need the MAC address of each device that must be statically set. The static (manually assigned) IP address and MAC address are entered into a table. When the device connects to the network and reveals its MAC address, the DHCP server will see that the IP address is reserved for the device and will enable it. The IP address cannot be given to any other device or MAC address.

The Domain Name System

The Domain Name System (DNS) is a hierarchical, distributed database that maps names to data such as IP addresses. A DNS server keeps track of all the equipment on the network and matches the equipment names so they can be located on the network easily or integrated into control and monitoring systems.

When a new device is connected to the network, the DNS server matches the name of the device to its IP address. Now a user can call devices by name instead of number. This is very useful, especially on the Internet. For example, it is easy to remember infocomm.org, but not so easy to remember 66.117.62.20 (go online and type the numbers into your browser's address field).

When DHCP and DNS servers are working together, you may never need to know the IP address of a device; you need only its name. This makes managing a network simpler, as the IP address does not need to be static, and the entire addressing scheme could change without affecting the communication between devices.

Another networking scheme, Windows Internet Name Service (WINS), behaves similarly to DNS, but the two are not exactly the same. Ask the IT manager which system the network is using. You will normally find two DNS or WINS servers per network.

This adds redundancy to the network, providing backup in the event that a server fails or allowing use during maintenance periods.

Network Switches and Routers

When working with networked AV systems, you should also know about the hardware building blocks of the network itself. For example, a network switch connects multiple devices together so they can communicate with other devices that are also connected to the switch. Some routers have switches built into them, but a router does much more than a switch.

Network Switches

As each device is connected to a network, a *switch* collects its MAC address and stores it in memory. When one device wants to communicate with a second device, the switch looks up the destination device's location in its memory. If the address is in the switch's memory, the switch will send the information to its destination.

Switches also have the ability to allow connected devices to operate at their maximum speeds. If a 100 Mbps device and a 1 Gbps device are connected to a 1 Gbps switch, both devices can use their maximum speeds. In other words, the 1 Gbps device will not need to slow down to accommodate the slower device.

There are two basic types of switches:

- **Unmanaged**   An unmanaged switch is one you simply plug in and connect devices—that's it. There are no adjustments. It just works.
- **Managed**   Managed switches give the IT manager the ability to adjust port speeds, set up VLANs, set up quality of service (QoS) settings, monitor traffic, and more. Managed switches are what you will find in most company networks.

Routers

A *router* works at the OSI layers above a switch's layer, which are the Network and Transport layers. A router reads the destination IP address of the packets being sent, so it can be used to send the packets to specific locations or via predefined paths on the network. The IT manager can use a router to change how a network works. Routers also allow for redundancy in a network.

Routers work together in a hierarchy, as one router can control the behavior of routers beneath it.

Gateways

A *gateway* is usually the topmost router in the hierarchy of routers. This top-level router connects to an outside network. All traffic must eventually travel through this router to get outside the local network.

A gateway has special duties. It passes traffic to the routers below, which look to the gateway to find names (DNS addresses) that are not found on the local network. Sometimes a videoconferencing device won't operate if it doesn't have a gateway address.

Bridges

A *bridge* connects two different types of networks together. It translates one network protocol to another protocol. One basic example of a bridge is a broadband modem. The modem converts, or bridges, the Ethernet protocol to a cable television protocol.

Network Security

In a digital world, it's hard to go a day without hearing or reading about network security issues. Often, it has to do with a network security breach, such as a hacker stealing financial information from a bank's computers, or some shadowy organization attacking a government website so that it can't function. To the extent your AV systems must run on IT networks, you, as an AV professional, must have a working knowledge of network security.

The purpose of network security is to prevent unauthorized users from accessing a LAN. It is implemented at many points in the network, but most defenses operate at the boundaries of a system. The first of these boundaries is the physical network interface. Think of the wall port as an extension of the network switch to the wall. One method of control is to limit the MAC addresses that may use a port. The port can be configured so only certain devices can access the network from that point.

AV professionals must understand the network architecture in order to identify the boundaries and possible points of attack. That said, AV professionals who are designing or installing networked AV systems will usually not be responsible for actual network security. In order to make the AV system functional, they will need to coordinate with a client's IT department or network security personnel so that the appropriate AV traffic can traverse firewalls and other security measures.

Network Access Control

Network access control (NAC) describes a group of technologies that secure a network. They are chosen to align with the security policy of an organization. In a nutshell, when an organization employs an NAC solution, any device that attempts to connect to that network must comply with the organization's security policy, which may dictate everything from how the device is configured to the software it runs. And once a device connects to a network, NAC can control what that device can access on the network, based on user actions and identities.

Your AV system may need to interface with the client's NAC technologies to work properly. Some AV devices have trouble with NAC because they are not fully functional computers.

Access Control Lists

When a network is broken into subnets or broadcast domains, routers may provide additional security. Routers may contain an access control list (ACL). An ACL controls what travels through a router based on the type of data traffic, source, and/or destination. If your AV system will require special access rights, be sure that IT creates an ACL for the system and adds the appropriate end users.

802.1x

One component of NAC may be the use of 802.1x, or port-based NAC. 802.1x is an IEEE standard that requires authentication before a device may connect to the network through a certain port. Prior to accessing the organization's network, the user must be verified by an authentication server. The process is analogous to an airport security agent verifying your ID and ticket before allowing you to enter the terminal.

Firewalls

Firewalls are among the most common network security technologies. They can be software programs (chances are your personal computer runs a software firewall) or dedicated hardware devices. Their primary job is to control incoming and outgoing network traffic by analyzing packets and determining whether they should be allowed through, based on a set of rules.

Most firewalls accomplish their goal through one or more of the following techniques:

- **Packet filtering**   This is a firewall technique that uses rules to determine whether a data packet will be allowed to pass through a firewall. Rules are configured by the network administrator and implemented based on the protocol header of each packet.

- **Network address translation (NAT)**   NAT is any method of altering IP address information in IP packet headers as the packet traverses a routing device. Firewalls often use NAT to hide the true addresses of protected network assets.

- **Port address translation (PAT)**   PAT is a method of NAT whereby devices with private, unregistered IP addresses can access the Internet through a device with a single registered IP address. PAT conserves address space, which is a concern in IPv4 implementations (though it's also used for IPv6 networks). PAT hides the original source of the data. From outside the network, all data appears to originate from a NAT server. Any data that arrives at the NAT server without being requested by a client has nowhere to go; it has the address of the building, but not the apartment number. Using PAT, all unrequested data is blocked by the firewall, and a malicious intruder can't trace the data's path beyond the edge of the network.

- **Port forwarding**   This method combines PAT and packet filtering. The firewall inspects the packet based on packet-filtering rules. It is also configured to translate certain ports to private addresses on the network.

Navigating Firewalls

One of the most important factors in whether AV system traffic reaches its intended destination is the configuration of network firewalls. The firewall is the ultimate arbiter of what kinds of traffic can access a network through what ports. Because the firewall must protect the entire network, firewall configuration is a crucial area of coordination between AV design and overall network management.

An AV designer must list the ports and protocols that need to pass through a firewall as specifically and narrowly as possible so that the network managers can support the system without endangering the enterprise network. If you are anticipating a specific data stream for an AV application, you need to ensure its port will be open.

You should be able to determine which ports and protocols are required for a given AV application from the manufacturers of the devices used. The manufacturer specifications of any networked device should list the protocols and port ranges the device will use.

Chapter Review

Today, more than ever before, it is important for CTS-certified AV professionals to have an in-depth knowledge of IT networking. The most current CTS exam incorporates new network-related questions because AV systems increasingly interface with or communicate directly over the same types of networks that organizations use to connect their computers. It is also important for AV professionals to understand the basics of network technology in order to collaborate with IT departments on the design and installation of networked AV systems.

Review Questions

The following review questions are not CTS exam questions, nor are they CTS practice exam questions. Material covered in Part II provides foundational knowledge of the technology behind AV systems, but it does not map directly to the domains/tasks covered on the CTS exam. These questions may resemble questions that could appear on the CTS exam, but may also cover material the exam does not. They are included here to help reinforce what you've learned in this chapter. For an official CTS practice exam, see the accompanying CD.

1. Which of the following networks connects devices in sequence along a linear path?
   A. Bus
   B. Star
   C. Application
   D. Ring

2. What type of network uses packets?
   A. IEEE
   B. Bus

C. Ring

D. Ethernet

3. What happens to the connection speed in a Wi-Fi connection if the signal strength declines?

A. Speeds up

B. Stops

C. Slows down

D. Remains constant

4. Which of the following is a type of fiber-optic cable identified by its yellow outer jacket?

A. ST

B. SC

C. Multimode

D. Single-mode

5. The _____ model is a guide that assists with conforming network communications and their processes to standards.

A. Network interface

B. OSI reference

C. Informal data link

D. IP

6. In the OSI model, cabling and patchbays are elements of _____.

A. Layer 2, the Data Link layer

B. Layer 4, the Transport layer

C. Layer 1, the Physical layer

D. Layer 3, the Network layer

7. A _____ address is unique to every device and identifies a network's equipment.

A. Transfer mode

B. Baseband

C. Digital subscriber line

D. MAC

8. IP deals with which of the following on a network?

A. Addresses

B. Names

C. Routes

D. Versions

9. An IPv6 address uses ____ groups of four hexadecimal numbers.

   A. Three

   B. Eight

   C. Six

   D. One

10. Subnet masks can indicate how many ___ are allowed on the network.

    A. Computers

    B. Gateways

    C. Devices

    D. Printers

11. What is required to set an IP address manually on a network?

    A. IP address and device name

    B. Subnet mask and gateway

    C. Subnet mask and DNS server

    D. IP address and subnet mask

12. Which type of server automatically assigns an IP address to the MAC address during the device's connection to a network?

    A. Gateway

    B. Virtual private network

    C. DNS

    D. DHCP

13. Which of the following switches just needs to be plugged in and connected to devices?

    A. Unmanaged

    B. LAN

    C. Addressing

    D. Managed

14. A _____ sends packets to different locations on a network and connects to outside networks.

    A. Switch

    B. Gateway

    C. Bridge

    D. Router

15. A _____ controls incoming and outgoing network traffic and determines what will be allowed through based on a set of security rules.

    A. Switch

    B. Gateway

    C. Firewall

    D. Router

Answers

1. **A.** A bus network connects devices in sequence along a linear path.

2. **D.** An Ethernet network uses packets.

3. **C.** The connection speed in a Wi-Fi connection slows down if the signal strength declines.

4. **D.** Single-mode fiber-optic cable can be identified by its yellow outer jacket.

5. **B.** The OSI reference model is a guide that assists with conforming network communications and their processes to standards.

6. **C.** In the OSI model, cabling and patchbays are elements of layer 1, the Physical layer.

7. **D.** A MAC address is unique to every device and identifies a network's equipment.

8. **A.** IP deals with addresses on a network.

9. **B.** An IPv6 address uses eight groups of four hexadecimal numbers.

10. **C.** Subnet masks can indicate how many devices are allowed on the network.

11. **D.** An IP address and subnet mask are required to set an IP address manually on a network.

12. **D.** A DHCP server automatically assigns an IP address to the MAC address during the device's connection to a network.

13. **A.** Unmanaged switches just need to be plugged in and connected to devices.

14. **B.** A gateway sends packets to different locations on a network and connects to outside networks.

15. **C.** A firewall controls incoming and outgoing network traffic and determines what will be allowed through.

Signal-Management Systems

In this chapter, you will learn about
- Wiring and cable for AV systems
- Ensuring signal integrity and distribution amplifiers
- Rack building

Think of a signal-management system as a number of individual elements that work together to achieve a goal. The individual elements include signal-management devices, cables, wires, and connectors. These elements work together to achieve effective signal transfer. If you properly assemble and configure the working parts of a signal-management system, you can preserve the quality of the signal throughout the system.

A *signal* is the desired information sent through an AV system. Some examples are video, audio, and control signals. Undesirable elements introduced into the system are considered noise and interference.

In most cases, signals are transported through an AV system by wires and cables. Each cable in a system has a specific purpose and physical destination. A coaxial cable can carry a single video signal. Other types of cables can carry multiple signals. A cable designed for use with a video camera may contain multiple wires for video, power, and other functions.

Signal types can be grouped together so that their path through the system is easier to trace. This path is referred to as *signal flow*. There are separate flows for video, audio, and control.

In this chapter, you will learn about the components of signal-management systems.

Wire and Cable

The terms *wire* and *cable* are often used interchangeably; however, each has its distinct characteristics.

Wire contains only one conductor. The conductor may either be solid or composed of strands of conductive material. In electronics, a conductor is made of a material that easily conducts an electrical signal, such as copper. Insulation around the conductor is composed of materials that do not efficiently conduct electrical signals. The insulation provides physical and electrical protection to the conductor.

Cable contains multiple insulated wires in a protected bundle, as shown in Figure 7-1. Some types of cables contain a shield that adds protection to the integrity of the signal carried by the conductors. A shield helps prevent noise from mixing with the signal. A jacket covers the internal components of a cable and provides protection from environmental factors. Here, you'll learn more about each of these elements.

Conductors

A *conductor* is a material that allows current and voltage to pass through it continuously. Signals are made up of current and voltage. Metals are excellent conductors because they are inexpensive and easy to work with. The conductor in a wire carries the signal by conducting current and voltage between a source, such as a battery, and a load, like a light bulb.

Conductors may be classified by the following:

- Size (gauge or circular mils)
- Construction
- Conductive material

A *solid conductor* is a single conductor and costs less than a stranded conductor. Solid conductor cable is assigned bend-radius limitations by the manufacturer, based on specific standards. The bend radius dictates how much you can bend the cable before it deforms and alters its electrical properties. This type of cable is relatively inflexible, but it can withstand more strain than a stranded conductor.

A *stranded conductor* uses multiple smaller, solid conductors that are wound or braided together, as shown in Figure 7-2. The finer the strands, the more flexible the conductor, which enables easier handling and installation.

Stranded conductors are more flexible and easier to handle than solid conductors, but they are more expensive. They are also slightly larger compared to solid conductors at a similar gauge.

Figure 7-1
A Category 5 cable with four unshielded twisted-pairs of wires

Figure 7-2

A stranded
conductor

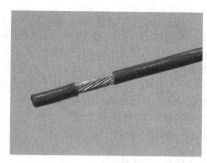

Stranded conductors are used where flexibility is required. One example is a live-event environment, where the cable needs to be flexed and moved around frequently. In a situation like this, the flex life of a cable becomes an important consideration. *Flex life* is a general term that describes how long a wire may last before it physically fails to conduct signals properly.

Insulation

The purpose of insulation, shown in Figure 7-3, is to prevent physical contact between multiple conductors, and to avoid voltage and current interactions between different conductors. If bare wires touch, they can create short circuits that prevent signal transmission. What's more, bare wires can injure and/or cause equipment to fail. To prevent such problems, insulation, which is a highly resistive material, surrounds the conductors.

In general, insulators are made of materials such as PVC or Teflon. Different types of insulation can impact the performance of wire. The performance quality of insulation materials can be measured in terms of *dielectric constant*.

Dielectric constant describes the ability of a material between two conductors to store an electrical charge. Dielectric strength is determined by the material's type and thickness, and reflects the amount of voltage the insulation can stand before it breaks down.

Temperature has an opposite impact on an insulator than it has on a conductor. For insulation, higher temperatures result in lower resistance.

Figure 7-3

Two conductors
with insulation

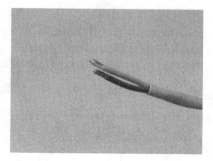

Shields

Shielding isolates and protects a signal from sources of electromagnetic (EM) and radio frequency (RF) interference. Cables may or may not have a shield.

Shields (sometimes called *screens*) can be implemented in a variety of ways. They can be used to provide overall coverage around a single insulated conductor or around individual insulated conductors in a multiconductor cable. A single shield can also be used to surround multiple insulated conductors.

There are three basic types of shielding:

- **Foil shield** This type of shielding uses a thin sheet of aluminum wrapped around an insulated conductor or conductors. For termination purposes, a bare-stranded conductor in continuous contact with the foil shield is used. The bare-stranded conductor is often called a *drain* or *ground wire*.

- **Braided shield** This type uses many tiny, interweaved wires covering an insulated conductor or conductors.

- **Combination shield** This type uses both foil and braid shielding, with the possible addition of a drain wire. Figure 7-4 shows an example of foil and braid shielding.

The shield may also provide a path for return current originating from sources of interference.

The main considerations in selecting a shield are coverage, flexibility, and frequency range. Coverage is expressed as a percentage and tells how much of the inner cable will be covered. Flexibility is a subjective measurement, which directly correlates to a cable's flex life. Shielding cannot protect a conductor from all sources of EM and RF interference. Because of this, a shield's effectiveness is narrowed to a certain frequency range.

Jackets

Jackets (also called *sheaths*) provide physical protection to a cable. They surround and protect the inner wire or wires and insulate the conductors from environmental factors. If the cable has a shield, the jacket is placed over the shield; otherwise, it is placed directly over the insulation.

Figure 7-4
A coaxial cable showing foil and braid shielding

Jackets are selected for their strength, integrity, abrasion resistance, and overall protection. Additional considerations are the environmental factors that can deteriorate the cable, color (for aesthetic or labeling concerns), flexibility, flammability, and electrical code requirements.

Most jacket materials are the same ones used for insulation. The choice is dictated by the application. For example, polyethylene is used outdoors only. It is never used inside because it is so flammable. Teflon is designed to withstand high heat, not support a fire, and produce very little smoke. It is, however, hard to work with and more expensive than other materials. Newer jacket materials that are less expensive and easier to work with have been developed; in some cases, you can even use them without conduit in certain spaces.

Cable Types

You should be familiar with three basic cable types: coaxial (coax), twisted-pair, and fiber-optic.

Coax Cable

Coax cable contains a single center conductor that is surrounded by a dielectric (see Figure 7-4, earlier in the chapter). Around the dielectric can be one or two types of shields and a jacket. The first shield is a solid-foil material; the second shield is made of a metal stranded or braided material. The term *coax* arose from the fact that the braid shielding and center conductor both have the same axis.

Coaxial cable is unbalanced. The single wire running down the center is the signal conductor, and the shield acts as the return. The signal conductor is a different impedance than the shield, which is at ground potential, so they are not balanced. This type of cable is used mainly for video signal connections, but it has other applications as well.

Twisted-Pair Cable

Twisted-pair cable is composed of two wires twisted together to form a single cable. Each wire is individually insulated, and each set of wires is twisted around each other and surrounded by an outer jacket. Twisting, when used in combination with balanced circuitry, helps reject interference.

Fiber-Optic Cable

Fiber-optic cable uses conductors that are made from transparent glass or plastic fibers, as shown in Figure 7-5. Because the conductors are transparent, light or infrared (IR) signals can be transmitted instead of voltage and current.

Fiber-optic cables are classified by what is called *mode*. Single mode and multimode are now more commonly used to transfer information by means of light pulses, rather than electrons.

Figure 7-5
Fiber-optic cable

Connectors

Connectors are terminated onto (attached to) cables and used on equipment to continue the pathway to the electronics inside. The type of connector used on the cable must be compatible with the connector type used on the equipment for a reliable connection. Connectors can be categorized by the way they mate:

- Twisting
- Twist and locking
- Threading
- Screw or snap-down lock
- Contact pressure

The following are some common connector types:

- **XLR**   Used for sending audio and control signals, and for supplying power. The connector is typically found on microphone mixers, amplifiers, and other audio-processing devices.

- **1/4-inch phone**   Used to transport audio signals from one audio-processing device to another. It is typically found on all types of audio-processing devices, such as audio mixers, musical instruments, loudspeakers, and amplifiers.

- **1/8-inch phone**   Used to carry audio or control signals. It is typically found on headphones, amplifiers, mixers, computer loudspeakers, laptop computers, cassette players, and a variety of other places.

- **RCA**   Carries video signals, audio signals, control signals, or power. It is found on video devices, audio devices, control systems, switching systems, and signal converters.

- **F type**   Carries audio and composite video signals. It is commonly found on VCRs, antennas, and televisions.

- **DB9A**   DB9 pin connector assembly, which is very common for control and video signal transport.

- **RJ-45 (8P8C or 8-pin, 8-conductor)**   Used for networking, control, power, and telephone purposes. It is often found on projectors, laptops, system-control devices, and network devices.

- **BNC**   Used to transport different types of signals such as RF, component video, time code, sync, and power. It is often found on wireless equipment that requires an antenna, and on some video equipment and projectors.

- **Speakon**   Commonly used to connect amplifiers to loudspeakers in temporary audio system setups.

- **Captive screw**   Designed with screws that secure the connector to the conductors within the cable.

- **Digital Visual Interface-Digital (DVI-D)**   Transmits only digital signals (24-pin).

- **Digital Visual Interface-Interlaced (DVI-I)**   Similar to DVI-D; however, it has a few more pins that allow analog and digital signals to be transmitted (29-pin).

- **HD15 (VGA)**   Commonly found on computers. The 15-pin assembly is used to transmit analog video signals.

- **DisplayPort and High-Definition Multimedia Interface (HDMI)**   Built on the success of DVI, this type adds audio and offers multiple channels. These interfaces are capable of sending either component or RGB formats. They are used for consumer and professional computer and video displays.

- **Serial Digital Interface (SDI) and High-Definition Serial Digital Interface (HD-SDI)**   Used for digital video. High-quality images for production and videoconference systems are sent using these standards. Coax cable and BNC connectors make using these digital signals very easy. The interface is capable of sending either component or RGB formats.

Signal Integrity

As a signal travels along conductors through an AV system, it is important to protect its quality. You need to arrange cables that carry signals in a way that minimizes interference from sources of noise. The goal is to preserve the signal's integrity until it arrives at its destination.

Just as resistance is a property of electrical energy, so are inductance and capacitance. *Inductance* is the property of a circuit that opposes any change in current. *Capacitance* is the property of a circuit that opposes any change in voltage.

Any two conductors separated by a nonconductive material exhibit some level of capacitance. This capacitance can store an electrical charge in an electrostatic field between the conductors. Capacitance in a cable can deteriorate the desired signal. In addition, each wire in a cable can act as an antenna, absorbing signals from other wires in the cable (crosstalk).

One way to protect signals is to group the cables together by signal types and then physically separate the groups. Cables that carry microphone-level signals should be bundled together only with other microphone-level cables. Similarly, cables that carry line-level signals should be bundled together only with other line-level cables. Placing cables that carry different voltages and frequencies near each other creates an environment where the individual signals can naturally interfere with each other.

Another way to protect signals is to avoid running cables near electrically noisy sources, such as motors, lighting, and the electrical power systems themselves. Other sources of noise include radio systems and fluorescent lights.

All of these situations can cause signal degradation and loss. Signal degradation can take many forms: voltage amplitude reductions, changes in the shape of the signal's waveform or phase, and power losses.

Generally, analog signals are much more susceptible to noise and distortion than digital signals. The best way for you to preserve signal integrity is through proper system design and installation, using industry best practices and established performance standards.

Distance Limits

A signal transmitted at the proper level will lose strength or amplitude over distance. Two important factors in determining how far a signal will travel are signal level and signal bandwidth. A third factor is the cable-loss properties. Every signal type has unique characteristics, and cable manufacturers provide documentation that indicates distance limits based on signal frequency.

In general, digital signals do not travel as far as analog signals without some form of conditioning or amplification. Each type may also require a unique cable configuration when splitting or looping them within a system.

Switchers

A *switcher* is a device that allows users to choose among two or more sources and send the information from that source to a particular destination. For example, you can connect an AM/FM tuner, a CD player, computer output, or an MP3 player to a switcher. Then you can connect the output of the switcher to a single-input amplifier. Choosing which source you want to listen to is as easy as selecting that source on the switcher. Switchers may be separate components, built into powered audio mixers, or included in a digital signal processor (DSP).

Switchers can be passive (not require power to operate), but many are active. Active switchers can provide some amplification on each input so that you can adjust the levels from each source. In that way, a listener is not subjected to a change in volume as different sources are selected.

Switcher Ins and Outs

Switchers are defined by the number of ins and outs they support. They are "either/or" devices, as opposed to mixers, which can send several sources to a destination simultaneously.

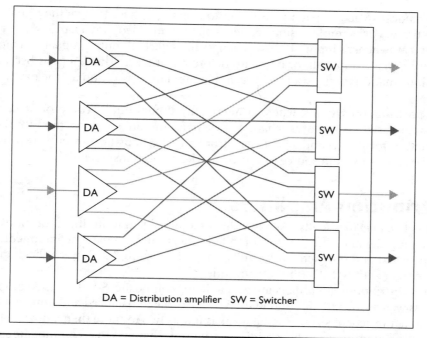

Figure 7-6 A 4-in 4-out matrix switcher diagram

A *matrix switcher* is a composite of several switchers, as illustrated in Figure 7-6. The amount of switchers is the actual number of outputs specified by the size of the matrix. For example, a 4 × 2 matrix switcher contains two switchers. The matrix is capable of selecting a particular source to go to one output and simultaneously selecting the same or a different source to be routed to the other output. To avoid loading at the inputs, each input contains a distribution amplifier, which distributes the input source to each input of every switcher contained in the matrix.

NOTE Video switchers are also defined by their control level. Can the switcher control audio and video separately? Can it control RF signals, component video, composite video, and so on?

Switcher Types

Two specific switcher types are mechanical and smooth. As you might guess, a *mechanical switcher* has a mechanical connection of the cables or circuit. It functions like a wall switch, meaning there is a mechanical connection or disconnection between two conductors.

Mechanical switchers pay no attention to a signal's phase (where the signal is at a certain time). For example, assume a media player and video camera are connected to a mechanical switcher's inputs. If a user changes the switcher's output signal to a monitor, the picture may jump, roll, or break up before it stabilizes. This happens because the signal information for the scan line was not in the starting position for the new source's image.

For presentation systems, where visual quality should not be compromised, *smooth switchers* are used instead of mechanical switchers. Smooth switchers monitor the sync information for the second signal, synchronizing the transition when it is in vertical retrace. Transitions from source to source are clean and transparent.

Distribution Amplifiers

A *distribution amplifier* sends a single video or audio signal to multiple locations while maintaining signal quality. Distribution amplifiers can be separate components, built in to powered audio mixers, or included in a DSP. Figure 7-7 shows a diagram of an audio system with two distribution amplifiers.

Video distribution amplifiers are available for composite, S-Video, and component video, as well as RGB, RGBS, and RGBHV. They come in balanced and unbalanced versions, and a variety of sizes, ranging from as few as two to more than a dozen outputs. Distribution amplifiers can have audio-distribution channels to accommodate simultaneous video and audio distribution. This prevents overloading system circuits.

For example, consider a speech-reinforcement system that requires press-box distribution to accommodate up to 24 press feeds. In this way, a speaker may make a presentation at a podium with one microphone, rather than being obscured from the camera by two dozen microphones. The press members simply connect their tape recorders to a receptacle on an output panel that provides a fixed, isolated signal from the output of a distribution amplifier. If one member of the press corps has a shorted cable, it would affect only that reporter's feed, rather than short the feeds to everyone else.

Figure 7-7
Audio system with two distribution amplifiers

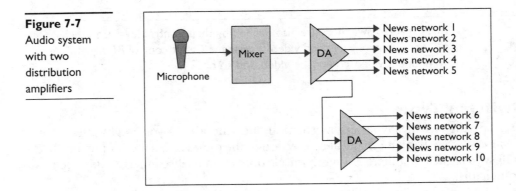

Rack-Building Considerations

One of the most common places you will find a signal-management system is in an AV rack. An *AV rack* is a housing unit that protects and organizes electronic equipment. Figure 7-8 shows examples of two AV racks.

The inside of a typical AV industry rack is 19 inches (482.6 mm) wide. Many of the technical specifications for a rack, including size and equipment height, are determined by standards that have been established by numerous standards-setting organizations.

The outside width of the rack varies from 21 to 25 inches (530 to 630 mm). Racks are also classified by their vertical mounting height. Rack height varies from about 1 foot (300 mm) to more than 7 feet (2,130 mm). The measurement specification is called a *rack unit* (RU). The height of one RU is equal to 1.75 inches (44 mm).

Most AV equipment is manufactured to be mounted within an AV rack, and therefore has mounting points that are an industry-standard 19 inches (482.6 mm) in width. Equipment height is also measured in RUs. An audio mixer might be 1 RU high, while an amplifier might be 2 RUs high. Some equipment, like switchers, can be 10 RUs high or even taller.

Designers and installers carefully plan for all the equipment that will go in a rack. One useful tool in the planning process is called a *rack elevation diagram*. This diagram can show the front and back of a rack and indicate the number of RUs required for each piece of gear. Special consideration is given to how much heat a piece of equipment emits, the types of signals carried to the equipment, and if the equipment has an interface component for a typical user.

Figure 7-8

Two AV racks

Installers use RUs to calculate necessary vertical-spacing requirements between electronic elements for ventilation and airflow paths in the rear of the rack. Electronics perform work, and work produces heat. The heat that builds up in and around equipment must be removed. Many electronic devices have vents on the top and bottom of the chassis so that the hot air can escape. When the warm air leaves the equipment, it does not flow easily out of the rack. This can cause equipment failure. To prevent this problem, vents are used to direct cool air in from the bottom of the rack, which forces the warm air out through the top.

Finally, equipment with user interface components, such as DVD players or CD-ROM drives, should be placed within convenient reach of users. There may be regulations in your location that determine proper placement of components. There are also diagrams that depict the back of a rack and the cables that connect each piece of equipment. These are called *signal separation diagrams*.

Chapter Review

Signal-management systems transmit AV signals from source to output device. The optimal path traverses various types of wire and cable, selected based not only on the signal itself, but also on a client's requirements.

Maintaining the integrity of a signal so that it provides the intended experience when it reaches, for example, a display or a loudspeaker, takes planning and an AV design that includes the appropriate signal-management devices.

Review Questions

The following review questions are not CTS exam questions, nor are they CTS practice exam questions. Material covered in Part II of this book provides foundational knowledge of the technology behind AV systems, but it does not map directly to the domains/tasks covered on the CTS exam. These questions may resemble questions that could appear on the CTS exam, but may also cover material the exam does not. They are included here to help reinforce what you've learned in this chapter. For an official CTS practice exam, see the accompanying CD.

1. The path on which signal types travel is called _____.

    A. Signal flow

    B. Signal transfer route

    C. Wires and cables

    D. Audio and video control

2. The purpose of shielding is to prevent _____ from mixing with the signal.

    A. Insulation

    B. Jackets

    C. Conductors

    D. Noise

3. Which of the following differentiates cable from wire?

   A. Cable contains a shield.

   B. Cable contains only one conductor.

   C. Cable contains multiple conductors.

   D. Conductors are insulated.

4. Twisted-pair cable using balanced circuitry can help in _____.

   A. Keeping noise from audio and video

   B. Blocking static

   C. Preserving the original transmission

   D. Rejecting interference

5. Unless amplified, digital signals generally do not travel as far as _____ signals.

   A. Wireless

   B. Analog

   C. Fiber

   D. Cable

6. Switchers _____.

   A. Must have power to operate

   B. Mix different inputs to a signal output

   C. Connect multiple inputs simultaneously to one output

   D. Allow the user to select one input from a number of inputs

7. A 4 × 2 matrix switcher _____.

   A. Can connect any of four inputs to any or both of two outputs

   B. Must have only one output connected at any given time

   C. Can connect any of two inputs to any or both of four outputs

   D. Has effectively eight outputs

8. An AV rack is a housing unit that _____ electronic equipment.

   A. Identifies

   B. Elevates

   C. Protects and organizes

   D. Cools

9. The inside of a typical AV rack is _____ wide, and the outside varies from _____.

   A. 1 foot; 2 to 7 feet

   B. 19 inches; 21 to 25 inches

   C. 25 inches; 19 to 21 inches

   D. 21 inches; 19 to 25 inches

Answers

1. **A.** The path on which signal types travel is called signal flow.
2. **D.** The purpose of shielding is to prevent noise from mixing with the signal.
3. **C.** Cable contains multiple conductors.
4. **D.** Twisted-pair cable using balanced circuitry can help in rejecting interference.
5. **B.** Unless amplified, digital signals generally do not travel as far as analog signals.
6. **D.** Switchers allow the user to select one input from a number of inputs.
7. **A.** A 4 × 2 matrix switcher can connect any of four inputs to any or both of two outputs.
8. **C.** An AV rack is a housing unit that protects and organizes electronic equipment.
9. **B.** The inside of a typical AV rack is 19 inches wide, and the outside varies from 21 to 25 inches.

Control Systems

In this chapter, you will learn about
- What a control system can do for an AV design
- The components and interfaces of a control system
- The various types of control systems you can employ

Control systems allow you and/or your client to operate complex AV systems using simple interfaces. Complex AV systems are composed of many individual components, integrated to perform like a single system. Ideally, however, you shouldn't need to know about each piece of equipment in order to operate the overall system. Control systems make the AV system more accessible because they don't require technical knowledge to operate the AV equipment.

A control system simplifies the complex individual functions and steps necessary to complete specific tasks. It is event-driven in that certain events, such as a button being pressed or a temperature setting crossing a threshold, initiate a change. The event causes the control system to react in a preprogrammed way.

There are a variety of different user interfaces for control systems, including hand-held remote controls, switches, rotary knobs, and touchpanels.

Control System Functions

What can a control system do? According to InfoComm best practice, a control system's behaviors are described in a "button-by-button" document. This document narrates the events that occur when someone presses a button on the control system interface. The events described are the button's functions and macros.

A *function* is an individual action. For example, if you select Lights On from a user interface, the lights in the room switch on. Common functions include the following:

- Raising or lowering a projection screen
- Commanding the switcher to change an input (see Chapter 7 for information about switchers)
- Setting volume levels
- Routing a signal from one device to another

- Powering on devices
- Activating various functions on a single device (play, pause, and stop)

In a more complex system, a user may execute several functions by pressing a single button. A set of functions activated by one button is called a *macro*. For example, a user might press a button marked Watch Film. After the user presses the button, the control system may execute the following series of functions, programmed as a macro:

- Power on the display
- Power on the audio amplifier
- Lower lighting levels in the room
- Turn on the Blu-ray player
- Switch the display input to the Blu-ray output
- Play the video

In general, as the client requests more functions, the design complexity of the control system increases. When a system has been designed correctly, anyone should be able to operate it seamlessly. The operators' level of knowledge and experience can vary widely, so accommodating a wide range of users makes control-system programming and design a challenge.

Control System Components and Interfaces

The primary component of a control system is the central processing unit (CPU). It is the "brain" of the entire system.

The CPU runs a custom program to accept the input or control requests from the operator, and then controls the individual devices as required by that program. A control system programmer makes changes to the software so that the CPU performs the correct functions.

A user may need to control devices in multiple rooms, across a college campus, or even in another part of the world. These advanced configurations may require multiple interconnected CPUs. These configurations are often referred to as *distributed intelligence*, *distributed processing*, or *edge appliances*. While these three terms have slightly different meanings, control system professionals use them for one-to-many processes.

Control systems require some form of human interface connected to the CPU. The CPU can't normally execute a function if it doesn't receive a command, though increasingly, control systems interface with sensors that can trigger functions based on what they sense in an environment, such as automatically turning off lights when no one is in a room.

An interface might be a multibutton panel installed at the entry door of a presentation room, allowing the user to select a lighting level for the room. Advanced graphical user interfaces (GUIs), touch-sensitive screens (as in Figure 8-1), and web-based control panels are also common. The needs of the user will dictate the appropriate user interface devices.

Figure 8-1
An example of
a control system
touchpanel

Control Signals

The CPU and the devices connected to it must be able to communicate in order to execute the desired command(s). The available types of communication, or *control signals*, vary depending on the capabilities of the CPU and the devices. Communication between devices can be bidirectional or unidirectional.

In a system with *bidirectional* communication, a CPU sends out an instruction to a device. The device executes the instruction, and then replies back to the CPU, "Okay, it's done." This is called *feedback*, and it allows the user to know the status of the device at all times.

Communication can also be in one direction only. This is called *unidirectional* communication. Unidirectional communication may be the only type of control available for a device. A command is sent to or sent out from a device, and there is no return acknowledgment to verify that the command was executed.

Types of Control Systems

Control systems come in many different configurations and use different control languages. In this section, we will look at the contact-closure, variable-voltage, IR, and RF types of control systems.

Contact-Closure Control

A *contact closure* is the simplest form of remote-control communication. It is a switch. This type of control point operates a device by opening or closing an electrical current or voltage loop. It has the most basic protocol language: on (closed circuit) or off (open circuit). Such a protocol language is the basis of binary logic systems represented by ones and zeros—on and off, respectively.

Mechanical devices that use contact-closure control include motors, projection screens, drapes, and shades. They typically require a low-voltage interface to allow the low voltage of the contact closure to operate the high or line voltage of the motor.

Use of contact closures is not limited to the control of a device. They can also provide input to the control system. For instance, a contact closure can provide the status of a room-dividing panel, or it can be installed in a door jamb to turn on the light when the door is opened.

Variable-Voltage Control

The *variable-voltage control*, also referred to as a *voltage ramp generator*, is an analog form of control communication. A voltage of specific parameters is applied to the control point of a device to adjust a level by a specific ratio of voltage value to device level.

Early versions of light-dimming system technology used this form of control to dim the lighting. A 0 to 24 VDC signal was applied to the dimming system zone or light circuit, and the lighting was illuminated at the same percentage. If a 12 VDC signal were applied, for example, the lights would be illuminated at 50 percent.

Another type of device that can be controlled via a variable-voltage signal is a video camera's pan/tilt head. When zero volts are applied to the unit, it is at rest. If a negative voltage is applied, it will move in one direction. A positive voltage will cause it to move in the other direction.

IR Control

IR control comes in two formats: optical and wired. The handheld remote that controls the television set in most homes uses the optical format. The wired format is generally called *serial communication*, or one of many other names created by individual manufacturers. Serial communication sends the same information pulses that would be relayed if it were an optical control, but bypasses the optical circuitry.

A pattern of light pulses is emitted from an LED in the light spectrum that is just beyond the view of the human eye. These light pulses form patterns that are recognized by the control point on the device. If the light patterns do not match the specific functions programmed into the controlled device, the function is ignored.

IR control is very inexpensive to manufacture and simple to integrate into many control technologies, from all-in-one handheld controls up to the most complex control system mainframe devices.

Optical IR control solutions do have disadvantages, however. This control type requires direct line of sight to the device's control point and is limited in range to 30 to 40 feet (9 to 12 meters). You can use optical repeaters and IR blasters to overcome the distance limitations. And to deal with line-of-sight issues, most integrated remote-control systems' IR ports employ IR LEDs installed directly over the controlled device's IR receiver or control point. This allows control of the device while it is installed in a cabinet or equipment rack.

Another disadvantage of optical IR control is that the signal is susceptible to interference from sunlight or fluorescent lighting. In environments where this lighting cannot be controlled, the wired serial format provides remote control of the device. The solution can be wired up to 250 feet (76 meters) from the control system.

Keep in mind that IR is a one-way communication path, meaning that the IR-controlled device has no way to provide feedback on its status or confirm that it has received a command from the control system.

RF Control

RF control is generally employed as a user interface to the control system. Some manufacturers' devices provide control links into their components using RF transmission, as this affords the ability to control devices when line of sight is not possible. RF control has a general limit of 100 feet (30 meters). Third-party remote-control system manufacturers do not easily emulate this type of control link because it requires specific frequencies, along with the protocol with which to communicate.

When using RF control, you must verify the frequencies that are already in use in the environment you intend to control. The RF spectrum is quite busy, and many electronic components release stray RF into the environment, which can interfere with the desired communication packets. By testing the space to be controlled, you can implement appropriate frequencies that will function properly.

Control System Cabling

Control system cabling used to interconnect components may include the following:

- **RS-232**   The most common form of digital data control is the Electronic Industries Association (EIA) standard RS-232. The RS-232 standard is an unbalanced circuit and thus susceptible to noise. The noise issue limits the length of cable you can use for RS-232 control. Earlier versions of the standard limited the maximum cable length to 50 feet (15 meters). Later revisions use a maximum cable capacitance load of 2500 pF. Therefore, depending on total cable capacitance, the maximum useful cable length can be about 50 to 65 feet (15 to 20 meters). RS-232 communication most often is connected using the DB9 or DB25 connector types and uses a minimum of two conductors plus a ground. RS-232 is a point-to-point protocol.

- **RS-422**   The EIA standards developers recognized that the unbalanced characteristics of the RS-232 protocol do not allow for long-distance cable runs. Therefore, they developed the RS-422, providing a balanced, four-wire solution allowing cable runs of up to 4,000 feet (1,200 meters). The length is dependent on data speed and cable quality. RS-422 uses a minimum of four wires plus (usually) a ground. RS-422 is a multidrop protocol that uses only one transmitter (driver) in most applications.

- **RS-485**   This EIA standard supports 32 transmitting and receiving devices or, in some applications, 256 transmitting and receiving devices. The balanced design allows cable distances up to 4,000 feet (1,200 meters) on unshielded twisted-pair cable. RS-485 is a multipoint protocol that allows multiple transmitters (drivers) and receivers.

Ethernet Control Systems

In recent years, more control systems have begun using standard Ethernet network topologies. This connection type could be defined as a digital data link, but the ramifications of its use are so extensive that it resides in a control-point classification of its own. This is because digital data communication is designed to communicate between devices, while the Ethernet solution allows communication among control components, applications, data, and the Internet.

Ethernet topologies are utilized mostly to extend the control environment at distances previously unimagined. Within a single building, a local-area network (LAN) allows a central control room to communicate with each presentation room, for example. LANs that are connected together between buildings on a corporate campus or in a city form a metropolitan-area network (MAN). The LANs and MANs of organizations that span multiple cities, when connected together, create a wide-area network (WAN).

From these IT-based topologies, control solutions can be created that allow effective, real-time room support, loss prevention, facility usage and scheduling, and web-centric presentation environments, to name only a few possibilities. Within this environment, the control scenario shifts from the needs of the facility to enterprise-wide application specifics.

Control Systems on Networks

You may think of control systems controlling AV equipment within individual rooms. But thanks to the addition of Ethernet networking capabilities to AV equipment, control systems take on new possibilities. By placing control systems on networks, users can control devices remotely, more efficiently maintain inventories of equipment, respond to customer-service issues more efficiently, and much more. In an enterprise-wide control environment, a user can turn on a projector from another time zone by using a smartphone. That said, placing control systems on networks may also require more planning and coordination among trades, particularly between AV professionals and IT departments.

Network-based control systems may mean more equipment to manage because the network enables users to glean new and more data from a wider range of devices and components. Collected data could help track projector bulb usage, for example. It may also show that AV-capable rooms in one building are more popular than rooms in another building. Such knowledge makes the control system more than just a "regular" control system. With the right software and IT integration, a control system can be used to adjust room scheduling or alert a facilities staff to update old equipment. In other words, when control systems ride a network, they consume and create actionable data that goes beyond just controlling AV devices.

Chapter Review

Professional AV systems are often the sum of various AV components, each integrated together to support a client's needs. A control system provides the single user interface for operating all those different integrated components as one.

Control systems come in various configurations and speak different control languages. Choosing the right solution depends on the AV technology you need to control, the space where you need to control it, and users' specific requirements.

Review Questions

The following review questions are not CTS exam questions, nor are they CTS practice exam questions. Material covered in Part II of this book provides foundational knowledge of the technology behind AV systems, but it does not map directly to the domains/tasks covered on the CTS exam. These questions may resemble questions that could appear on the CTS exam, but may also cover material the exam does not. They are included here to help reinforce what you've learned in this chapter. For an official CTS practice exam, see the accompanying CD.

1. Remote control systems _____ the operation of an AV system.
   A. Complicate
   B. Automate
   C. Reconfigure
   D. Simplify

2. Which of these is *not* a common method of interfacing with a control system?
   A. GUIs
   B. Flip charts
   C. Wireless touchpanels
   D. Wall switches

3. Which of the following is most likely to be a function rather than a macro?
   A. Starting a videoconference
   B. Playing a video
   C. Dimming the lights and starting a Blu-ray player
   D. Setting volume levels

4. Which of the following would most likely be a macro?
   A. Turning on the lights
   B. Activating a single function on a device
   C. Powering on the audio amplifier and display
   D. Setting a volume level

5. The purpose of a control system interface is to send instructions to the _____.
   A. Internet gateway
   B. Network printers
   C. Hard drive
   D. Control system CPU

6. Communication that allows a return message is called _____.

    A. Omnidirectional

    B. Bidirectional

    C. Multidirectional

    D. Unidirectional

7. Contact-closure control communication provides device operation by _____.

    A. Increasing the power level of the electrical circuit

    B. Reducing the wattage in a current loop

    C. Increasing resistance in a voltage loop

    D. Closing or opening an electrical current or voltage loop

8. Which of the following can interfere with an IR control?

    A. Quality of the device

    B. Strength of the power source

    C. Shadows and darkness

    D. Sunlight or fluorescent lighting

9. Ethernet is mainly used in control systems _____.

    A. To control environments at greater distances

    B. For one-way device communication

    C. To create IP addresses

    D. For analog devices

Answers

1. D. Control systems simplify the operation of an AV system.

2. B. Flip charts are not a common method of interfacing with a control system.

3. D. Setting volume levels is more likely to be a function than a macro.

4. C. Powering on the audio amplifier and display would most likely be a macro.

5. D. The purpose of a control system interface is to send instructions to the control system CPU.

6. B. Communication that allows a return message is called bidirectional.

7. D. Contact-closure control communication provides device operation by closing or opening an electrical current or voltage loop.

8. D. Sunlight or fluorescent lighting can interfere with an IR control.

9. A. Ethernet is mainly used in control systems to control environments at greater distances.

CHAPTER 9

Electrical Systems

In this chapter, you will learn about
- The basics of electrical power and distribution
- The characteristics of electricity
- How an electrical circuit works
- How electricity is managed in an AV system
- Electrical safety issues

At some point, using an AV system will require a source of power. Most elements in an AV system use a connector, cable, or plug. Electrical engineers and electricians plan and install power-distribution systems to handle these power requirements.

Different countries have different electrical systems. In most countries, electrical wiring systems are subject to strict government codes and regulations. Always check with the authority having jurisdiction (AHJ) for the requirements that apply to your system's location.

Electrical Power Basics

To understand electrical systems, you need to become familiar with concepts and terms such as voltage, current, resistance, impedance, circuits, Ohm's law, power, and grounding.

Voltage and Current

Voltage is the measure of electrical pressure. Voltage reflects the potential force in an electrical signal. It is the force that causes current (described next) to flow through an electrical conductor. Voltage represents the greatest difference of potential between any two conductors of a circuit. In calculations, it is symbolized as V, for electrical pressure, or in some circumstances, as E, for electromotive force.

The flow of electrons through a conductor, like a wire, is called *current*. It is symbolized by I, for intensity. Current is measured in amperes (amps). Devices that require electrical power are said to "draw" current from a circuit. Note that an increase in current is an increase in the *quantity* of electricity, not the speed of the electricity. The speed of electricity is constant.

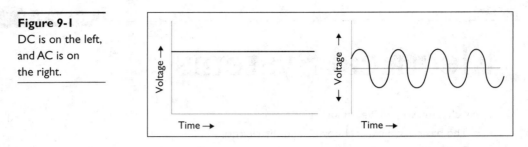

Figure 9-1
DC is on the left,
and AC is on
the right.

There are two types of electrical current: direct current (DC) and alternating current (AC), as illustrated in Figure 9-1.

Direct Current

DC flows in one direction. A battery is a good example of a DC device; it has both a positive terminal and a negative terminal. In a DC circuit, the negative terminal sends electrons into the wire, where the electrons in the wire continue to process toward the positive terminal, which is attracting the electrons.

Here are some facts about DC power:

- The charge stays at a constant, regulated flow and does not reverse direction.
- The power can be either positive or negative.
- It is extremely common for computer signals, batteries, and power supplies of AV equipment to use DC power.

Alternating Current

AC refers to an electrical signal or current that reverses its polarity and its direction at regular intervals. The voltage of the AC power in a wall outlet is typically 120 volts (V) in the United States and 230V in Europe. The voltage level in most of the hardware that is installed is usually substantially lower than this. References to low voltage may include anything under 70V AC.

Here are some facts about AC power:

- The signal (current) alternates back and forth.
- Current travels in cycles and can be thought of as oscillating.
- It is quantified by frequency, measured in cycles per second or in hertz (Hz).
- The electricity in the United States is typically 60 Hz. In Europe, it is 50 Hz.

About Current

Here are two important things to know about electricity and current:

- Electrical current is the flow of electrons from the power source, through the circuit, and back to the source.
- Electrical current will take all available pathways in its return to the source.

While we have all heard that electrical current takes the path of least resistance, that is not quite true. Again, it will take all available pathways. The amount of current depends on the resistance. Obviously, the majority of the current will take the pathway of lesser resistance, rather than the pathway of greater resistance, but all pathways will have current flow. If a conductive path is created and that path bypasses a portion of the circuit, you get a short circuit. Potentially, this can be very dangerous.

Resistance and Impedance

The amperage at which electrons can flow depends on resistance, which is measured in ohms. *Resistance*, symbolized R, is the opposition to the flow of electrons. It refers to the property of a substance that impedes current and results in the dissipation of power in the form of heat. Here's an analogy: If you try to push a heavy boulder uphill on a hot day, the rock will resist you, and the work you must do to make it move will make you sweat. The rock symbolizes resistance, and the sweat from working hard symbolizes the dissipation of power.

The measure of resistance in a conductor is based on the size of the wire used and increases as the cable increases. Size is determined by gauge or cross-sectional mils.

One way to overcome resistance is to use thicker wire. With a thicker wire, there is less resistance and better electrical transfer. Resistance increases with cable length; therefore, a longer cable run will result in a weaker signal at the end of the cable. When resistance is too large, it will degrade the signal or prevent it from reaching its destination.

Impedance, also measured in ohms, must not be confused with resistance. Although both resistance and impedance are measured in ohms, impedance uses the symbol Z. Impedance includes resistance and reactance, which varies with the frequency of the AC. If the current is direct, or one-way, the reactive component is zero, and the impedance equals the resistance.

Power

Energy expended in one form manifests itself in another form—motion, heat, or light, for example. In physics terms, energy is dissipated or consumed by the work being done. This is power (P), or the rate at which work is done. The result in a measurement called watts (W). For example, an audio power amplifier's output is rated in watts. A watt is the power expended when 1 ampere of DC flows through a resistance of 1 ohm.

Table 9-1 summarizes voltage, current, resistance, impedance, and power.

Ohm's Law

Voltage, current, and resistance/impedance are all related. The current in an electrical circuit is proportional to the applied voltage. An increase in voltage means an increase in current, if resistance stays the same. The relationship between current and resistance is inversely proportional, meaning that as one increases, the other decreases. An increase in resistance produces a decrease in current, if voltage stays the same.

As an analogy of the relationship between current and resistance at work in DC, think of a sink with a stopper. The water pressure in the sink basin is like voltage. It has the

Characteristic	Symbol	Definition	Measurement
Voltage	V	Electrical pressure	Volts (V)
Current	I (for intensity)	The flow of electrons through a conductor	Amperes (amps)
Resistance	R	The opposition to the flow of electrons in DC circuits	Ohms
Impedance	Z	The opposition to the flow of electrons in AC circuits	Ohms
Power	P	The rate at which work is done	Watts (W)

Table 9-1   Characteristics of Voltage, Current, Resistance, Impedance, and Power

potential to go down the drain, but it has not done so yet. (Voltage has the potential to become power, but it isn't power yet.) If you remove the stopper, the amount of water flowing through the drainpipe represents current. The pipe size is resistance. A larger pipe will drain water more quickly than a narrower pipe.

This relationship among voltage, current, and resistance is defined in mathematical form by Ohm's law:

$I = V/R$
Current = Voltage/Resistance

So, 1 amp is equal to the steady current produced by 1V applied across a resistance of 1 ohm.

Here are some similar forms of Ohm's law:

$V = I*R$
$R = V/I$
$P = IV$

Electrical Circuits

The flow of electricity requires a continuous (closed) circuit. A circuit always has a source and a load. The *source* is the supplier of power, information, or a signal. The *load* is the reactive component, or the receiver of information.

One of the simplest examples of a circuit involves a battery (the source) and a light bulb (the load). A battery uses chemicals to produce electricity. The light bulb turns the flow of current through it into heat and light. The electricity goes from the battery to the light bulb through a conductor—typically a wire. A switch is used to turn the electrical flow on and off, and subsequently, control the light bulb.

The complexities of circuits found in AV systems are far greater than the light bulb example. Managing the power properly protects the equipment, the signals, and the users. An important technique for protecting electronic systems is grounding, which we'll cover next.

A key concept of electrical theory is that electrons flow from the power source, through the circuit, and return to the power source. A common misconception is that all current is seeking to go to ground (meaning the earth). This is not true. Current is seeking to return to the source, whatever that source may be. You can see how this works by studying basic series and parallel circuits, as shown in Figure 9-2.

In the series circuit, all of the current supplied by the source will flow through the entire circuit. You can think of it this way: All of the electrons leaving the source of power (the negative terminal of the battery in this case) go through each component of the circuit and return to the source (the positive terminal of the battery). While all of the current flows through all of the circuit, the voltage divides across the three resistors (loads), as well as the wire that connects them. (Wire has resistance; therefore, it is considered to be a load in the circuit.)

However, in the parallel circuit, it is the voltage that remains the same across the loads, but the current divides and takes all the paths available to it in order to return to the source.

Grounding

Grounding (*earthing*) refers to the techniques and hardware that connect electronic equipment and wiring to a reference point, which is then connected to the ground—literally, the earth. Grounding ensures that systems are stable, safe, and reliable. The primary objective of grounding is to limit human exposure to high-voltage electricity.

In agreement with industry best practices, a grounded power system is one that makes use of both system grounding and equipment grounding. *System grounding* requires connecting one side of an AC power circuit to the ground by physically running a conductor into the earth.

Equipment grounding is the practice of connecting to the ground system all metal units used to carry and house power cables, connectors, or electronics. Equipment that may be grounded includes metal conduits, outlet boxes, and metal cabinets and enclosures. Because the elements are metal, they are conductive. In the event of a fault where a live wire contacts conductive elements, they will attract an electrical charge. Unless this charge has somewhere to go, it can increase the possibility of electrical fire, destroy the electrical circuits in the equipment, or possibly even present a fatal situation (if a person comes in contact with a charged conductive object).

To provide an escape route for this charge, ground wires connect the metal elements to the ground wires in the infrastructure, which are connected to the system ground. These ground wires serve as a backup and, under usual circumstances, carry no current.

Figure 9-2
Series circuit
(left) and parallel
circuit (right)

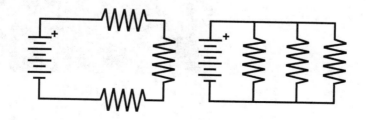

Electrical Power and Distribution

To convert the AC power that comes into a building into DC power used by most electronics, a power supply is required. The power utility service brings electrical power to a building. You don't need to know how the power utility company harnesses and distributes electricity. You do need to know what is going on once it is in the building.

Many facilities have a dedicated power system that is used solely for computer and electronic systems. Formerly known as a *technical power system*, current practice identifies this as an *isolated ground system*. A building typically has all its power coming from one central location. The main distribution panel distributes the power to all necessary locations within the facility via feeders and subpanels. In a presentation theater or training facility, for example, the subpanels for the AV system are often located in or near a control room or equipment closet.

From the subpanels, or *panelboards*, branch circuits will run to the various electronics, racks, communication equipment, and wall outlets. Figure 9-3 shows an example of a panelboard. Circuit breakers in these panelboards help limit the amount of current to any one circuit. These branch circuits must be used only for the AV system or computers to be used with the AV system.

If the branch circuits are used to power other equipment—such as a coffee machine, vacuum cleaner, or heater—electromagnetic interference may be introduced into the system. To avoid this, circuits that are accessible to nontechnical users should be labeled "For use by AV equipment only."

Figure 9-3
A panelboard

Electrical Safety

As you have learned, the primary purpose of electrical codes, including grounding, is the practical safeguarding of persons and property from hazards arising from the use of electricity. There are different safety issues to be aware of when you are working with permanent installations and portable installations.

When working with permanent installations, keep the following in mind:

- Recognize that permanent installations can exist in adverse or hazardous environments. Therefore, you may notice specialized hardware for extreme temperature variations or wet environments.

- Do not exceed conduit area fill allowances.

- Maintain applicable code standards for electrical supply systems.

- Follow proper procedures for grounding of equipment racks.

- Look for proper current protection in the form of circuit breakers.

- Know where the supply cutoff is located. This may be a labeled main breaker in a local panel.

- Have proper lighting and safe access to electrical system service rooms.

When working with temporary installations, note the following:

- Test power from portable distribution systems before connecting equipment.

- Balance the loads when using multiple circuits.

- Use extension cables that are designed for heavy-duty use.

- Use the proper wire size for the current required.

- Dispose of frayed or damaged cables.

- Use cable covers and ramps, or tape all cables securely to the floor, to prevent a tripping hazard.

- Use weather-resistant equipment for outdoor events.

Whether permanent or temporary, the following points apply to safety:

- Do not exceed 80 percent of the capacity of any circuit.

- Specify or use equipment that uses equipment grounding conductors.

- Discard any cord or equipment where the equipment grounding conductor or pin has been removed. Also remove any ground lift adapters that have been used.

- Do not daisy-chain power strips together.

Chapter Review

In an AV system, power-distribution systems are necessary. In this chapter, you became familiar with the concepts and terms associated with these systems. You also learned about the basics of electrical power and distribution, the characteristics of electricity, how an electrical circuit works, how electricity is managed in an AV system, and electrical safety issues.

Review Questions

The following review questions are not CTS exam questions, nor are they CTS practice exam questions. Material covered in Part II of this book provides foundational knowledge of the technology behind AV systems, but it does not map directly to the domains/tasks covered on the CTS exam. These questions may resemble questions that could appear on the CTS exam, but may also cover material the exam does not. They are included here to help reinforce what you've learned in this chapter. For an official CTS practice exam, see the accompanying CD.

1. Voltage is the force that causes _____ to flow through a conductor.
    A. Ohms
    B. Neutrons
    C. Watts
    D. Current

2. Current is measured in _____.
    A. Amperes
    B. Volts
    C. Ohms
    D. Watts

3. The opposition to the flow of electrons in AC is called _____.
    A. Ohms
    B. Resistance
    C. Impedance
    D. Dissipation

4. Power is measured in _____.
    A. Voltage
    B. Impedance
    C. Resistance
    D. Watts

5. The relationship among voltage, current, and resistance is defined by the formula _____.

   A. $V = I/R$

   B. $I = R/V$

   C. $I = V/R$

   D. $R = V*I$

6. A current is always seeking to go to the _____.

   A. Battery

   B. Load

   C. Ground

   D. Source

7. The main panel distributes power using _____.

   A. Panelboards and service entrances

   B. Feeders and subpanels

   C. Branch circuits and feeders

   D. Service entrances and subpanels

8. At what point in the AC system are the branch circuits that power wall outlets and AV equipment connected?

   A. Lateral feed

   B. Feeders

   C. Main distribution

   D. Subpanel (panelboard)

9. When planning an electrical system, do not exceed ____ of the capacity of any circuit.

   A. 95 percent

   B. 80 percent

   C. 65 percent

   D. 75 percent

Answers

1. D. Voltage is the force that causes current to flow through a conductor.
2. A. Current is measured in amperes.
3. C. The opposition to the flow of electrons in AC is called impedance.
4. D. Power is measured in watts.

5. C. The relationship among voltage, current, and resistance is defined by the formula $I = V/R$.

6. D. A current is always seeking to go to the source.

7. B. The main panel distributes power using feeders and subpanels.

8. D. The branch circuits that power wall outlets and AV equipment are connected to the subpanel (panelboard).

9. B. When planning an electrical system, do not exceed 80 percent of the capacity of any circuit.

CHAPTER 10

Radio Waves

In this chapter, you will learn about
- The basics of radio waves
- What is required to transmit wireless RF signals
- How the right antenna setup makes a difference
- What goes into established RF video systems

Every time people connect to the Internet from a computer without using a cable, they are using radio waves to transmit data. When a presenter communicates a message using a wireless microphone, radio waves transmit the audio information. A global positioning system (GPS) uses radio waves to transmit location information. Many handheld remote controls for AV equipment use radio waves. Digital television signals can be broadcast using radio waves from antenna towers.

What are radio waves? Radio waves are AC signals in the form of electrons or photons (wireless), and are classified as electromagnetic energy. These signals, or range of radio frequencies, make up a small section of the electromagnetic spectrum. Any AC signal in the range from 3 kHz to 300 GHz falls into the RF spectrum (although, in the United States, the FCC has not allocated any spectrum below 9 kHz).

Transmitting and Receiving RF

Any communication system that uses RF signals requires at least one transmitting device and one receiving device. The role of the transmitter is to take audio, video, and/ or data information and convert it for transmission via an antenna. The information could be audio from a microphone, control information from a remote control, or data from an Ethernet network. Increasingly, AV system manufacturers are coming up with new, reliable ways of transmitting video wirelessly.

For transmittal and reception, the information must be modulated into an RF carrier, whether it is analog or digital. The RF carrier is the "envelope" that contains the information. Demodulating the RF carrier extracts the information from the envelope. So, modulation and demodulation are essential for conversion.

Modulation is the most important part of the conversion process. The transmitter modulates information into an RF carrier and sends it through the air from an antenna.

During the process of modulation, a special circuit adds the program signal(s) to a carrier frequency. After modulation is complete, the information is transmitted into the air using the antenna.

The receiver uses an antenna to receive the RF signal sent from the transmitter and sends it to the next device in the AV system. The receiver tunes in the modulated RF signal and separates the program signal from the carrier frequency by demodulating it. Once the carrier is demodulated, the original audio, video, and/or data information it contained is passed on to the next device in the AV system, using conventional cabling.

Allocations of RFs

Due to the heavy demand for RFs, the range of available frequencies is carefully regulated and coordinated throughout most of the world. The allocations will vary depending on your location.

The RF spectrum is enormous and is divided into several smaller groupings, or bands, for descriptive purposes. Two commonly used bands are very high frequency (VHF) and ultra-high frequency (UHF). VHF channels start at 30 to 300 MHz, and UHF is in the range of 300 MHz to 3 GHz (3,000 MHz).

Most wireless microphone systems and television broadcasting stations operate from 500 to 800 MHz. Consumer mobile devices, such as cell phones, use frequencies near 800 and 1900 MHz. Most wireless devices that are used in data networks and mobile applications operate in the range of 2.4 to 5 GHz, but even that is changing.

Recently, in the United States, television broadcasting transitioned from analog to digital, freeing large amounts of the RF spectrum, much of it in the 700 MHz to 800 MHz range, which held analog TV channels 52 to 69. As a result of a series of decisions by the Federal Communications Commission (FCC), the 700-MHz spectrum, in which wireless microphone systems commonly operated, was cleared out and either auctioned off or set aside for public-safety applications. Wireless microphone users, therefore, are generally confined to 470 to 698 MHz in the United States

To complicate things, the FCC has allowed a new generation of smartphones, tablets, and other wireless consumer devices to also operate in the spectrum that wireless microphones might otherwise use. One solution from the FCC has been to reserve at least 12 MHz (two 6-MHz channels) for wireless audio systems in the U.S. markets where AV professionals operate. Users who need more spectrum—whether for live events or applications that require many wireless microphones—can reserve more channels, either through the FCC or an FCC-approved database of available frequencies.

As you can tell, a working knowledge of governmental spectrum allocations within the VHF and UHF bands is extremely helpful prior to using a wireless microphone system. Randomly selecting frequencies may result in interference. By identifying the operating range of frequencies of your systems, you can select open frequencies to prevent problems. Your wireless microphone manufacturer can provide you with a chart of these frequencies for your area.

The Importance of Antennas

Antennas are vital to the transmission and reception of RF signals. If antennas are incorrectly matched to the transmitting and/or receiving equipment, or are set up in the wrong position or orientation, it will degrade the overall performance of the RF equipment. An understanding of antenna types can help you make better-informed decisions when purchasing wireless microphone systems or other RF-based systems.

Antenna Lengths

The dimensions of antennas are directly related to the carrier frequency of the desired RF transmission. Antenna elements are usually cut to lengths that match 1/4, 1/2, 3/4, or 1 wavelength of the carrier frequency. When using an antenna cut to the proper length, virtually all of the RF energy is transferred to the antenna.

When the wavelength of the carrier frequency and the antenna length are equivalent, the antenna is *resonant*. Resonance is important on the receiving end, as the antenna will be most sensitive to wavelengths closest to its physical length.

An easy way to determine the wavelength of a signal, in meters, is to divide the frequency in megahertz into 300. To determine frequency, divide the wavelength into 300. If you have a signal that is 100 MHz (which puts it in the VHF frequency band of FM radio and TV broadcasts), its wavelength is 3 meters. Conversely, if you have a signal with a wavelength of 1 meter, its frequency is 300 MHz (right on the border of VHF and UHF). When you get into gigahertz frequencies, where technologies such as Wi-Fi operate, you're talking about wavelength measurements in millimeters.

Ground Plane

Because all RF signals are AC signals, they have a positive and negative swing through one complete wavelength. For 1/4-, 1/2-, or 3/4-wavelength antennas to radiate efficiently, they require a *counterpoise*, also known as a *ground plane*. The angle of the ground plane affects both the impedance of the antenna and its angle of radiation. For an AM radio station, the counterpoise can be a large system of ground radials, often in watery soil. For an FM or television station, a small reflector or screen will be used.

For many AV systems, the ground plane could be the equipment case or metal mounting rack. Ideally, the ground plane should measure to the same fractional wavelength as the antenna element. Increasing the size of the ground plane beyond this length will have minimal effect on the antenna's performance. However, a smaller ground plane will reduce antenna gain and performance.

Antenna Orientation

It is important to properly orient, or *polarize*, the transmitting and receiving antennas when setting up an RF system. Proper orientation of the antennas maximizes signal transmission and reception.

For example, wireless microphone systems usually employ antennas that are perpendicular to the floor. This orientation is known as *vertical polarization*. A vertically polarized antenna radiates RF energy from the sides of the radiating element, parallel to the floor. If the receiving antenna is also vertically polarized, it will receive the maximum RF energy from the transmitter. If the antennas are not oriented in the same direction (*cross-polarized*), the received signal will be much weaker, degrading system performance.

Diversity Systems

When RF energy is transmitted from an omnidirectional antenna, it radiates in all directions. Some of the energy may go straight toward the receiving antenna (this is called *line of sight*); other RF energy may bounce off walls, floors, and other surfaces before it reaches the receiving antenna. The best RF signal will be the one that goes straight from the transmitter to the receiver, with minimal or no reflection from other nearby surfaces.

If the incident signal and one or more reflections (echoes) arrive at the receiving antenna at different times, they will be out of phase, and the signals will cancel out. To overcome this problem, some RF receivers use a pair of antennas to receive a transmitted RF signal. These devices are called *diversity receivers*. Diversity receivers are constantly calculating phase differences between signals to dynamically shift between the two antennas and avoid cancellation. It's possible that both the incident and reflected signal could arrive in phase and sound just fine.

A diversity receiver will dynamically compare the intensities of direct and reflected RF signals between the two antennas. Then the receiver will select the strongest signal and process it while ignoring echoes. Any RF signal that is not line of sight is categorized as a multipath signal.

Figure 10-1 shows an example of several wireless microphone receivers in an AV rack.

Figure 10-1
Several
wireless
microphone
receivers in
an AV rack

RF Video Systems

Not all RF signals are received directly through the air. Some video systems receive RF signals from one central antenna, and then retransmit those RF signals through coaxial cables to multiple displays. Figure 10-2 illustrates an RF distribution system in a facility.

Master antenna television (MATV) systems receive broadcast programs from a single master antenna array, and then redistribute programs over coaxial cable to a small number of users. Community antenna television (CATV) systems broadcast signals are received by one or more centrally located antenna towers and distributed by coaxial or fiber-optic cable through a larger geographic region. CATV systems are generally known as *cable*. This type of system tends to have more services and features than an MATV system.

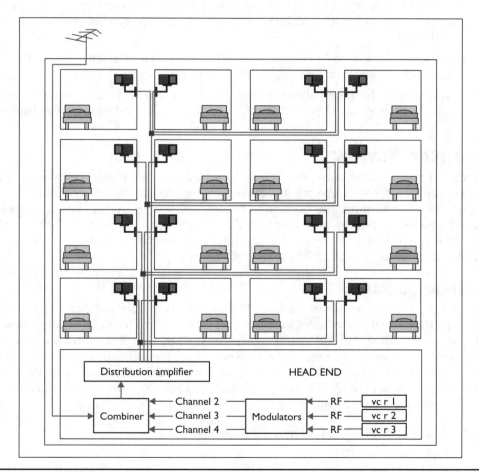

Figure 10-2　An RF distribution system in a facility

The following components are commonly found in RF video systems:

- A *modulator* converts analog composite or digital component video signals, along with corresponding audio signals, into modulated signals on a specific carrier frequency.

- A *combiner*, using a process called *multiplexing*, combines multiple modulated video and audio signals (called *channels*) together into one cable that constitutes a broadband signal. This process usually takes place at a central location in a building known as the *headend room*.

- A *splitter* is a passive device that can divide a signal's strength equally between two or more destinations.

- A *tap* is a passive device (*transformer*) that can extract, or "tap," some of the RF energy passing through it. While a reduced RF signal is present on the tapped output, the original RF signal proceeds through the tap with little reduction in strength.

- *Line amplifiers* are used to boost or amplify the signal for transmission to the next part of the system and overcome system losses.

- A *demodulator* (or tuner) allows the user to select a desired channel. It extracts the selected channel signal from the carrier, providing video and audio to the display.

Chapter Review

Though RF systems have been around for a long time, the AV industry is realizing new ways to use them to come up with designs that exploit the installation flexibility inherent to wireless communications. In this chapter, you learned about radio waves and should now be able to identify them on the electromagnetic spectrum. You should also be able to differentiate between a transmitter and a receiver, explain the RF signal process, and identify common RF bands and RF video systems.

Review Questions

The following review questions are not CTS exam questions, nor are they CTS practice exam questions. Material covered in Part II of this book provides foundational knowledge of the technology behind AV systems, but it does not map directly to the domains/tasks covered on the CTS exam. These questions may resemble questions that could appear on the CTS exam, but may also cover material the exam does not. They are included here to help reinforce what you've learned in this chapter. For an official CTS practice exam, see the accompanying CD.

1. An AC signal of _____ falls into the RF spectrum.
   A. 9 kHz to 300 GHz
   B. No less than 9 kHz
   C. 300 GHz and below
   D. 3 kHz to 300 GHz

2. What does the transmitter do with its audio, video, and/or data?

    A. Converts it to be sent out via antenna

    B. Stores and reads it

    C. Demodulates it

    D. Converts it so that it can be translated and received

3. _____ is the most important step in converting data in a transmitter.

    A. Transmission

    B. Demodulation

    C. Extraction

    D. Modulation

4. The range of RFs between 300 MHz and 3 GHz is the ___ band.

    A. VHF

    B. HF

    C. VLF

    D. UHF

5. An antenna will be most sensitive to transmissions that _____.

    A. Have a wavelength closest to its physical length

    B. Have a wavelength one half the size of the receiving antenna

    C. Have proper orientation

    D. Are transmitted within proximity

6. The main function of a diversity receiver is to _____.

    A. Amplify echoes so that the main signal is strengthened

    B. Find incident and reflected signals that arrive in phase

    C. Calculate phase differences between signals in order to avoid cancellation

    D. Eliminate multipath signals that come from the transmitter

7. Which of the following receives broadcast programs from multiple antennas and is redistributed by coaxial or fiber-optic cable?

    A. HDTV

    B. CATV

    C. MATV

    D. RFTV

Answers

1. D. An AC signal of 3 kHz to 300 GHz falls into the RF spectrum.

2. A. A transmitter converts audio, video, and/or data to be sent out via antenna.

3. D. Modulation is the most important step in converting data in a transmitter.

4. D. The range of RFs between 300 MHz and 3 GHz is the UHF band.

5. A. An antenna will be most sensitive to transmissions that have a wavelength closest to its physical length.

6. C. The main function of a diversity receiver is to calculate phase differences between signals in order to avoid cancellation.

7. B. CATV systems receive broadcast programs from multiple antennas and redistribute them by coaxial or fiber-optic cables.

PART III

Preparing for AV Solutions

Part III of the *CTS Exam Guide* deals with preparation. It covers the steps a CTS-certified AV professional should take before proceeding to the technical steps of specifying, designing, and installing an AV system. These steps are critical to ensuring that CTS holders understand exactly what customers are looking for in an AV system, and that customers ultimately receive exactly the system they want.

Domain Check

Chapters 11 through 15 address tasks within Domain A (Creating AV Solutions). Although Domain A includes several other tasks, those covered in this part of the book focus on preparatory steps. They account for about 32% of the CTS exam.

CHAPTER 11

Gathering Customer Information

In this chapter, you will learn about
- Communicating with clients, designers, and contractors
- Identifying client needs
- Interpreting scale drawings of the client site
- Identifying client contact information
- Evaluating possible constraints at the client site

This chapter covers the various types of information required to design and/or install an AV solution at a client's site. To obtain this information, you generally need knowledge of customer service and negotiation techniques, and an ability to interpret scale drawings and other project documentation. You also will need skills in basic math and listening techniques.

Domain Check

Questions addressing the knowledge and skills related to gathering customer information account for about 6% of your final score on the CTS exam (about six questions).

Communicating with Clients

CTS-certified AV personnel are able to communicate effectively and accurately with clients and the project teams. They know how to present themselves and to correspond in a professional manner. Repeat business from satisfied clients and word-of-mouth referrals are important to the professional AV business. If clients and business partners feel you've established a clear, open, and straightforward relationship, they're more likely to do business with you again or recommend your work to others.

In communicating with clients and the project team, you will need to determine AV requirements by listening carefully and presenting AV-related questions, issues, and options. This section reviews strategies to improve your ability to communicate with clients, coworkers, and members of allied trades.

Obtaining Client Contact Information

It may sound obvious, but it is critical in the early stages of the project—and throughout the design and installation—that you have the correct and proper contact information to ensure that you can communicate with the client in a timely manner. The contact information should include any participants or stakeholders that you may need to contact throughout the project, such as the architect, building manager, AV manager, information technology (IT) manager, and so on.

The following is the typical contact information necessary to collect at initial meetings:

- **Telephone number**   This is the main telephone number and extension.
- **Fax number**   This is a fax number for transmitting signed contracts or other documents.
- **Mailing address**   The full mailing address should be recorded for sending important written communications, contracts, plans, and other documents.
- **E-mail address**   E-mail is excellent for quick communication, especially when you need to retain the content of the messages for your records.
- **Mobile phone number**   This is an important number to have in the event that you need to contact a client team member immediately to address an issue in a timely manner. For example, if the installation team needs access to a specific building space, calling the appropriate team member's mobile phone would likely provide the timeliest response. Mobile phones also allow you to transmit and receive text messages in a timely manner.
- **Chat or instant message ID**   A brief communication can often be handled through chat or instant messages transmitted via a computer or mobile phone.

Face-to-Face Communication

It almost goes without saying, but face-to-face communication is the most direct method of communicating on a project. Looking at a person helps you gauge that person's competence, establish a rapport, and assess the reaction to your message. By sitting down and talking directly to clients and the project team, you may be able to assess the following:

- Is this person knowledgeable?
- Is this person trustworthy?
- Is this person listening?
- Does this person care about your message?

Also, in a face-to-face conversation, you can use a number of strategies to communicate that you yourself are competent, trustworthy, and sincere.

Personal Appearance

In face-to-face communication, your overall appearance sends the first message to the client. Clients usually feel more confident in your abilities if you appear professional. Make sure you dress appropriately. Wear neat, professional-looking clothing. Tuck in your shirt and wear a belt. Wear modest clothes that cover you up when you are working. And avoid clothing with logos or advertising (with the exception of your organization's branded apparel).

Many companies provide uniforms or have a dress code that you should follow. If you are unsure what to wear, talk to your manager.

Body Language

Your body language also sends a message. Nonverbal communication can sometimes be as important as verbal communication when speaking to a client. Keep the following tips in mind when meeting with clients:

- Smile when you greet clients and when you close your business. A smile conveys a positive and friendly attitude that will help establish rapport.
- Introduce yourself and shake clients' hands. Do not wait for clients to make the first move.
- Stand up straight and don't slouch. Slouching can make you appear like you lack confidence, energy, or even competence.
- When clients are speaking, face them, smile, and listen to what they are saying.
- Observe clients' body language. Do they seem uncomfortable or nervous? If so, adjust your responses based on that impression. If they appear confused, simplify your explanation or statement, and then ask if they have questions.

Active Listening

Active listening is critical to effective communications. When someone is speaking, it is easy to fall into the pattern of thinking about what you will say once that person is finished talking. Unfortunately, that can prevent you from hearing what the other person is actually saying. Here are some tips to improve your client listening skills:

- Stop what you are doing and look at the person who is speaking to you.
- Maintain eye contact with the client. If you are wearing sunglasses, remove them so the other person can see your eyes.
- Focus on what the client is saying, instead of planning what you will say when the client is finished speaking.
- If you understand what the client is saying, nod your head to indicate so.

- When appropriate, repeat or summarize the main points of what the client is saying using your own words (paraphrase), and provide some feedback to confirm that you understand.

- Ask questions, when necessary, to clarify your understanding of what the client is saying.

- Let the client speak freely. Do not dominate the conversation, interrupt, or jump in to finish people's statements.

Sensitive Topics

Handle sensitive topics carefully. How you express what you need to say can be just as important as the content of your statement. There can be cognitive effects from the logic you use to phrase your message, emotional effects from the empathy you show while delivering your message, and legal effects from the implications your message leaves. Consider the following strategies for sensitive communication when responding to client statements or questions:

- Preface questions with statements such as, "To make sure I understand . . . ," which tells clients that you are taking responsibility for understanding their needs.

- State your question as concisely as possible.

- Consider past experiences with other clients and formulate questions to clear up any common areas of misconception.

- Phrase your comments and questions carefully to avoid confusion or insult. Vague phrasing can make the client question your abilities to help. Seemingly confrontational or accusatory wording can put the client on the defensive.

- Be conscious of your tone of voice and facial expressions.

- Address the client's concerns in a straightforward way and keep to the point.

- Explain pertinent terms in plain language and avoid jargon. Be tactful when the client doesn't seem to understand. No one likes to feel stupid, so protect the client's feelings. This is especially important if your client is a presenter about to have an audience.

- Answer every question asked.

Conflicts

Managing conflict or working with challenging clients can be difficult. Challenging clients are not always negative clients; rather, they are clients who may not be easy to serve. When conflicts arise, the following may be helpful:

- Be calm and polite.

- Listen and be supportive.

- Try saying, "I can see that you are upset. I apologize . . . "

- Remember that clients are not allowed to verbally assault you.

Generally, you may encounter argumentative, overly talkative, and "know-it-all" types of clients, who require special handling. With argumentative clients, keep these points in mind:

- Do not make the situation worse with a negative comment.
- Listen to what they are saying and try to figure out why they are rejecting your suggestions.

Overly talkative clients can also cause problems. You might deal with them as follows:

- Never assume you know what they need and interrupt.
- Details can be very important when you are trying to solve a problem. Try to listen for the exact nature of the problem.

"Know-it-all" clients can be another source of conflict. Keep these points in mind:

- These clients may react to your assistance with a solution of their own.
- They may interrupt you, telling you what they know.
- Sometimes their suggestion is a great solution, so be open to possibilities.
- Listen to the client before responding.

Don't forget that you can contact your supervisor for advice when a situation becomes unmanageable.

Language Barriers

Language barriers may also impact your ability to communicate effectively. If you are providing assistance to someone who speaks a language other than your own, you may need to do one or more of the following:

- Speak slowly and clearly.
- Use plain language.
- Avoid slang. Remember that it is hard enough to comprehend an unfamiliar language, so do not complicate it with slang.
- Avoid jargon and acronyms unless the person is technically proficient.
- If the client asks you to repeat something, do not respond with increased volume.
- Write, draw, gesture, and demonstrate to show clients that it is important to you that they understand.
- Diagrams can be a helpful form of communication. Math can help as well. If you can figure out how to demonstrate a concept visually or mathematically, it may overcome a language barrier.

Speaking on the Phone

Your initial contact with a potential client will usually be on the telephone. During that first conversation, people will draw conclusions and make judgments about your business skills. If you are professional and knowledgeable, clients will likely return for your services. It is essential to the image of your company that you understand how to communicate over the phone successfully. Here are some suggestions to improve your telephone communication skills:

- Speak in a natural, friendly tone of voice.
- Smile. Your smile will not be seen, but it will make your voice sound friendlier.
- If you initiate the call, start off with your first and last name, the name of your company, and the reason for your call.
- No matter who placed the call, use the client's name frequently. Listen to the client's general tone and the way the name is given to determine how to address the client, such as whether to use just a first name or be more formal.
- Practice active listening. Give the call your full concentration.
- Strive to be clear in your communication, because you won't have the advantage of visual aids to help communicate your message.
- Keep your tone positive even if the caller has a complaint. Offer solutions, never excuses.
- Take notes when the discussion is detailed.
- Verify the client's factual information by reading from your notes or from an existing file.
- Keep in mind that some procedures are too complicated to explain over the phone. If this is the case, suggest meeting in person so that you can provide the details.
- If you feel that there is a language barrier, find someone who is familiar with the client's language and the technology to help interpret. Alternatively, suggest a different form of communication.

One challenge of telephone communication is that clients' descriptions of problems or needs are often subject to their own interpretations. Too often, clients assume that they understand the nature of the issue and leave important details out of the discussion. To overcome this, you must carefully word your questions in a manner that provides an objective description of the issue or problem, or arrange a meeting at the site to be able to personally examine the issue.

NOTE   In communicating with clients, remember that your role is to help the client accurately describe the problem or need. It is not the clients' responsibility to correctly identify the source of the problem or what they require of their AV system.

Business Writing

In pro AV, as in any business, written communication can take many different forms, such as e-mail messages, online chats, letters, and faxes. To write effectively using any of these forms, it is important to organize your writing, keep your audience's needs in mind, and follow basic grammatical rules. To successfully communicate in all these forms of writing, you should follow some basic guidelines:

- Begin by informally listing the points you want to make for your own notes, and then write the content.

- In business writing, it is appropriate to begin, "The purpose of this [e-mail/letter/ fax] is to [assist/inform/report] . . . "

- Creative, verbose, or academic writing styles are not ideal for business communication.

- After writing, compare the message to your initial list of important points.

- Eliminate any information that does not pertain to these points.

- Use meaningful details and do not ramble.

- Check your spelling, sentence structure, grammar, and punctuation.

- If you are delivering bad news, focus on what you are doing to fix the problem.

- Do not use technical jargon or acronyms that your audience will not understand.

- Do not type in all capital letters.

- Use an easily legible font.

- What you write is a reflection of you and your company. Remember that it may be reproduced and sent to anyone—prospective clients, competitors, superiors within your company, other people at the client's company, or even the media.

Some forms of writing require a more formal style than others. However, to write effectively in any form, you should follow the guidelines presented here.

Choice of Communication Form

With so many forms of communication to choose from, how do you decide which one to use? You obviously want to use the form of communication that allows you to communicate both accurately and efficiently with your clients. Be aware of your own preferences, and don't allow your personal feelings to influence your choice. Practice with different forms of communication and think of your client's preferences before your own. Consider these questions when selecting a form of communication for your client:

- With which forms of communication does your client contact you?

- With which type of communication will you fulfill your client's need?

- Which form of communication is most efficient for your client and you?

If you have several options of communicating, choose a form that your client is comfortable using. This will decrease the likelihood of clients missing messages, and it will make the interaction more comfortable for them.

E-mail

E-mail messages should briefly cover only a few main ideas. Information requiring long, detailed paragraphs, including multiple topics, should be conveyed in a formal letter. The more details you put into an e-mail message, the greater chance you have of losing your audience.

People typically do not like sending or receiving lengthy, technical e-mail messages with excessive information. These types of messages can be intimidating to clients and may portray the author in an unflattering manner. Instead of an overly detailed message, send specific directions or information. If you feel details may be valuable, then attach a document containing a thorough explanation.

When sending e-mail messages, keep the following in mind:

- Verify that the e-mail address is correct.
- Ensure that the subject line reflects the main topic of your message.
- Simply address the recipient using the client's name. It is not necessary to use "dear" in the greeting.
- Respond to e-mail in a reasonable amount of time, even if only to inform the client that an answer to the request is in progress.
- Do not conduct personal business using a company e-mail account.
- Do not gossip in e-mail. That is unprofessional, and your message could accidentally be forwarded to an undesired recipient.
- Do not include date or return e-mail addresses; they are automatically included.
- Do not make jokes. Even the most well-intentioned jokes can be misunderstood.
- Know when to stop. If your conversation goes longer than three e-mail exchanges, call the client for the sake of time efficiency.
- Take care when sending e-mail—you never know who may read it and how it will affect your company, your reputation, and your career. The best advice is to always assume that all e-mail could become public with no warning. If you must provide confidential company information, include only information that is required to communicate clearly. It may be appropriate to mark the e-mail "Confidential," or set a "Do Not Forward" flag if it is available in your e-mail system.
- While there are many benefits to communicating online, not everyone is comfortable with this form of communication. Consider the method your intended recipient has used to contact you. If the customer typically uses a telephone to contact you, then you should use the phone to reply. If you decide the most accurate way to convey your message is using e-mail, call the client first and explain that you will reply in e-mail form.

Online Chat

Online chatting and instant messaging are excellent ways to conduct brief, impromptu discussions internally and with clients. Online chat is intended to be very informal, but there are still a few guidelines that apply:

- Conversations should consist of only a few words and quick responses. If the client is asking a question that requires a longer explanation, choose a different form of communication.

- For those who have trouble with proper grammar, there is less expectation for accuracy due to the fast-paced nature of chatting.

- The nature of online chat has generated a need to abbreviate. Emoticons and acronyms are commonplace in online chat sessions, but be aware that they may not be understood by all of your clients.

- When answering questions, be accurate. Incorrect information can be recorded and repeated.

Faxes

Facsimile messages (faxes) have their own etiquette and protocol. Fax communication can be used to quickly obtain a legal signature on documents, to send graphics or diagrams that complement an explanation of a procedure or point of confusion, and to send messages when electronic files are too large or the recipient is not near a computer.

Use a fax cover sheet as the first page of every fax communication. Many companies have cover sheets professionally designed. If your company doesn't provide a template or cover sheet, you should create your own that includes the following:

- The contact information of the sender and receiver

- The number of pages you plan to send and if this count includes or excludes the cover sheet

- The purpose of the fax and what response you expect from the recipient

- The date and time you sent the fax

 NOTE Many fax machines automatically print the date and time on the received document, but it is good information to have for your records. This practice will become important if the receiving fax machine is set incorrectly or not functioning.

Keep in mind the characteristics of a fax when sending documents, and how these characteristics might affect your message. Here are some examples:

- The quality of the received fax will be poor compared with the original. When creating a fax, use a large font that is easy to read.

- It is acceptable to fax an AV design drawing to explain a point or procedure, but never use these versions as an official diagram. The process of faxing the document typically distorts the image detail and scale.

- Information that is faxed is legal and binding. You will be expected to honor all faxed agreements, estimates, and bids.

- Do not fax any personal information without the consent of the person receiving the information.

- Faxes are not completely confidential. Misdialed numbers or fax machines in common areas can compromise the confidentiality of the document being sent.

Legal Ramifications

There may be legal implications to your client communications, so be careful, even with your coworkers. Restrictions on disclosure of confidential or proprietary information may lead to legal action against individuals who disclose information about a company. Even repeating simple gossip can lead to rumors that become public, and you may find yourself named as the source of sensitive information.

Obtaining Information About Client Needs

Meetings with clients can happen at various times during a project, depending on many variables. One of the most important variables is when in the process the AV professional joins the project team.

AV designers and installers may be brought into a project at any stage of the design and construction process. In some cases, an AV company is contacted after an architect has already designed the building and/or room. When certain aspects of the building cannot be modified to create a suitable AV environment, it can compromise the effectiveness of the AV systems. Moreover, requiring changes to existing designs in order to support an AV system may result in additional cost.

In other cases, AV integrators or designers are hired at the beginning stages of a project. This allows for including the required infrastructure into the original building design, typically leading to a more successful AV environment.

In either case, the AV team should know at what point in the project it is beginning its role, because that will help determine the agenda for early client meetings. During those initial meetings with the client, the AV team must determine exactly what the client needs to address their AV requirements, including the services, infrastructure, and equipment.

Conducting Initial Meetings with the Client and End Users

The main objective of the initial client meetings is to gain an understanding of the end users' needs. The goal is to develop an AV design document that will include descriptions of the needed systems, along with preliminary budgets.

Depending on the scope of the project and the number of users to be interviewed, client meetings can take many forms. For example, you may have a single meeting with a few individuals that lasts for a few hours. Alternatively, you may have a series of meetings with different groups over several days. Regardless of the size and duration of the meetings, the issues to be addressed are the same: the functions that the system will need to support.

Best Practice: Focus on Functionality, Not Equipment

All participants in AV design meetings should understand that the goal of the meeting is to determine *functionality*, not equipment. While some discussion of equipment is inevitable (especially in short-term projects), the meeting should address the functionality required to address the users' needs.

For example, projectors often come up in discussions, because they can impact both the budget and the room design. But a discussion of projectors during an initial client meeting should focus more on what users want to do with the projectors than on projectors themselves. How many images do users need to support their tasks? What image sources need to be displayed? At what aspect ratio and resolution should the images appear to support users' tasks? The answers to questions like these will determine the parameters of the projectors you ultimately specify.

Client meetings to identify needs should typically be attended by the following people:

- Administrative representatives/stakeholders, including owner executives, facility managers, department heads, and funding organizations

- End users, including representatives from each organization or department that will use the new systems

- AV technology managers, representing the technical side of the owner's AV operational needs

- IT representatives, to address issues in utilizing or accessing the IT system and how IT relates to the users' activities

- Architects, to provide input regarding space allocation and planning

- Program or construction managers, who may be onboard as stakeholders in the overall project outcome

If people outside the client's immediate organization will be using the AV systems (as in videoconferencing facilities for hire, or a classroom building serving several departments), their perspectives should be represented whenever possible.

Conducting a Needs Analysis

The needs analysis is the most important part of the AV design process because the results determine the nature of the AV systems, their infrastructure, and the budget, including the impact of AV-related expenses on the building.

The goal of the needs analysis is to define the functional requirements of the AV systems based on the users' needs and desires and on how the systems will be applied to perform specific tasks. Merely developing an equipment list is not enough. While the equipment list is an essential part of the design and installation process, it is not part of the needs analysis. It is tempting—particularly from a sales perspective—to go straight to an equipment list. However, only by analyzing the users' needs can you ensure you address them fully through the selected equipment and system design.

A formal needs analysis includes identifying the activities that end users must perform and developing the functional descriptions of the systems that support those needs. During the needs analysis phase (which is known as the *programming phase* in the architectural trades), the AV professional determines the end users' needs by examining the following:

- The specific activities the end users want to accomplish within the space/room
- The required AV presentation application(s) based on the users' needs
- The tasks and functions that support the application(s)

The result of this needs analysis is a thorough understanding of which types of presentation system capabilities the users require to support their desired activities. These needs are translated into a detailed document, often called the *program report* or *brief*, that delineates the overall functional needs of the AV system (including budgets), based on the tasks the owner requires for operation. (We will cover the program report in more detail in Chapter 15.)

Understand a Need When You Hear It

The following are some typical examples of end users' needs:

- To communicate with staff within a company
- To save money on travel while enhancing face-to-face communication
- To support and operate a conference center
- To host many different types of events in the same venue
- To provide adult education
- To market or sell a product
- To improve situational awareness for a federal, state, or local government agency
- To monitor and operate an area-wide, nationwide, or worldwide network
- To provide entertainment in a performing arts, sports, or themed entertainment venue
- To enhance the worship experience
- To provide primary and secondary education in multiple locations

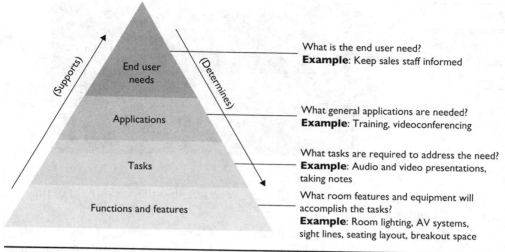

Figure 11-1 The needs analysis pyramid

The needs analysis pyramid, shown in Figure 11-1, illustrates both top-down and bottom-up approaches to examining end users' needs. The needs determine the applications that are required to support them, which in turn determine the functionality of the required AV systems (the equipment). Based on this information, you can determine some of the necessary infrastructure for the overall space (architectural, electrical, and mechanical).

Room Function

In some cases, the needs that the AV system must address are determined by the room. The facility type or room function is often known before the first client meetings. For example, maybe the client is building a new classroom, distance-learning center, corporate-training center, headquarters, boardroom, or network operations center. The facility or room type often has well-established AV system requirements that can address the typical users' applications within the room.

The following are some examples of standard rooms that AV designers and integrators work on:

- **Boardroom or conference room**   These are usually richly appointed rooms where interactive meetings are held, generally attended by the executives of a company. Most clients want these spaces to look impressive, which usually requires that all equipment be aesthetically integrated into the space. Seating is around a conference table, typically positioned in the center of the room. The presentation is made from one of three locations:

  - At the head of the table, with the projection screen behind the presenter

Figure 11-2
Rear-projection
system installation in a typical
boardroom

- From a lectern, with the projection screen to either the left or right of the presenter
- From the end of the table that faces the projection screen (see Figure 11-2)

- **Training room or classroom** This type of room tends to be multifunctional, meaning it needs to handle various types of presentations and AV applications. People are typically seated classroom-style at desks or theater-style in rows of seats. Presentations are given at the front of the room, commonly from a lectern, with the presentation screen to either the right or the left of the presenter.

- **Auditorium** This is a large room, commonly used for noninteractive presentations (usually informative or entertaining) to large groups. Presentations are made from the stage at the front of the room, with the presentation screens typically at center stage.

- **Divisible room or zone** This refers to a large area that can function as one large room or be divided into several rooms, each hosting its own event. For example, hotels and convention centers have large multipurpose spaces with sliding walls that are used to define room spaces. The shape and size of these rooms can be changed easily and quickly to accommodate the size of the group.

- **Videoconferencing or "telepresence" room** This is a space designed for meetings via two-way video and audio transmissions. Presentations are made from specific locations in the room, which have been preprogrammed into the system. Depending on need, these rooms can be designed so that the parties on either end appear as if they were in the same physical location.

Knowing the intended function of a room (or rooms, on larger projects) gives the AV team a starting point for determining client needs. Each of these types of rooms has standard technical requirements that the AV designer will be expected to address. Each room also presents specific design challenges. For example, certain videoconferencing rooms require that the space provide specific sight lines, lighting, and acoustical treatments. Large, divisible rooms are often subject to excessive sound reverberation and echoes, requiring the designer to pay special attention to room acoustics and sound reinforcement.

A Word About Client Politics

At times, relationships within the client's organization may affect how meetings are scheduled, attended, and managed. It is important to identify the differences among end users, decision makers, and funding groups. If they are separate entities, conflicts may arise among what end users say they need/want, what decision makers think end users need, and what the finance group is willing to spend money on. If the design team can decipher these relationships before any meetings are held, the meeting leader can be sensitive to potential problems in addressing each group's issues.

AV Tasks and Parameters

Once you have determined what activities will take place in a room and which AV applications are required to support them, you need to get more granular and identify the specific AV tasks that the system will handle, as well as the parameters of those tasks. This information helps establish system configurations and budgets.

One common AV task that supports many AV applications is image display. For this task, the following parameters might be identified:

- Sources/signal types to be displayed
- Aspect ratio of sources
- Number of simultaneous images
- Source resolutions
- Display distance from viewers

These parameters feed into the design process to inform the team about both facility requirements (display size, room size, room configuration, lighting, etc.) and system requirements (type and brightness of projectors or displays, number of inputs, video-switching requirements, etc.).

Another common task is audio playback, which could have the following parameters:

- Audio signal types
- Number of audio sources
- Area to be covered by loudspeakers
- Distribution of audio to other locations

You should also seek out information about infrastructure issues that might impact the AV design and installation. The following are some of the issues to cover:

- **Space allocation**   If space has already been allocated to the AV systems, part of your assessment should be to verify that the space will accommodate the application.
    - Is there adequate seating and workspace area?
    - Is there adequate accommodation for a presenter and lectern or podium, if needed?
    - Is there adequate space for equipment, including rear-projection, control, or equipment rooms, if needed?
    - What are the sizes and locations of the required images, and can they be accommodated?
    - What issues related to acoustics; lighting; and heating, ventilation, and air conditioning (HVAC) might have an impact on the overall project budget and space allocation?
- **Security**   What are the hours of operation for each area of the facility that will house the AV systems?
- **Electrical/lighting**   There are several issues related to electrical and lighting systems. Ask yourself and the client:
    - Is a light-dimming system required?
    - What areas require zoned lighting?
    - Are control interfaces required for lighting, drapes, shades, or other systems that are not included under the AV scope of work?
    - Are there nonstandard lighting systems required (such as for videoconferencing or performing-arts functions)?
    - Will there be additional or special electrical systems dedicated to the AV systems?
- **Data/telecommunications**   Are special or additional data/telecom outlets and services required for audio, video, or control systems?

Table 11-1 shows possible AV tasks and parameters to support end users' applications and needs. This gives you an idea of the type and level of detail you need to create the AV system design specification and budget.

AV Task	Parameters of AV Task
Image display	Number of simultaneous images Source resolutions Sources/signal types to be displayed Aspect ratio of sources
Audio playback	Number of audio sources Audio signal types Area to be covered by loudspeakers Distribution of audio to other locations
Speech reinforcement	Number of talkers to be reinforced Location of talkers Area to be covered by loudspeakers Interface to other systems
Audioconferencing	Stand-alone or part of videoconferencing? Concurrent with speech reinforcement? Local bridging for multiparty audio
Videoconferencing	Dedicated function or incorporated with presentation system? Number of participants Single-axis or dual-axis (participants in audience only, or is there also a presenter)? Number of images required Resolution of conferencing images Type of connections to be supported (ISDN, IP, satellite, fiber-optic, broadcast, etc.)
Overflow/interconnection of spaces	Identification of potential source spaces Interconnection of spaces Identification of potential destination spaces Number, type, resolution, and format of audio and video signals to be connected One-way or two-way connections required? Are sites within or outside the project facility? Type of connections to be supported (ISDN, IP, satellite, fiber-optic, broadcast, etc.)
Recorded media	Types of recorded media playback Audio and video parameters of media content Accessed by technicians and end users?
Recording of events	Audio and/or video Audio and video parameters of recorded signals Set up and controlled by technicians and end users?
System control	Locations of control within each space What do the control systems need to do? Who will be controlling the system (end users, assistants, technicians, etc.)? Local and/or remote control required? System-wide monitoring Help desk functions Interfacing to other devices or systems required (lighting, drapes, building automation system, etc.)?
Ancillary systems	Any ancillary systems required to support the users' activities or spaces (audience response systems, background music, sound masking, intercom, nurse call, security, etc.)?

Table 11-1 AV Task Parameters

AV Needs Analysis Meeting Agenda

The following example of an AV program meeting agenda shows some typical issues and items for discussion. Note that the specification process may require several meetings to interact with different stakeholder and end-user groups.

AV System Needs Analysis Meeting Agenda

Overview

  A. Project overview

    1. Identify project type.

    2. Identify project schedule.

    3. Identify major project applications required for end users' operations.

    4. Identify user groups and functions.

  B. Identification of owner and end-user vision and style

  C. Differentiation of AV system functionality vs. AV equipment

    1. Define expectations of the process.

      (a) What is needed/expected from the design team?

      (b) What is needed/expected from the owner/user?

      (c) How will the design team and owner/user communicate?

      (d) System quality

  D. Technology trends

Review Existing Documents, Facilities, and Infrastructure

  A. Review pertinent parts of architectural program

  B. Review any existing AV program information

  C. Tour existing facilities

Identify User Functions

  A. Existing functions

  B. Anticipated functions

Identify Overall User Standards and Requirements

  A. Standards

  B. Benchmarks

  C. Known connectivity requirements

  D. Known basic AV requirements

E. Internal tech support availability

F. Accessibility considerations for disabled persons as required by AHJ

Discuss Each Space or Area

A. Identify each area that requires a system

B. Space-by-space functional review

1. Functions required for each space

2. Operational requirements (day, night, remote monitoring, etc.)

C. Identify AV tasks and parameters for each area

1. Identify major equipment requirements (number of images required, room size and seating, conferencing required, audio and video sources, etc.).

2. Identify potential impact on infrastructure.

(a) HVAC

(b) Security

(c) Electrical

(d) Lighting

(e) Data/telecom

D. Owner-furnished equipment

E. Budget issues and priorities

Conclusion

A. Identify key individuals and contact information for follow-up

B. Identify follow-up meetings

C. Discuss schedule for completion and distribution of report

Figure 11-3 shows an overview of how the end users' needs ultimately define the design team's ability to create the required AV systems, and how the AV systems have an impact on the architectural, electrical, and mechanical infrastructure design.

Benchmarking

Benchmarking refers to the process of examining methods, techniques, and principles from peer organizations and facilities. AV designers and integrators can use these benchmarks as a basis for designing a new or renovated facility. For example, a client may identify a specific existing facility that provides capabilities or design elements that the client would like included in the facility your company will design.

Figure 11-3
Translating needs
into a design

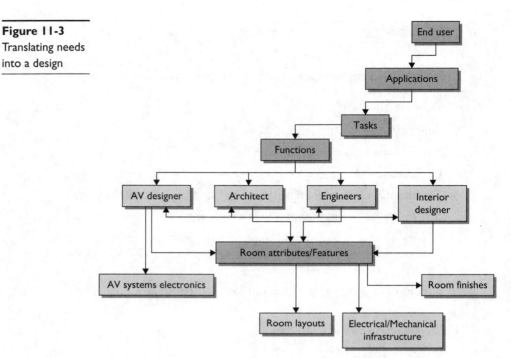

Benchmarking Benefits

Benchmarking offers the owner and the design team a common (and sometimes expanded) vision of what the client wants and needs. Seeing a number of locations of similar size, scope, and functionality can help guide a new facility design. Benchmarking offers the following benefits:

- It provides an opportunity to see varying approaches to design-versus-budget decisions.
- It allows project stakeholders to open a line of communication with other building managers and end users to evaluate what they learned through the design and construction process, and to discuss what they would do the same or differently if they did it over again.
- It may inspire new design ideas.
- The team can identify successful (and unsuccessful) designs and installations that include elements that relate to their project.
- It can help to determine which functions and designs are most applicable to the current project.

Benchmarking Site Visits

During a benchmarking site visit, the AV team and the client should collect information about site characteristics they intend to consider during design and/or installation. The main objective is to identify desirable AV features and functionality.

Arranging a benchmarking trip involves the following steps:

- Determine appropriate facility types to visit. These may be a precise match to the owner's operation, or they may be facilities with similar functions and operational needs.

- Create a list of potential facilities to visit.

- Check whether the benchmark sites allow such visits. Some benchmark visits may only require a user's perspective of a public facility, but often you will want to visit a private facility, for which you need permission to enter. If you are planning to benchmark a secure facility—in support of a government or health-care project, for instance—you may also need to gain security clearance.

- Most benchmarking visits benefit from a behind-the-scenes tour, which may require coordination with the technical staff, including those who support the facilities computer networks.

- Narrow down the options to a final list.

- Determine who will go. The benchmark group may draw from the end users, the owner's technical staff, the owner's administrative managers, and the architectural design team, as well as the AV provider.

- Schedule and make the visits.

- Write a benchmarking report summarizing the sites visited, the pros and cons of each site, what impact there is on the client's anticipated needs, and the resulting AV systems that will support those needs.

The following checklist provides examples of useful information you may want to collect during a benchmarking site visit. Once the benchmarking tour is set up, use this list to make the most of the visit and glean information that will be useful to the design team, owner, and end users.

Benchmarking Checklist

Facility Information

- ☐ Organization, facility, and location
- ☐ Contacts at the benchmark organization
- ☐ AV project installation date
- ☐ AV budget at time of installation
- ☐ Delivery method used for the project
- ☐ Style of the project: low-, mid-, or high-end
- ☐ Is there any facility or system documentation that can be shared with the design team?
- ☐ Is photography allowed?

Benchmark Information

- ☐ What is the benchmark facility's focus?
- ☐ How is it like the facility you are working on?
- ☐ How is it different?
- ☐ What technologies are used and how do they support the end users' activities?

Features As They Relate to the Project Under Design

- ☐ Audio systems
- ☐ Video systems
- ☐ Control systems
- ☐ Local-area and wide-area networks
- ☐ Integration with IT
- ☐ Lighting
- ☐ Acoustics
- ☐ What technologies, design approaches, or criteria were used at the benchmark facility?
- ☐ How do the AV systems integrate with (or not integrate with) the facility's IT network?
- ☐ What are the benefits and drawbacks of the design approach taken?
- ☐ What accommodations were made for upgrades and additions to the systems?

Facility Management

- ☐ How does the benchmark facility organization manage its AV technology?
- ☐ How is the benchmark project serviced in terms of operations, maintenance, and help desk?
- ☐ How much staffing is required to operate and manage the AV systems, and what are the qualifications of the staff members?
- ☐ What are the benefits or drawbacks of their technology management approach?
- ☐ How do these approaches compare to the existing and/or planned facilities?

Owner and User Feedback

- ☐ What do the benchmark facility end users have to say about the technology and the facility?
- ☐ What do the benchmark facility technology managers and technicians have to say about it?
- ☐ What do the benchmark administrators have to say about it?

Conclusions

- ☐ Which features of the benchmark facility should be included in your project design?
- ☐ Which features of the benchmark facility should be avoided in your project design?
- ☐ What aspects of the benchmark facility's technology management approach should be developed or avoided in the current project owner's organization, and what impact does that have on the facility and its systems?

Obtaining Scale Drawings of the Customer Space

In Chapter 12, you will examine further some of the standards and formats for site-plan documentation. Whether the facility you will be working on is already built or is being built, you will want to acquire scale drawings of the space. For projects that are under construction, a range of documentation about the existing physical, organizational, and technical characteristics of the site should be available for review. If the AV systems are to be installed in an existing facility, inquire about scale drawings, and tour the area during the design process to document the physical aspects of the space and how it is currently being used.

Depending on the building system you want to understand, there may be several sources of documentation. Each of a building's systems is defined by its own subset of drawings, but even those may not include all the information you require, so it is important to work with building owners and/or architects and contractors to round up all the necessary information. A particular system's functions are usually defined by discipline drawings, while the locations of system devices are shown on the architectural set of drawings. For example, the subset of drawings for an electrical system may define the electrical system itself, but not its physical relationship to the building. That information may be found in architectural drawings and would be important to the AV system design and installation.

In addition, you will want scale drawings and information about so-called "hidden" systems, which may not strike other members of the project team as critical to the AV design, but could actually play an important role in how AV systems will be installed or mounted on ceilings and walls. These hidden systems include structural, mechanical, fire protection, and others that an AV system designer and installer must consider when incorporating an AV system in the room. Figure 11-4 shows an example of a plan for speaker installation in a ceiling.

Ultimately, when locating an AV device based on information in architectural drawings, it is important to consider various elements and dimensions provided in the drawings. When examining drawings of a space, consider the following:

- Mechanical drawings show the ductwork that goes through the building.
- Electrical drawings show locations of power and lighting.
- Reflected ceiling plans depict the ceiling grid, diffusers, fluorescent lights, down lights, sprinklers, projection system, and loudspeakers.
- Site drawings locate geographically where you will need to go to perform the installation.

The most important thing to remember is that you are seeking the fullest set of drawings available for the rooms that your AV design and installation tasks will address. And be sure to examine all of the elements that may potentially affect the layout, mounting, installation, and operation of the AV system components.

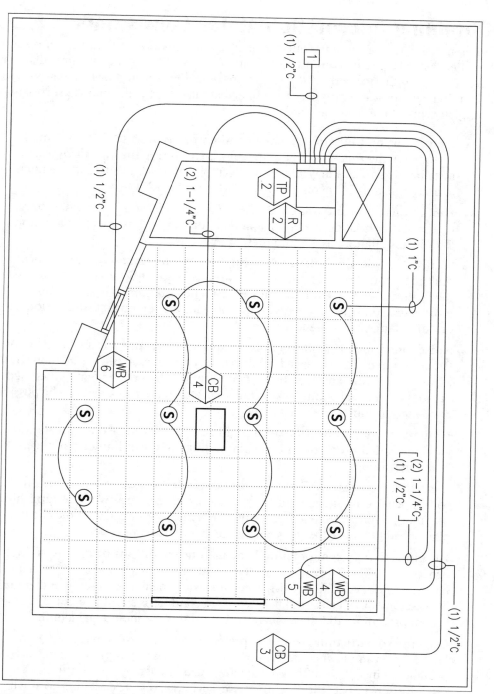

Figure 11-4 Example of plan for speaker installation within a ceiling

Obtaining Information About Constraints

When meeting with a client to collect general site information and information about the client's needs, it is also important to learn about any issues or constraints that may impact your ability to work at the client site once the design and installation tasks begin. Here are some examples of constraints that may affect your ability to successfully complete work at the customer site:

- **Limitations on times of day when on-site work is acceptable** This could include restrictions on working in some areas during the day, since your installation activities may adversely affect the ability of on-site staff members to accomplish their ongoing work.

- **Limitations to on-site work activities** For example, you may need to store equipment in backroom areas of a hotel in order to minimize the disruption that the installation tasks could present to guests.

- **Security** If you are working in a secure facility—for example, a government building—you may be required to obtain special identification to gain access to the facility. Certain facilities require even higher levels of security clearance. In most cases, you will not have won the job without demonstrating that you have or can obtain such clearances.

- **Noise limitations** Limits on noise in specific areas may affect what types of tasks you can perform at specific times of the day. For example, installation within a broadcast studio area may be limited to times when there are no scheduled broadcasts, or installation work within a cafeteria dining area may not be possible until after the lunchtime rush is over.

- **Limitations on parking or loading vehicles** These can be due to rush-hour parking restrictions on the streets surrounding the site or limited parking in the building parking lot.

- **Areas that require special care or attention** These areas might include rooms with expensive and fragile finishes or features, expensive artwork, or furnishings that must be protected. Note these types of issues, and work with the client to plan ahead to ensure that these items are safe and secure.

- **Locations that present potentially dangerous conditions or require special procedures** These might include areas within manufacturing facilities that may have hazardous conditions, such as moving equipment, the presence of flammable gases, or special electrical hazards. Such conditions should be noted, and any special procedures to follow when working at the site should be identified.

- **Ongoing construction** This may present a range of problems for an installation. For example, incomplete portions of the site might present dangers or prevent the installation from proceeding. Also, there may be conditions that can adversely impact the AV equipment, such as excessive dust or exposure to weather.

- **Cultural issues**   These issues can affect how and when your team performs its work. For example, when conducting work activities within a religious institution, it is important to understand which days are considered nonwork days for that religion. You should also be aware of any behavioral issues that may offend the adherents to the particular religion. Language barriers are another potential issue at specific sites.

When collecting the client and site information, make sure that you note any of these types of issues and plan ahead to ensure that they do not adversely impact the design and installation tasks. In many cases, the team can identify a method to work around constraints, such as working during the evening or night in areas where daytime disruptions are prohibited.

Chapter Review

In this chapter, you examined gathering customer information. You can expect about six exam questions addressing the knowledge and skills related to material in this chapter. You should understand the following:

- How to communicate effectively with clients, from ascertaining contact information to ensuring you use the best communication medium in the best possible way
- How to obtain information about client needs, including specific AV-related tasks that the AV system should be designed to support—that is, defining the end users' needs and using that information to determine the AV system functions necessary to address those needs
- Issues associated with obtaining scale drawings of the client space, including both existing spaces and any new buildings or rooms that remain to be completed
- What type of information you should collect on site-related constraints that may affect your ability to successfully complete the work at the client site, including the times of the day in which you are authorized to work at the site, any limitations on noise levels that could affect installation tasks, and cultural issues that you should be aware of when working at the site

Review Questions

The following review questions are not CTS exam questions, nor are they CTS practice exam questions. These questions may resemble questions that could appear on the CTS exam, but may also cover material the exam does not. They are included here to help reinforce what you've learned in this chapter. For an official CTS practice exam, see the accompanying CD.

1. Which of the following is *not* considered part of "active listening"?

   A. Focusing on what the person is saying

   B. Summarizing and paraphrasing the person's statements

C. Asking frequent questions to guide the conversation to the topic you are interested in discussing

D. Maintaining eye contact with the person

2. What is usually the best approach for communicating detailed technical AV plan information to a client?

A. Sending the information within an e-mail message

B. Sending the client a document containing the information

C. Sending the client a fax

D. Calling the client on the phone

3. What is the most valuable source of information when defining the needs for an AV system?

A. Architectural drawings of the building

B. Feedback from benchmarking site visits

C. End-user descriptions of the tasks and applications the AV system will support

D. Client/building owner preferences for AV system equipment

4. What is the main purpose of an initial needs analysis?

A. Identify the specific equipment needs for the desired AV system

B. Determine the overall design of the AV system

C. Obtain the client's vision of the AV system design

D. Identify the activities that the end users will perform and the functions that the AV system should provide to support these activities

5. How does knowledge of the overall room function assist the AV system designer in defining the client needs?

A. Provides a starting point from which to determine client needs

B. Defines the functionality that the system elements should provide

C. Provides a standard design that can be used for most clients

D. Provides a standard design template that can be given to the building architect

6. How are task parameters used when defining user needs?

A. To provide overall information about the desired general AV needs

B. To define the layout of AV components within a room

C. To define the specific AV functions that the components must support

D. To define the specific AV components

7. What is the purpose of benchmarking?

   A. To demonstrate specific AV equipment in operation

   B. To give the client an opportunity to experience a number of AV system designs that address similar needs

   C. To test AV system designs

   D. To evaluate AV vendors prior to final selection of a vendor

8. Why is it important for the AV professional to obtain a full set of building plans?

   A. To understand the full range of room features to ensure that the AV system design takes other building systems and components into account

   B. To evaluate and approve plans to ensure that room elements are compatible with the AV system installation needs

   C. To be able to use general site plans to determine the layout of the selected AV components

   D. Detailed building plans for the rooms in which the systems will be installed are typically not required; only a general floor plan is necessary

9. What client contact information should be collected during initial client meetings?

   A. The main client contact

   B. The client technical representative

   C. Contacts that may be required for site inspection and installation, including the architect, building manager, construction manager, security manager, IT manager, and so on

   D. All of the above

10. How does the AV team use information about any identified constraints to the AV design and installation tasks?

    A. To select alternative AV system components that are not affected by the identified constraints

    B. To inform the client that these constraints must be removed prior to installation tasks

    C. To develop a work-around plan when these constraints will affect the design or installation tasks

    D. To eliminate specific tasks that may be adversely impacted by constraints

Answers

1. **C.** Active listening includes focusing on what the person is saying, summarizing and paraphrasing the person's statements, and maintaining eye contact with a person. It does not include asking frequent questions to guide the conversation to the topic you are interested in discussing, because the point of active listening is to actually hear what the client is trying to communicate.

2. **B.** Usually, the best approach for communicating detailed technical AV plan information to a client is to send the client a document containing the information. Other methods, such as e-mail, fax, or telephone conversations, are likely too informal to enable the client to review the detailed information.

3. **C.** The most valuable source of information when defining the needs for an AV system is usually end-user descriptions of the tasks and applications the AV system will support.

4. **D.** The main purpose of an initial needs analysis is to identify the activities that the end users will perform and the functions that the AV system should provide to support these activities.

5. **A.** Knowledge of the overall room function assists the AV system designer by providing a starting point from which to determine client needs.

6. **A.** Task parameters are used to provide overall information about the desired general AV needs.

7. **B.** The purpose of benchmarking is to give the client an opportunity to experience a number of AV system designs that address similar needs.

8. **A.** It is important for the AV vendor to obtain a full set of building plans because the AV designer must understand the full range of room features to ensure that the AV system design takes other building systems and components into account.

9. **D.** The AV team should collect the full range of client contact information, including the main client contact and the client technical representative. Also, the AV team should collect information about any contacts that may be required for site inspection and installation, including the architect, building manager, construction manager, security manager, IT manager, and so on.

10. **C.** The AV team uses information about any identified constraints to the AV design and installation tasks to develop a work-around plan when these constraints will affect the design or installation tasks.

Conducting a Site Survey

In this chapter, you will learn about
- Visiting construction sites
- Observing and documenting site conditions
- Interpreting blueprints, site plans, and CAD drawings
- Calculating areas and volumes of installation spaces

Visiting the client site is a key step in the AV system design and installation process. A site survey is your opportunity to meet with clients, discuss their AV support needs, and review the characteristics of the various locations where AV systems will be installed. Your objective is to collect as much information as possible about the space you will be working in to provide a basis for creating your system proposal.

You will need to examine the site and document its relevant features, such as room dimensions, layout, and any issues that may impact the AV system design and installation. You will also need to interpret blueprints or other drawings that depict how the room will appear once construction is complete. In many cases, the client site may be under construction, so it is important that you understand safety requirements and follow proper procedures when visiting the site.

Domain Check

Questions addressing the knowledge and skills related to conducting a site survey account for about 6% of your final score on the CTS exam (about six questions).

Health and Safety Requirements at a Work Site

Safety at a work site is everyone's responsibility. From falling objects at a construction site to electrical dangers at a client-occupied facility, anywhere you install AV systems could potentially present hazards that might cause serious injury or even death.

While an AV technician will not be performing installation tasks during an initial site visit, other construction activities that pose safety hazards may be underway at that time. In some cases, a technician inspecting a site may be required to climb ladders to examine ceilings or enter potentially hazardous areas.

The following are some examples of potential safety issues at construction sites and event locations:

- Improperly taped, frayed, or otherwise damaged cables
- Blocked egress or covered exit signage
- Improper overhead rigging or hanging equipment
- Hazardous equipment operation
- Trip hazards

Prior to visiting the work site, be sure to ask the client about any safety requirements or personal protective equipment (PPE) that may be required to enter the site. Also, your company should have guidelines concerning safety. Make sure you read those guidelines and follow them.

Safety Regulations

When present at a work site, AV technicians must comply with all applicable safety regulations. These include those set out by administrative rules in the project documentation package, as well as local, regional, and national safety codes and regulations. Applicable rules are defined by what is known as the authority having jurisdiction (AHJ).

 NOTE In construction and contracting, the AHJ is often—but not always—a government agency (state, local, or other) that regulates various aspects of the process in that location. For example, the local fire marshal in a town where you're working may be the AHJ over fire codes. You will often be referred to the AHJ because rules and regulations can vary significantly from one town, city, or state to another.

In most cases, compliance with safety rules is also a requirement of the client's insurance company. Depending on conditions at the site, anyone entering the work site area may be required to use PPE, such as hard hats, steel-toed shoes, hearing protection, and safety glasses. In some specialized locations, specific safety training and/or equipment may be required for all persons working at the site.

Best Practice

AV professionals should follow the most restrictive code for the region in which they are working. If you find the interpretations of applicable codes and standards are in conflict, follow the requirements of the local AHJ.

Safety Equipment

The following are examples of standard safety equipment that an AV technician may be required to use at a work site:

- **PPE**   This includes equipment such as the following:
  - Hard hats help protect your head from unexpected injuries. For example, when inspecting a ceiling, a hard hat can protect you from sharp support elements that may be hard to see. Hard hats also protect your head if you fall from the access equipment.
  - Work boots or steel-toed footwear that meets applicable standards can help protect your feet from being crushed or cut.
  - Work gloves protect your hands from cuts, scrapes, burns, or contact with harmful materials.
  - Safety glasses help protect your eyes from dust and other hard-to-spot debris.
- **Approved ladders**   Make sure you use the correct type of ladder at a work site. Never use electrically conductive ladders, even if you are not working with electrical components. Someone else at the site may make a mistake that results in an electrically charged surface, which could cause electrocution of anyone standing on a conductive ladder. Fiberglass ladders are preferred, and wooden ladders are generally acceptable.
- **Fall protection**   Equipment for fall protection is a requirement for many site inspection and installation tasks. Since falls are the leading cause of worker fatalities at work sites, your AHJ may require that workers use approved fall-protection harnesses whenever working at a height greater than 6 feet, or 2 meters.

Always use the proper tools for the task you are about to perform. You should inspect your equipment regularly to make sure it is in good operating condition.

Gathering General Site Information

CTS-certified professionals visit a client site in order to collect information about its layout and the client's needs. The goals are to ensure that the designer can create an appropriate AV system design and that the installation team can install it properly.

Site information can include items such as building location and characteristics, locations and sizes of rooms, construction styles and features, electrical capacity, and HVAC component locations. You want to collect information about the current state of site conditions as they relate to client needs (Chapter 11 covers gathering client information), which may lead to changes and additions to the infrastructure.

Prior to visiting the site, think about what information is necessary to support the needs of the AV design and installation. The following sample checklist illustrates the type of general site information you should collect to support AV tasks.

Onsite Survey Checklist

Use this checklist to collect general information about the characteristics of the site.

☐ Site contact name _____

☐ Site contact phone number _____

☐ Site contact mobile number _____

☐ Exact address of job site _____

☐ Best route to job site _____

_____

☐ Travel time from warehouse to site, taking into account normal traffic conditions for your scheduled trip _____

☐ Type of loading/unloading/parking access (hour or time restrictions)

_____

☐ Location of loading/unloading access _____

☐ Access route from delivery dock to storage area

_____

☐ Elevator dimensions _____

☐ Security concerns (such as whether the area can be locked and who has a key)

_____

_____

☐ The primary function of the work site room(s)

_____

☐ The room's proximity to other functions in the same building and area

_____

_____

☐ Name and contact number of facility owner's representative, site's maintenance chief, or AV technician (who will know where power connections and ducts are?)

_____

_____

☐ Potential for electrical interference with other equipment _____

_____

_____

☐ Potential for any other problems regarding room location (for example, ambient noise from outside the room) _____

_____

☐ Potential for any problems regarding traffic patterns during installation

_____

_____

☐ Dimensions of the room(s): ceiling height, room length, and width

_____

☐ Ceiling type (drywall, drop ceiling, location of joists)

_____

☐ Wall material (drywall, block, etc.) _____

☐ Ambient noise measurement (using a meter) _____

☐ Acoustical properties of the room (echoes, loud mechanical noise, outside noise, voices or sounds from adjacent room, etc.)

_____

_____

☐ Existing sound system (if there is one, and if so, what kind is it?)

_____

☐ Ambient light measurement (using a meter)

_____

☐ Natural light from windows (can it be masked if necessary?)

_____

☐ Existing lighting _____

☐ Existing security lights that might make lighting difficult

_____

☐ Electrical capacity of the room _____

☐ Location(s) of electrical panelboard(s) serving the room _____

☐ Existing AV features or equipment _____

☐ Existing IT network and location of network ports in the room

_____

_____

☐ Location of network data center or server/equipment room

_____

_____

(Continued)

☐ Possible obstructions to audience view (such as chandeliers, sliding walls that are not completely retractable, and pillars)

_____

☐ Suitability of room(s) to accommodate the audience size and the type of equipment being considered

_____

☐ Seating capacity of room(s) according to requested setup

_____

☐ Room shape(s) and orientation of the requested setup

_____

_____

While collecting site information, also make note of the following:

- Any extra gear that you may need during the installation, such as extension ladders, lifts, coring tools, and hole saws.
- Potential mounting positions and boxes required for equipment within the room. For example, if any mounting must be performed in lecterns or on walls, it may be necessary to acquire the services of a carpenter for professional finishing details.

It is a good idea to bring a camera to the site to take pictures for future reference, in case questions come up later in the project.

Interpreting Site Layout Drawings

It is extremely important that CTS-certified professionals involved in designing and installing AV systems be able to read and interpret architectural site drawings and AV system plans. Through these plans, you understand the vision and scope of the client's project, as well as the AV system installation requirements.

Two primary groups of drawings are of interest to the AV system designer and installer:

- **Architectural drawing package** The drawings in this package illustrate the overall design of the building and rooms. These include the room layouts; features and dimensions; and locations of building systems such as HVAC, plumbing, lighting, and electricity. Small jobs may have only one or two drawings.

Bigger jobs can have entire sets, divided into different groups based on the construction process, such as the following:

- Mechanical drawings to show heating, plumbing, and ventilation
- Electrical drawings to show power and lighting systems
- Structural drawings to show the wood, concrete, and steel structure of the building support system
- **AV project drawing package**   This package provides the overall picture for an AV installation. These plans, created by the AV system designer, depict the equipment, where each component will be located, and how each component should be physically installed. The package typically contains functional diagrams, connection details, plate and panel details, patch panel details, equipment diagrams, rack elevation diagrams, control panel layout, and more.

This section focuses on architectural diagrams, because these provide the information that the CTS-certified professional should collect when making an initial site visit. You will learn more about AV project drawings in Chapter 16.

Drawing Views

The first thing to notice when reading a drawing is the perspective from which it is drawn. The perspective from which a drawing looks at the space is referred to as a *view*. For example, a drawing could be depicted from a ceiling view or a side view.

Architectural drawings provide a number of different views of the site, each from a different reference point, to convey design and layout concepts. Views used for drawings depict the site from the top, front, side, and back. Here, we'll look at a few of the different types and subsets of drawings you will typically encounter during a project.

Plan Drawings

The top view, also called a *plan drawing*, is a view taken from directly above, showing the floor plan or site plan, as shown in Figure 12-1. The floor plan identifies the room locations and layout dimensions; locations of walls, doors, and windows; and more. A floor plan view may also include indicators to other (more detailed) views or drawings.

Reflected Ceiling Plans

To illustrate elements in the ceiling with respect to the floor, a *reflected ceiling plan* is used, as shown in Figure 12-2. It is called *reflected* because it is intended to be interpreted as though the floor was a mirrored surface, reflecting the ceiling plan. The reflected ceiling plan shows the locations of elements such as ventilation diffusers and returns, sprinkler heads, and lights.

These drawings are of interest to the AV designer and installer because locating AV equipment such as loudspeakers and projectors may require coordination with allied trades. For example, if you are designing and installing AV systems to meet performance standards, the placement of lights and HVAC diffusers may conflict with the necessary location of projection screens or loudspeakers.

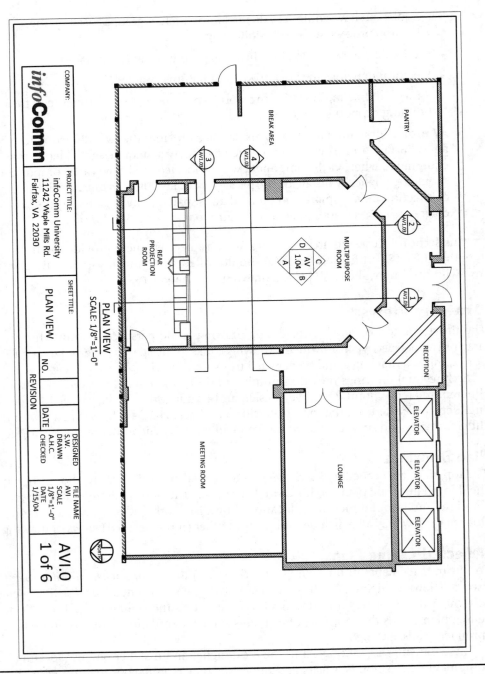

Figure 12-1   Example of a plan view

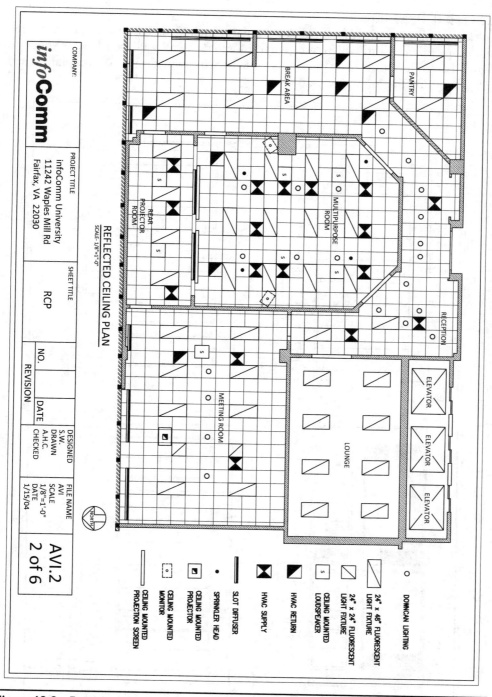

Figure 12-2 Example of a reflected ceiling plan (shows the position of loudspeakers and rear-projection system screens)

Elevation Drawings

An *elevation* is a drawing that looks at the environment from a front, side, or back view. Elevations provide a true picture of what the interior wall will look like, as shown in Figure 12-3. Elevations show electrical outlets, windows, doors, AV faceplates, chair rails—anything that might be on the wall.

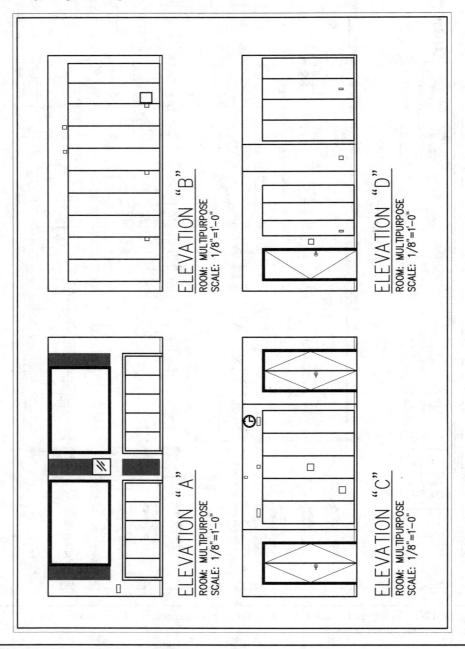

Figure 12-3 Examples of elevation drawings

Section Drawings

A view of the interior of a building in the vertical plane is called a *section*. A section drawing shows the space as if it were cut apart, and the direction you are looking is indicated by an arrow in the plan drawing, as shown in Figure 12-1. Section drawings show walls bisected, which allows you to view what is behind the wall and the internal height of the infrastructure. A section drawing may be rendered at an angle, so study it carefully. Figure 12-4 shows examples of section drawings.

AV professionals use section drawings to plan for installation needs, such as mounting locations, cable runs, and more.

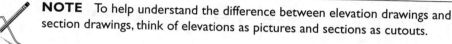

NOTE To help understand the difference between elevation drawings and section drawings, think of elevations as pictures and sections as cutouts.

Detail Drawings

Detail drawings depict small items that need to be magnified in order to show how they must be installed. Detail drawings show items that are too small to see at the project's typical drawing scale, as shown in Figure 12-5. Details may show how very small items are put together or illustrate mounting requirements for a specific hardware item, such as a ceiling projector mount.

Title Blocks and Drawing Schedules

In the *title block* of a drawing, you will find valuable information to help ensure that you are working with the proper and most current drawing. For orientation, you need to identify the drawing numbers, compass points, scale, drawing date, revision date, and sheet title. Figure 12-6 shows an example of an architectural drawing title block.

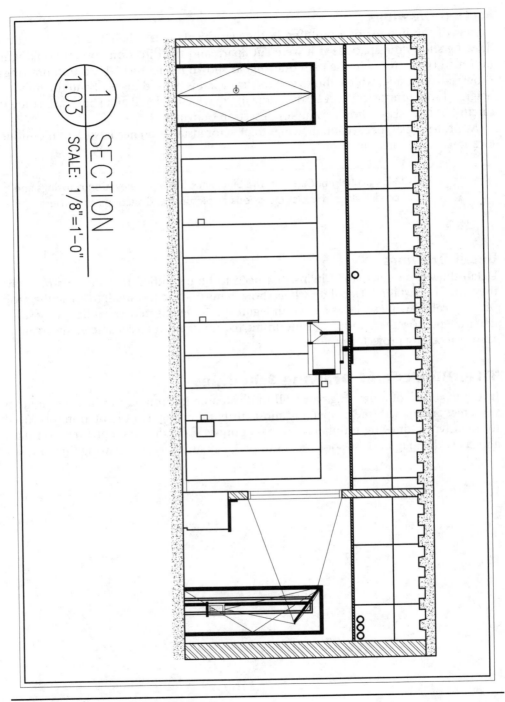

Figure 12-4 Examples of section drawings

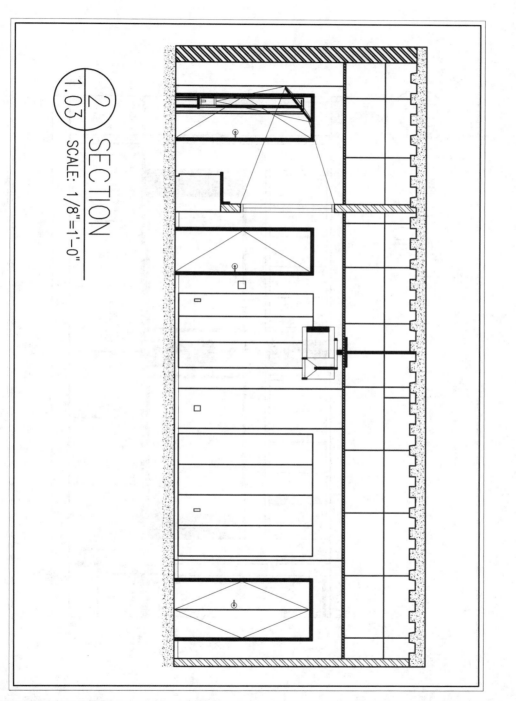

Figure 12-4 Examples of section drawings *(Continued)*

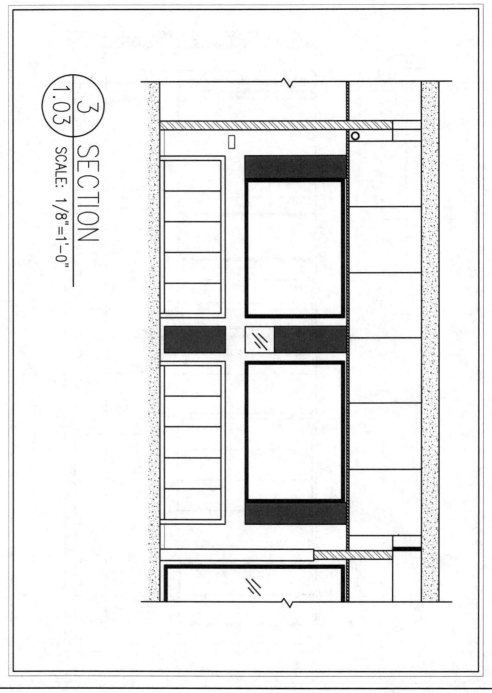

Figure 12-4 Examples of section drawings *(Continued)*

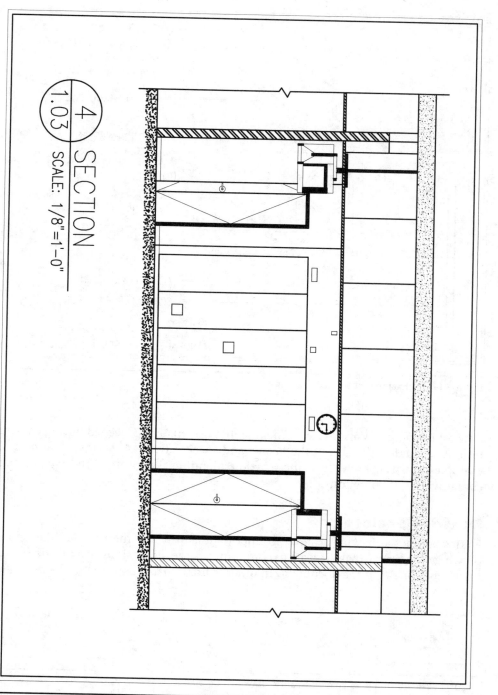

Figure 12-4 Examples of section drawings

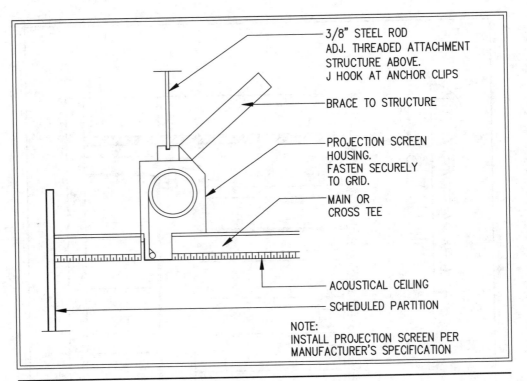

Figure 12-5 Example of a detail drawing

Most drawing sets also have a schedule that lists materials needed during construction. A schedule will call out the details and specifications of the installation. In the example shown in Table 12-1, the schedule identifies the location and type of boxes containing AV terminations.

Drawing Scales

Understanding a drawing scale is an important skill when interpreting dimensioned drawings. Scale drawings are used to communicate the dimensions of a full-size project on a paper or electronic document. In other words, what is depicted in the drawing is

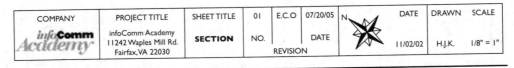

Figure 12-6 Example of an architectural drawing title block

Box	Description	Type	Mount	Location	Cover Type	Source	Note
AV1	Presenter input	4-gang	Flush	@ 18" AFF	Custom	AV contractor	
AV2	Presenter input	4-gang	Flush	@ 18" AFF	Custom	AV contractor	
CT1	Conduit termination point	12×12	Surface	@ 48" AFF	Blank	Electrical contractor	Verify location with AV contractor
V1	Projector	3-gang	Flush	In ceiling	Custom	AV contractor	Verify location with AV contractor
V2	Document camera	2-gang	Flush	@ 18" AFF	Custom	AV contractor	
V3	Auxiliary input	1-gang	Flush	@ 18" AFF	Custom	AV contractor	

Table 12-1   Example of a drawing schedule

proportionately the same as the much larger room/space/system, based on the specified scale.

The selected scale is usually in the title block in the lower-right corner of the drawing, but may be located anywhere on the plans. A set of plans may include more than one scale ratio, and you must check each drawing page for its scale. In situations where the plans also include detail drawings (described in the previous section), you may find more than one scale listed on a single page.

Scale Measurements

Drawings using U.S. customary measurements will often state a scale using a particular fraction of an inch to represent a foot. For example, you might see that 1/4 inch equals 1 foot, as shown in Figure 12-7. Drawings done using metric measurements often state the scale as a single ratio, such as 1:50, which means that 1 unit on paper is equal to 50 of the same units in the real space.

Figure 12-7
Example of a
scale key in U.S.
customary units
that you may find
on a drawing set

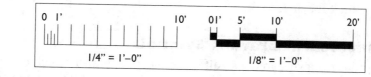

The following are typical U.S. unit scale measurements:

$3/32$ inch = 1 foot	¼ inch = 1 foot	¾ inch = 1 foot
$3/16$ inch = 1 foot	⅜ inch = 1 foot	1 inch = 1 foot
⅛ inch = 1 foot	½ inch = 1 foot	1½ inches = 1 foot

The following are some common metric (SI) units:

1:1/1:10	1:2/1:20	1:5/1:50
1:100/1:1000	1:500/1:5000	1:1250/1:25000

Using a Scale

The word *scale* is used to represent the size reduction (ratio) in the drawing, as well as the tool commonly used to interpret scaled drawings. The scale tool provides a quick method for measuring the object on paper and interpreting its true size in the actual space.

Traditional scales are prism-shaped tools that look similar to rulers. Architectural scales have numbers that are read from both left to right and right to left. A whole or fractional number to the left or right edge of the measurement tool indicates the scale those numbers represent.

A scale marked 16 is a standard U.S. customary ruler. For metric architectural scales, there are paired ratios, so each side of the scale can be used to measure two different ratios, such as 1:5/1:50.

Using your architectural scale, select the face of the tool that matches the indicated scale. Lay the 0 point at the extreme end of a line and read the corresponding value at the other end of the line. If the distance read is dramatically different, you may have the wrong scale selected. If the scale is marginally different, it could be that your drawing has not been printed to scale. It is important to remember that if dimensions are written on the document, they take precedence over the scale measurements.

Common Architectural Drawing Abbreviations

Table 12-2 shows some common abbreviations you may encounter when looking at an architectural drawing.

Architectural Drawing Symbols

Each project drawing may use different symbols and/or icons to depict specific elements of the project design or relationships among multiple drawings, such as where a room depiction in one drawing is continued on another drawing. Symbols do not necessarily follow a single standard, so the meanings of specific symbols are defined in the drawing legend.

Abbreviation	Definition
AFF	Above finished floor
AS	Above slab
CL	Center line
CM	Construction manager
DIA	Diameter
E.	East
E.C.	Empty conduit
EC	Electrical contractor
(E) or EXG	Existing
EL	Elevation
ELEC	Electrical
FUT	Future
GC	General contractor
MISC	Miscellaneous
NIC	Not in contract
NTS	Not to scale
OC	On center
OD	Outer diameter
OFCI	Owner-furnished, contractor-installed
OFE	Owner-furnished equipment
OFOI	Owner-furnished, owner-installed
PM	Project manager
RCP	Reflected ceiling plan
SECT	Section
VIF	Verify in field

Table 12-2 Common Site Drawing Abbreviations

Grid System Symbols

A grid system is used to indicate the locations of columns, load-bearing walls, and other structural elements. Grid lines are used for dimensioning and to reference the schedule. Vertical grid lines should have designators at the top and be numbered from left to right. Horizontal grid lines should have designators at the left and be alphabetized from bottom to top. Figure 12-8 shows an example of this system.

Figure 12-8

Example of grid system symbols

You can use the grid system to find your way around a work site in the early construction phase. Contractors will refer to a point in the space with respect to where it exists on the drawing, as in "4 meters west of B6." Because the space may not yet be divided into rooms, identifying a point by the grid system ensures everyone at the site understands the location.

Match Lines

Technicians use match lines to line up, or rebuild, a drawing that has been cut into separate drawings because it does not fit onto one page. For example, it may take multiple pages to depict one area of a building. To assemble the separate drawings so you can see the big picture, the individual pieces are aligned using the match lines as a guide.

Reference Flags

Some symbols on a drawing tell you to reference another drawing, such as an elevation, a section, or a detail drawing.

Elevation flags, such as the one shown in Figure 12-9, are used on plan drawings to indicate related elevation drawings. The center number indicates the page number where the elevation drawings can be found. The letters in the triangle portions of the figure indicate the elevation drawing on that page. Each elevation view for a room

Figure 12-9

Example of an elevation flag

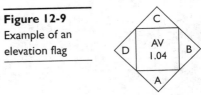

may be drawn in order and presented in a clockwise manner. This way, you can "look around the room" as if you were standing on the elevation flag and viewing the walls.

Section cut flags indicate which section drawing depicts a section of a master drawing in more detail, as shown in Figure 12-10. The bottom number indicates the page number where the section drawing can be found. The top number is the section drawing identification, because multiple

Figure 12-10

Example of a section cut flag

section drawings may be on the same page. A line extends from the center of the symbol indicating the path of the section cut. The right angle of the triangle is an arrow indicating the view direction of the section drawing.

Detail flags indicate which detail drawing depicts a specific feature of a larger drawing in more detail, as shown in Figure 12-11. In a large system, there may be hundreds of areas requiring more detail to read accurately. Each instance is numbered. The

Figure 12-11

Example of a detail flag

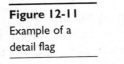

bottom number indicates the page number where the detail drawing can be found. The top number is the detail drawing identification, because multiple detail drawings may be on the same page. In the example in Figure 12-11, the symbol refers to detail number 1, where you can find more information about drawing AV1.06.

Converting Measurements

The AV industry exists around the world. For measurements, most areas of the world use either the metric system or the U.S. customary system. When collecting site information, the CTS-certified professional must be able to convert from one system to another to maintain accuracy.

Converting measurements from one system begins with establishing base units of measure. Common base units for metric (SI) measurements include the meter (m), centimeter (cm), and millimeter (mm). Common base units for U.S. customary measurements include the foot (ft) and inch (in).

 NOTE What does the "SI" mean when referring to metric measurements? It's an abbreviation of the French term *Système International d'Unités*, which is the modern form of the metric system.

To perform the conversion from one system to another, use Table 12-3. In the left column of the table, locate the unit of measure from which you are going to convert. Follow straight across the row to the middle column for the desired unit of measure (for example, from feet to meters). Finally, locate the conversion factor (such as 0.3048).

 NOTE While proper math rules state that any rounding should be delayed until the final step to avoid rounding errors, it is not always necessary to use a multiplier out to the hundred-millionth place. For example, 3.28084 is often rounded to 3.28, while 0.03937008 is rounded to 0.03937.

Convert From	To	Multiply By
Meters (m)	Feet (ft)	3.28084
Centimeters (cm)	Inches (in)	0.3937008
Millimeters (mm)	Inches (in)	0.03937008
feet (ft)	Meters (m)	0.3048
Inches (in)	Centimeters (cm)	2.54
Inches (in)	Millimeters (mm)	25.4

Table 12-3 Converting Between U.S. Customary and Metric Measurements

Figure 12-12
Determine the area of a room by multiplying the length by the width. In this example, the area is 600 square feet.

Figure 12-13
Determine the volume of the room space by multiplying the length by the width and then by the height. In this example, the volume is 162 cubic meters.

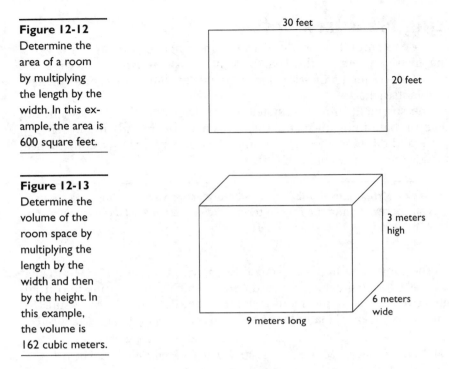

30 feet

20 feet

3 meters high

6 meters wide

9 meters long

Calculating Area and Volume

Typically, during a site visit, you will want to collect measurements of rooms and other spaces, and then calculate the area of the room or the total volume of a space in order to determine the required AV system capabilities. These are two simple calculations that all AV professionals should be able to handle.

To determine the area of a room, multiply the room length by the width. The result is the square footage of the room (or the square meters of a room). For example, Figure 12-12 shows a room that is 20 feet wide by 30 feet long.

To determine the volume of a room, multiply the room length by the width, and then by the height. The result is the cubic footage (or cubic meters) of the room. For example, Figure 12-13 shows a room that is 6 meters wide by 9 meters long by 3 meters high.

Chapter Review

A site survey provides much of the information required to design or install an AV solution at a client site. This chapter covered some of the knowledge and skills that CTS-certified professionals must apply to conduct such a survey, including the following:

- Awareness of the safety requirements for visiting construction sites, including health and safety regulations, use of PPE, and how to identify and follow site-specific safety requirements

- The process for observing site conditions in a consistent and comprehensive manner during a site visit, and documenting these observations in a clear manner

- A basic knowledge of how to interpret architectural drawings and site plans

- Knowledge of basic math required to calculate areas and volumes of the spaces, and how to convert between various measurement standards and formats (such as U.S. customary to metric)

Review Questions

The following review questions are not CTS exam questions, nor are they CTS practice exam questions. These questions may resemble questions that could appear on the CTS exam, but may also cover material the exam does not. They are included here to help reinforce what you've learned in this chapter. For an official CTS practice exam, see the accompanying CD.

1. When visiting a client site, the primary work site regulations governing the use of PPE are typically defined by which of the following?

   A. The client

   B. The AHJ

   C. The client's insurance company

   D. Your company

2. Which of the following is *not* considered part of the safety equipment that AV technicians should use when working at a work site?

   A. PPE, such as hard hats, work boots, gloves, and safety glasses

   B. Tools with nonslip handles

   C. Approved nonconductive ladders

   D. Fall protection

3. What is a site survey checklist typically used to do?

   A. Document the required AV system components for a client

   B. Document the AV, electrical, and mechanical systems

   C. Document information about client satisfaction with the AV system

   D. Document general information about a client and the site that may be relevant for the AV design and installation tasks

4. What type of plan drawing depicts the layout of items on the ceiling?

   A. Reflected ceiling plan

   B. Detail ceiling drawing

   C. Reflected floor plan

   D. Elevation drawing

5. What type of drawing depicts the ductwork that goes through the building?

   A. Mechanical drawing

   B. Reflected ceiling plan

   C. Detail drawing

   D. Section drawing

6. What type of drawing would you use to determine the characteristics of a wall to ascertain the appropriate height and location of a video display?

   A. Elevation drawing

   B. Reflected ceiling plan

   C. Detail drawing

   D. Mechanical drawing

7. What type of drawing would you use to determine what is behind a wall or above a ceiling that could interfere with your system installation plan?

   A. Elevation drawing

   B. Reflected ceiling plan

   C. Section drawing

   D. Mechanical drawing

8. What drawing would you review to find out exactly how the projector should be mounted to the ceiling?

   A. Reflected ceiling plan

   B. Detail drawing

   C. AV system schematic

   D. Mechanical drawing

9. Which abbreviation on a drawing tells you something on the drawing is *not* consistent with the scale?

   A. NTS

   B. NCS

   C. SCN

   D. NIC

10. Which abbreviation, along with a measurement given on a drawing, tells you how high on a wall an interface plate should be installed?

    A. AFF

    B. FUT

    C. SCN

    D. NIC

11. Which type of drawing provides the best information about the ducts and piping in a facility?

    A. Mechanical drawing

    B. Section drawing

    C. Electrical drawing

    D. Structural drawing

12. On a drawing with a 1:50 scale, 2 centimeters equal how many meters?

    A. 1

    B. 2

    C. 1/2

    D. None of the above

13. On a drawing using a 1/4 scale, 6 inches equal how many feet?

    A. 24

    B. 8

    C. 16

    D. None of the above

14. What does a section cut symbol on a drawing indicate?

    A. An interior view of a wall or ceiling structure

    B. The number of a more detailed drawing that depicts a specific portion of a master drawing

    C. A drawing that depicts items in great detail, such as equipment mounting plans

    D. A view of a wall from an angle

15. Which of the following equations allows you to convert a measurement of 10 inches into an equivalent measurement in millimeters?

    A. $25.4 \times 10 = 254$ mm

    B. $3.94 \times 10 = 39.4$ mm

    C. $25.4 / 10 = 2.54$ mm

    D. $3.94 / 10 = .394$ mm

16. Which of the following equations allows you to convert a measurement of 100 millimeters into an equivalent measurement in inches?

    A. $100 \times 25.4 = 2540$ in

    B. $100 \times 2.54 = 254$ in

    C. $100 / 25.4 = 3.94$ in

    D. $100 / 2.54 = .394$ in

17. What is the square footage of a space measuring 15 feet wide by 20 feet long by 10 feet high?

   A. 30 square feet

   B. 35 square feet

   C. 300 square feet

   D. 3000 square feet

18. What is the cubic footage of a space measuring 15 meters wide by 20 meters long by 3 meters high?

   A. 39 cubic meters

   B. 95 cubic meters

   C. 900 cubic meters

   D. 9000 cubic meters

Answers

1. **B.** When visiting a client site, the primary work site regulations governing the use of PPE are typically defined by the AHJ. However, keep in mind that the client, the client's insurance company, and your company may also have safety requirements that you should follow.

2. **B.** The safety equipment that AV technicians should use when working at a site include PPE (such as hard hats, work boots, gloves, and safety glasses), approved nonconductive ladders, and fall protection. Nonslip handle tools are not typically considered required safety equipment.

3. **D.** A site survey checklist is typically used to capture general information about a client and site that may be relevant for the AV system design and installation tasks.

4. **A.** Reflected ceiling plans depict the layout of items on the ceiling.

5. **A.** Mechanical drawings depict the ductwork that goes through the building.

6. **A.** Elevation drawings depict the characteristics of a wall using a side view, which would allow you to determine the appropriate height and location of a video display.

7. **C.** Section drawings depict the structures behind a wall or above a ceiling that could interfere with your system installation plan.

8. **B.** A detail drawing would depict exactly how the designer wants the projector mounted to the ceiling.

9. **A.** The abbreviation NTS refers to "not to scale" and is used to tell you something on the drawing is not consistent with the scale.

10. **A.** The abbreviation AFF refers to "above finished floor" and would indicate how high on a wall the interface plate would be installed.

11. **A.** Mechanical drawings provide the best information about the ducts and piping in a facility.

12. **A.** On a drawing with a 1:50 scale, 2 centimeters equal 100 centimeters, or 1 meter.

13. **A.** On a drawing using a 1/4 scale, 6 inches equal 24 feet.

14. **A.** A section cut symbol on a drawing indicates an interior view of a wall or ceiling structure.

15. **A.** The equation $25.4 \times 10 = 254$ mm allows you to convert a measurement of 10 inches into an equivalent measurement in millimeters.

16. **C.** The equation $100 / 25.4 = 3.94$ in allows you to convert a measurement of 100 millimeters into an equivalent measurement in inches.

17. **C.** The square footage of a space measuring 15 feet wide by 20 feet long by 10 feet high is 300 square feet.

18. **C.** The cubic footage of a space measuring 15 meters wide by 20 meters long by 3 meters high is 900 cubic meters.

Evaluating a Site Environment

In this chapter, you will learn about
- Identifying the room needs
- Assessing the room environment
- Assessing the services at the site

One of the first issues that AV designers should address when evaluating a client space is the intended use (or uses) of the space. For example, a videoconference room has very different requirements than a gymnasium, and the space should be able to support it.

For new construction, you may assess the proposed environment by reviewing project documentation, drawings, and construction schedules. When renovating an existing space, the assessment should also include a site visit to confirm the actual site conditions, both within the room and in adjoining areas.

This chapter reviews key characteristics of a room environment that CTS-certified professionals should evaluate before proposing an AV solution. For such evaluations, you'll need to collect typical site construction information, such as structure plans, historical/heritage considerations, and building regulations (code) considerations, as discussed in Chapter 12.

The knowledge and skills required to accomplish the site evaluation task typically include the following:

- Knowledge of customer service and negotiation techniques
- An understanding of how to read scale drawings and other project documentation
- Skills in basic math
- Skills in listening techniques

Identifying Room Needs

When preparing to design and install an AV system, it's important to understand that client need largely determines the overall room design, layout, and AV system characteristics. If you've identified the client's needs, you can begin planning how the room should be set up.

Many types of rooms use AV equipment, and despite how different they may seem, the needs they address are often similar. For example, an AV-enabled room may need to foster communications among participants, which means the people who use it must be able to see and hear each other clearly.

Regardless of the type of room, an AV professional should take all necessary steps to understand the following:

- **Client's requirements**   At least initially, rooms should be designed to focus on the client's needs, not the system's needs. Designers should not begin to design the AV systems until they have collected all of the facts and understand these needs.

- **Client's processes**   Designers should learn the end users' processes for activities they will conduct in the room. Each client may have a different approach to doing things, whether it's giving presentations or holding videoconferences. The AV system should become an integral part of the client's processes.

- **Client awareness**   End users and/or clients do not always understand clearly what they want or need. Clients often have a general idea of what they want an AV system to do, but they may not actually know what is possible or practical to implement. Because AV designers typically have experience addressing the needs of other clients, as well as in-depth knowledge of AV system applications, they may need to educate the client on the capabilities, technology, and techniques of using AV systems.

Types of Rooms

AV designers are accustomed to working with many different types of rooms, each with its own purposes and requirements.

- **Boardrooms**   These are richly appointed spaces for holding interactive meetings, typically with executives of a company. Seating is often arranged around a specialized conference table, positioned in the center of the room. Boardroom presentations are usually made from one of three locations:

- From the head of the table, with the projection screen behind the presenter
- From a lectern, with the projection screen either to the left or right of the presenter
- From the end of the table, facing the projection screen

  When designing a boardroom, appearances mean a lot. Executives expect their space to look impressive, with AV equipment integrated aesthetically and/or located out of sight. The challenge in boardrooms is balancing technical practicality, ease of use, and appearance.

- **Conference and meeting rooms**   Like boardrooms, these rooms host interactive discussions among small to large groups. Their AV requirements are similar to those of boardrooms, but the rooms themselves are generally not as elegant.

- **Training rooms and classrooms**   These rooms tend to be multifunctional spaces. Participants typically sit classroom-style at desks, theater-style in rows of seats, or grouped together to promote collaboration. Presentations are usually given from the front of the room, commonly from a lectern, with the presenter or teacher to one side of one or more displays or screens.

- **Auditoriums**   Clients use auditoriums for many purposes, such as lectures, theatrical productions, and movie screenings. They are large rooms, commonly used for one-way presentations to large groups, rather than interactive meetings. Presentations are typically made from the stage at the front of the room, with a large screen at center stage.

- **Divisible rooms**   These multifunctional spaces often amount to one large area or room that clients can partition into more than one usable room. It is common to find divisible rooms at hotels and convention centers, where folding walls slide into place, creating several, temporary spaces for separate events, each with its own AV requirements. Clients or their facility managers can quickly and easily change the shape and size of each temporary room to match the size of an event or group. Reverberation, echoes, and sound isolation are significant concerns in these types of rooms, so acoustics and sound-reinforcement principles must be carefully considered and applied.

- **Videoconference rooms**   These spaces are designed for face-to-face meetings via displays or screens and typically set up like a standard conference room. Cameras and microphones are strategically positioned within the room to capture what presenters say and do, and then to transmit the audio and video to a remote location. Audio and video of participants at remote locations are similarly captured and transmitted back to the videoconference room, with images displayed on a large screen and sound routed to the room's speakers. Videoconferencing rooms may be dedicated to such applications, but many corporations are also adding videoconferencing capabilities to existing spaces, such as boardrooms and conference rooms.

- **Courtrooms** Courtrooms require AV technology to help legal teams make effective presentations on behalf of their clients. Many courtrooms are equipped with sophisticated trial-presentation systems, including videoconferencing for remote witnesses. Audio and video recording, as well as speech privacy systems, are often integrated into courtroom systems.

- **Arenas, sports complexes, and stadiums** Most large arenas and stadiums seat thousands of people. They must be equipped with large-scale AV systems, from line array speakers, to public address systems, to digital signage networks, to large high-definition (HD) video screens. Although the screen is often used to display video of the event as it is taking place (a performance, sporting event, or live theater), it may also be used for other visuals that enhance the show.

- **Houses of worship and religious education centers** Commonly known by their abbreviation, HOW, houses of worship and related venues have been an important pro AV market for years. Many religious services have evolved into productions that feature sophisticated video, audio, and lighting to enhance the experience. Even smaller worship rooms may use displays for song lyrics, announcements, and religious text.

General Room Features and Characteristics

Although each room will be different, there are a number of standard room features you should consider when evaluating the site environment:

- **Accessibility** Entry and egress must be taken into account. Entry and egress for standard events, accessibility by disabled persons, and emergency egress are some considerations.

- **Doors** Consider locations, finish, type, and so on. For example, placing a door at the rear of a videoconference room creates more on-screen movement, which could be a distraction to remote participants and increase transmission bandwidth requirements.

- **Ceiling height** Pay close attention to issues such as how high the ceiling is above the floor and how much room for installation is available above the ceiling.

- **Windows** Plan for blinds or curtains to block light that can interfere with projected images.

- **Interior finishes** While the interior designer is responsible for the color and texture of the finishes, the AV designer should brief the interior designer on issues that affect acoustics and the appearance of projected images. For example, the AV designer should coordinate color, materials, and patterns with the architect and interior designer to ensure compatibility with the overall interior design and to minimize strong visual patterns in the vicinity of projection screens and/or behind people in a videoconference room.

- **Static display systems** Specify the type and location of any static display systems, such as bulletin boards, notice boards, and easels within the room.

- **Lighting and light levels** Determine the required light levels by correlating them with tasks that users will perform in the room. Specify the appropriate number, type, and location of luminaires, as well as how the lighting will be controlled.

- **Sightlines** Carefully analyze sightlines to ensure that the audience is able to see the presenter and any projected images.

- **Electrical power** Specify each location that will require electrical power in order to operate the AV system and related equipment, such as computers. You will need to calculate the total amount of power requested for each location. Obviously, you will need to locate and calculate power for presenter and control room areas, but you also need to consider electrical power required for audience members. The electrical branch circuits installed for the AV and related components should not be shared with non-AV devices.

- **Grounding** Ensure that a suitable electrical grounding scheme, appropriate for the type of system, exists.

- **Phone and networks** Plan communication needs for presenters, audience members, and AV components. The location of network and telephone receptacles is critical.

- **Ventilation** Although the HVAC engineer will provide for audience comfort, there will be additional heat loads related to AV systems and other devices. You will need to calculate these loads and communicate the information to the HVAC engineers. Additionally, if equipment racks are located in closets or small rooms, or if AV components will be housed in credenzas or other enclosed spaces, you will need to communicate those ventilation needs.

- **Structural mountings** Identify the locations of mounted devices (such as wall-mounted flat-panel displays) to ensure that adequate reinforcement is installed within the wall structure to provide support.

- **Acoustics** Ensure the design offers good acoustic characteristics and specify acoustic control features where necessary.

When evaluating a site environment, it is important to consider all of the room features and characteristics identified here. Each of these elements may play a part in defining elements of the final facility and AV system design.

The Technical Functions of a Room

Many AV systems are used to communicate information. Therefore, this function should guide all design decisions. It is also important to consider a room's other features—such as the HVAC system, lighting, windows, etc.—as they relate to the needs of the AV system so that they do not detract from the primary function of the room.

A typical room layout can be broken down into the following areas:

- Audience area
- Presenter area
- Control and/or projection area

Let's take a closer look at each.

Audience Area

Looking over the space where your AV design will reside, you want to consider the needs of people who will be listening to and/or participating in the meetings or presentations that the space must host. Following are some concerns you should address when planning the audience area:

- **Visual** Can everyone in the audience see the presenter and images? Ensuring that the answer is "yes" helps define the required screen size, image resolution, sightlines, task requirements, and other aspects of the system.

- **Sound** Can everyone in the audience hear the presentation? The AV designer must consider the acoustical properties of the room and determine the appropriate configuration of microphones and loudspeakers.

- **Ease of movement throughout the room** The designer should plan a seating layout to ensure that all members of the audience can enter and exit the room and seating area with ease.

- **Audience comfort** There should be sufficient room between seats and in the aisles. You may need to design for task lighting, so the audience can comfortably see notes and meeting materials, as well as system connectivity—wired or wireless—so that audience members can plug into a network or transmit their own presentation materials without moving too far from their seats, if they must move at all.

Presenter Area

With the client's input, consider where presenters will normally stand in the room. This area typically includes a lectern, microphones, and an interface with the presentation control system. Here are some of the issues you will need to address:

- **Presenter workstation** Is the lectern or console laid out to support the range of presentation needs that the client has identified?

- **Presenter and equipment** Is there enough room for the presenter and the presenter's equipment, which may include notebook computers or presentation notes? Also keep in mind that the presenter may want to be able to move around during a presentation and not be anchored to the lectern.

- **Power, voice, and data** Are there adequate electrical power, data, and phone connections in the front of the room to meet the needs of presenters?

- **Sightlines** Can the audience see both the presenter and the projected images clearly? For example, if the presenter will be too close to the screen, it may be necessary to limit the amount of light cast on the presenter so that it doesn't interfere with the projected image. Doing so, however, may make it difficult to see the presenter. Or the presenter may be positioned too far off axis to allow the audience to view both the presenter and the projected image at the same time.

- **Versatility**   Have all the needs of the typical client presenter been identified and met? As with the other room elements, this initial discovery phase is the opportunity for the AV designer to determine what is required to meet the client's specific needs. For example, the client may not want a lectern permanently attached to a specific location in the room. The client may need to remove the lectern, depending on how the room will be used or reoriented to accommodate different audiences, events, or presentation types.

Control and/or Projection Area

Depending on the type and size of a room, you may need to plan for a projection room or other control area. In this case, consider the following:

- **Equipment space**   Under certain conditions, AV operators may need to support a presentation's technical needs. The operator may need to remain in a control area for long periods. Is there enough room for both the projection equipment (if applicable) and the AV operator? Will it be comfortable?

- **Heating and cooling**   Can the heat generated by the AV equipment be removed from the room? The AV designer must coordinate this requirement with the architect and HVAC designer to ensure adequate ventilation.

- **Electrical power requirements**   Is there sufficient power for the AV and related equipment?

- **Voice and data**   Does the projection/control area include appropriate telephone and data communication connections? For example, a network cable may be required to ensure that computer-based presentations can access the required data. Phone or intercom communication between the control area and the presenter area may also be desirable.

- **Task lighting**   There should be sufficient task lighting to ensure the AV operator can see the system controls without adversely affecting the quality of the projected image or otherwise distracting from the presentation in the main room/space.

- **Sound isolation**   Noise from AV components in the projection/control area, such as projector fans, must be controlled and minimized so that it does not distract the audience.

- **Monitoring**   Operators must be able to hear what is going on in the main room during a presentation in order to make adjustments and/or queue the proper AV systems at the proper times. The operators should not be tucked away in a closet.

Functions of Associated Spaces

Many facilities also include other associated spaces—such as lounge areas, restrooms, and storage rooms—that support the room where the AV system will be located, its participants, and other activities in the building. The AV designer should know how the other spaces in the facility will be used in order to ensure they're compatible with the room where the AV system will be located.

Associated spaces can impact the success of the room for which the AV designer is responsible. For example, a reception lounge adjacent to the presenter area can create a potential source of noise that might interfere with the presentations taking place in the room. Consider these associated spaces and how they might impact your design:

- **Reception area** Is the reception area properly designed and situated to direct persons to the presentation room?

- **Phones/network access** Is there nearby access to communications to support applications in the room?

- **Break area or lounge** Are the location and layout of this area compatible with the intended use of the room? Is there a method of informing the audience that the break is over and they should return to the presentation room?

- **Restrooms** Are the restrooms the correct size for the anticipated audience? Are they conveniently located? If participants must walk through the front of the room to go to the restroom, it may disrupt the presentation.

- **Beverage and food service** Are the location and layout of these areas compatible with the intended use of the room? Food service should be located at the rear of the room to minimize disruption.

- **Kitchen and vending area** Are the locations of these areas convenient? Are the facilities adequate for the anticipated needs of the participants?

- **Breakout room** Are the location, size, and layout of this room compatible with its intended use? Which AV systems might the breakout room require to support its functions?

The overall design scheme is ultimately the responsibility of the architect. However, understanding how the design scheme will affect the success of your AV system will help you determine the types of issues and questions that you should address during the program phase (discussed in Chapter 15). In addition, the more the architect fully understands the function of the room for your AV system, the more closely the overall room design will address the specific functions planned for the room.

 NOTE The program phase of an AV project describes all the steps an AV designer takes before actually beginning the design, including reviewing documentation, conducting site visits, meeting with end users, and developing a needs analysis.

Assessing the Room Environment

Each room has specific environmental characteristics that affect the performance of an AV system. The AV designer must evaluate the acoustic, visual, and lighting environment characteristics of each room to determine their impact on design and installation. Here, we will review each of these areas.

Assessing the Acoustic Environment

Acoustics is the study of sound energy, its transmission, and its interaction with the materials within the environment. The manner in which sound energy moves through the space and is reflected or absorbed can determine whether the audience can hear the message being presented. Acoustics affect intelligibility, and good intelligibility is important for effective communication.

AV designers and integrators should evaluate acoustical conditions to determine the following:

- Room treatments needed
- Maximum appropriate background noise levels
- Reinforcement system devices
- How the devices will perform together once integrated within the space

Your assessment should uncover any adverse acoustical conditions that might detract from the presentation so they can be addressed. Even during the initial construction stages, the acoustic characteristics of a room can be modeled and evaluated based on construction plans. With some experience, the AV designer should be able to identify and solve potential acoustical problems before the meeting space or display environment is built.

The acoustical environment should be appropriate for the type of venue. The acoustic criteria for a performing arts center will be quite different from those for a gymnasium, an auditorium, or a classroom. In a venue such as an auditorium, classroom, or meeting room, the acoustical environment should be suited for speech intelligibility, so that the presenter or participants can be heard clearly.

There are a couple common acoustical issues you should pay attention to, particularly as they relate to a building's HVAC system: ambient noise and reverberation.

Ambient Noise

Ambient noise, sometimes called *background noise*, can be a major factor in the effectiveness of an AV system. High ambient noise levels can lead to listener fatigue, which adversely affects communication.

Ambient noise can come from outside a room. For example, it may be the result of sounds from adjacent rooms, foot traffic from spaces above, structure-borne vibrations, or traffic on the street. Ambient noise can also come from sources within a room, particularly from the HVAC system, meeting participants, and/or AV equipment (such as projectors, displays, and power amplifiers).

You should evaluate room walls and partitions to determine how well they prevent unwanted sound from transmitting from one space to another. Moreover, consider the sound transmission characteristics of the floors and ceilings, and note where they are deficient for supporting effective AV experiences.

Reverberation

Reverberation and *echo* refer to reflections of sound energy within a room, which may reduce the intelligibility. The amount of reverberation in a room depends, in part, on the following factors:

- **The shape of the room**   Shapes such as parallel walls, domed ceilings, and other concave walls and surfaces can create excessive sound reflections within the space.

- **The composition of walls and other materials in the room**   Some surfaces create minimal sound reflections. Other surfaces, such as marble (commonly used in lobby and reception areas), can be highly reflective.

Obtaining a good understanding of the acoustic characteristics of the space will enable the AV designer to create an audio system that best meets the client's needs.

Assessing the Visual Environment

When evaluating rooms that are intended for visual presentations—such as boardrooms, videoconference rooms, and auditoriums—you must consider how environmental issues will impact the audience's ability to clearly see the presentation. How well people can see images within a room may be affected by a number of factors—environmental and otherwise—including viewing distance, viewing angle, resolution, and aspect ratio. Your assessment should be methodical and cover the following aspects of the visual environment.

Viewing Tasks

The actual tasks users intend to perform using a display help define the required image size. For example, viewers who need to read bullet points have different image-size requirements than those who must inspect detailed engineering drawings.

There are three main categories of viewing tasks:

- **Viewing**   General viewing that does not require close inspection. This is the least demanding of these tasks.

- **Reading**   Detail viewing with clues. Reading requires more detail and "viewability."

- **Inspecting**   Detail viewing without clues. Inspecting requires the viewer to resolve the highest level of detail within the image.

Visual acuity is the ability of your eye to see details. There is a certain distance at which a given height of text can no longer be read clearly, which is a point beyond your eye's acuity. In AV design, you want the text size to be comfortable to read; the audience members should not strain their eyes. Logically, as the image is set farther away from the viewer, the screen size must be larger to accommodate comfortable viewing parameters.

For determining the legibility of text characters, viewers should be no farther away than 150 times the height of the character. For example, if you determine that the text will display 1-inch (25-mm) tall characters, the viewers should be no farther than 150 inches (3,810 mm) from the screen.

Determining Image Size

The required image size is based on viewers' positions and the level of detail they must discern to accomplish their viewing tasks. Keeping these criteria in mind, industry experts have agreed on a best practice for determining the recommended image height. This is based on the requirement for the farthest viewer of the AV presentation for each of the following task categories:

- **Viewing (general)**   The farthest viewer should not be any farther away from the screen than eight times the image height. This applies to general viewing of video, film, and simple graphics.

- **Reading (detailed)**   The farthest viewer should not be any farther from the screen than six times the image height. This applies when the audience is reading detailed graphics or data images.

- **Inspecting**   If meeting participants will be inspecting graphics, they should not be seated any farther from the screen than four times the screen height.

To determine the appropriate size of a viewing area within a room, measure the distance from the screen to the farthest viewer or the length of the room, and divide as follows for each viewing task area:

- Divide by eight to calculate image height for general viewing ($H = D/8$)
- Divide by six for reading ($H = D/6$)
- Divide by four for inspection ($H = D/4$)

For example, if you have a room in which the farthest viewer will be 24 feet (7.3 meters) from the screen, here are the required image heights for each category:

- **General viewing** $H = D/8$, or 3 feet (0.9 meter) high
- **Detailed viewing** $H = D/6$, or 4 feet (1.2 meters) high
- **Inspection** $H = D/4$, or 6 feet (1.8 meters) high

Determining Horizontal Screen Size

The correlation between image height and width can be expressed by a number referred to as the image's *aspect ratio*. If you know the required screen height and the aspect ratio of the image to be displayed, you can use the aspect ratio to determine the required screen width. Here are some examples:

- For standard 35 mm film, with an aspect ratio of 3:2, multiply the height by 1.5 (3 divided by 2) to obtain the width.

- For standard video, with an aspect ratio of 4:3, multiply the height by 1.33 (4 divided by 3) to obtain the width.

- For HD video, with an aspect ratio of 16:9, multiply the height by 1.78 (16 divided by 9) to obtain the width.

So, a screen with a 6-foot (1.8-meter) height, displaying HD video with a 16:9 aspect ratio, needs to be 10.27 feet (3.13 meters) wide.

Applying these formulas will help you determine the screen size appropriate for the viewing environment.

Projected Image and Flat-Screen Display Considerations

When assessing the visual display environment of a site, the AV team should determine whether to use a projected image or a flat-screen display.

Projection systems present a range of considerations:

- Front projection or rear projection?
  - In a front-projection system, the image is viewed from the same side of the screen as the projector is placed. A front screen is simply a reflecting device. It will reflect both wanted (projected) and unwanted (ambient) light back into the viewing zone, so achieving proper contrast with a front-screen system requires controlling ambient light.
  - In a rear-projection system, the projected light is transmitted through the back of the screen. Portable rear-projection screens use a fabric screen; there are also optical glass and acrylic rear-projection screens. Rear projection often provides better contrast and color saturation than front projection when operating within environments of high ambient light. Rear projection requires additional space for a dedicated projection environment to contain the projection system.
- Where should you locate the projector? Projectors can be floor- or ceiling-mounted to provide flexibility of integration into the display environment, depending on the application and the space. In order to save space in a rear-projection environment, mirror assemblies may be used.
- Is there a potential problem with ambient light washing out the displayed image? If so, how should it be dealt with?
- Where are the windows and how does the sun track in relation to the optimum audience placement and screen location? Are blackout blinds or curtains to cover the windows possible and/or appropriate?
- Is the structure (the ceiling) capable of supporting the weight of a projector and its associated mounting equipment? Is structural reinforcement needed?
- Will the projector be hanging in the sightline of any other equipment, or will it create an obstruction for persons within the room? Is a projector lift necessary?
- Will hanging or pendant light fixtures interfere with the light path of the projector?

When assessing the ability to use a projected image, the AV designer should review these considerations and check for any limitations. These issues will determine the position of the audience and the equipment.

Flat-panel displays, similar in shape and size to today's flat-panel TVs, are also affected by ambient light and present challenges similar to those for projected images. The main considerations with flat-panel displays are screen size and display resolution. Also, you need to ensure they can be safely mounted in the proper location. Structural reinforcement behind walls may be necessary to support the wall mounting of large flat-panel displays.

Audience Sightlines

Sightlines affect screen size and placement. The AV team should assess the proposed audience seating, and determine whether it is arranged so that the line of sight of each participant is minimally obstructed. This includes each participant's ability to see the screen, and in the cases where videoconferencing is used, ensuring that the cameras have unobstructed views of each of the participants.

In placing the screen, you need to consider the optimal viewing angle characteristics of the screen selected. Position the screen at a comfortable viewing height, just above eye level, so that all participants have minimally obstructed sightlines.

The resulting AV room design based on this information should use proper screen size and placement, and/or staggered seating and other techniques, to ensure clear sightlines.

Quality of the Projected Image

The AV team should understand how the visual environment affects perception of the image. The goal is to ensure a site environment that does the following:

- Establishes an acceptable "system black" image level by controlling the ambient light in the environment

- Controls the amount of glare from lights in the room

- Achieves the most accurate color rendering on the screen, taking into account room finishes, the color temperature of light sources within the room, and how humans are influenced by juxtaposed colors

The contrast ratio of the system and the image uniformity of the projected image are two of the key factors in judging image quality. The difference between system black and white in an image is the system *contrast ratio*. The greater the contrast ratio, the better the image is perceived to be as it creates the impression of high resolution.

 NOTE In June 2011, InfoComm International published the ANSI/INFOCOMM 3M-2011: Projected Image System Contrast Ratio standard, which provides metrics for establishing a minimum contrast ratio for rear- and front-projection systems. It takes into account the viewers' requirements and the presence of ambient light. The standard is available for purchase online. Visit InfoComm's website for more information (www.infocomm.org/standards).

When assessing a room as a projection-system environment, keep in mind that black is not created by the display device. A black image cannot be any blacker than when the display is turned off. Control over ambient light within the projection environment is therefore critical to establishing a good system black level. Ambient light may come from windows, exit signs, light from the screen reflected back by the walls or ceiling, or the fixtures in the room.

> **NOTE** Some degree of ambient light is not only appropriate, but expected in AV installations. For example, most presentation spaces, with the exception of certain entertainment theaters, require task lighting so that participants can see the presenter, each other, and their notes or printed text. The use of task lighting will necessarily create ambient light, which will then reduce the image contrast ratio.

Glare is one source of ambient light that can affect the quality of a projected image. Glare occurs as the result of excessive luminance in the field of view. This can be any bright, distracting light that makes it difficult to see the intended image. Direct glare is caused by excessively bright light sources in the field of view that shine directly into the eyes; indirect glare results from light reflected into the eyes off surfaces or work areas. The sun is a direct source; an automobile may indirectly reflect the sun to your eyes. The design goal is to avoid creating an environment where glare is an issue.

Room finishes also impact the quality of a projected image. For example, lightly colored walls reflect the light from projection screens and lighting fixtures. When such reflected light bounces back toward the displayed image, it reduces image contrast.

AV designers should strive for rooms where the color of walls, ceilings, and floors skew toward neutral, nonreflective shades. The area surrounding a screen should be a darker, neutral color that creates the perception that the image is much brighter, increasing the perceived contrast. Wall colors should be neutral and dark. Even with all the lights in a room turned off, the light reflected off the screen will illuminate the walls and ultimately reflect light back onto the screen, reducing the contrast ratio. Walls with bright colors reflect back more light onto the screen, due to the nature of the paint or color.

Interior design and finishes, such as wall colors, are also a consideration when the room is to be used for video capture or recording, such as videoconferencing. In these cases, the walls will be within the field of view of the camera and can have a significant impact on the receiver-end perception of the room video quality.

Certain colors are better suited to video rooms than others. The automatic settings of a camera try to adjust for a reference "white" within the room. Walls that are painted a hue just off white are sometimes mistaken as a pure white by the camera. This results in incorrect colors from the video camera. Consider the use of neutral gray color tones, or perhaps bold colors that will not be mistaken for white. In keeping with these color recommendations, any acoustic panels defined for the space that will appear within the camera field of view should be ordered in colors such as silver-gray, quartz, or champagne. For aesthetic variety, you can alternate different-colored panels along the wall.

Wall coverings should not include too many patterns. The video codec will spend more processing time on the background image if it is highly detailed. This extraneous processing will take away from processing that could be spent on making the presenter look better.

Furniture selection and placement should also be assessed for their effect on image quality. For example, furniture in a videoconferencing room should be selected to meet the needs of the local participants, but also to broadcast acceptable images to remote participants. Any tables in the environment should have a light top surface. Glossy tops should be avoided, as should strong colors or bold wood grains.

Consider suggesting fixed chairs so presenters are not swaying back and forth. And remember that sunlight poses a major problem to the camera environment. If you must design a videoconferencing room—or any room—with windows, you need to include appropriate window treatments, such as blackout curtains.

Assessing Room Lighting

The lighting system in a room that's designed to include an AV display must maintain a balance between controlling ambient light and allowing for an adequate contrast ratio. *And* it should be accomplished while providing sufficient task lighting for participants

Lighting Terminology

The lighting industry has its own terminology. Here are a few terms you should know:

- **Luminaire**   A complete lighting instrument composed of a lamp and housing.

- **Lighting fixture**   An installed luminaire.

- **Lamp**   The source of radiated light. This is sometimes referred as a *light bulb*, but it could be any of many different technologies.

- **Lumen**   The unit of luminous flux emitted from a source of light. Lumens, abbreviated lm, are used to rate a light source such as a lamp.

- **Lux**   The metric unit of measure for light that falls onto an object. We use lux to state how much light is falling onto a table surface for reading or on a projection screen as ambient light. One lux is equal to $1 \ lm/m^2$.

- **Foot-candles**   A U.S. customary unit of measure for light that falls onto an object. A foot-candle is equal to $1 \ lm/ft^2$, or 10.76 lux.

- **Color temperature**   Light sources have a color to them and are rated in a numerical color temperature, expressed in kelvins (K). Lower color temperatures (2,000K) radiate a reddish hue. Higher color temperatures (6,000K) radiate a bluish hue.

- **Zone**   Grouping of lighting fixtures that are controlled together.

- **Scene**   Preset of lighting levels for one or more zones. With multiple zones, the scene settings can be quite complex.

to view documents, take notes, or produce an acceptable video image in a videoconference environment.

Each type of AV display application will require different lighting approaches to accomplish specific goals. Properly applied lighting can enhance the environment. Improper lighting can distract from the intended message. Your site environment evaluation should consider the range of AV applications that will be used in each room to determine the most appropriate types of lighting fixtures and lamps.

Types of Lighting

Several types of light sources are available. Each has its own unique qualities.

- **Incandescent lamps** Sometimes referred to as *Edison lamps*, these produce light by heating a filament in a sealed glass housing. Because incandescent lights become hot when illuminated, designers must account for the temperature to ensure that the heat does not damage other components. Generally, these types of lights have a color temperature of about 3,000K to 3,200K, resulting in a slightly yellow hue. Incandescent lights are being replaced by more efficient devices in most business environments.

- **Fluorescent lamps** These are found in many commercial sites. They are energy-efficient and create widely dispersed illumination. Different color temperature options are available. Fluorescent lights may cause IR and RF interference that could affect an AV system. For years, fluorescent lights could not be dimmed without special ballasts, but that is changing.

- **Halogen lamps** Halogen lamps are often used as small spotlights to illuminate specific elements in a room. They are considered highly efficient, luminous, and long-lasting.

- **LED lamps** These are low power and highly efficient. LEDs have a long lamp life (more than 100,000 hours). Multicolor LED lamps are also available. These types of lamps are currently being used as primary fixtures within a room.

Control of Light

The light within an AV presentation space will often need to be altered. For example, the presenter may need to turn off or dim specific lights during a video presentation to achieve the proper image contrast on a display. The lights are often grouped together into a zone controlled by a switch or dimmer.

The most basic lighting design for an AV room is two zones of switched lights. Typically, the lights in front of the projection screen can be turned off separately from the rest of the room lighting. Adding dimmers and more zones gives greater flexibility in the level of light within a space.

Assessing Services

The site evaluation should also provide a general understanding of the configuration of the building services that will impact the rooms in which AV systems will be installed. These services include electric power, data, cable, and HVAC systems. AV professionals should know how these elements are depicted on building plans and how they operate in general.

As discussed in Chapter 12, AV staff conducting a site survey should collect a wide range of information about the site. The documentation that can help with assessing services includes the following:

- Building plans
- Historical/heritage considerations
- Zoning and planning considerations
- Building regulations (code) considerations
- Architectural plans (which are essential for any site design or installation task)

Assessing Electric Power Availability

Assessing the available electric power at the client site is critical to an AV design. AV system designers should have a general understanding of standard commercial AC power systems and how they are documented in project drawings.

Here are some issues to consider when evaluating a site for power:

- How much power will be necessary for the system, and how does it compare to what is available or planned? Adding power will require more of the electrician's services.
- Where will the panelboards (distribution boards) that power the AV equipment be located?
- Is any non-AV equipment on the same panelboard as the AV equipment? Some devices can affect the operation of an AV system. For example, vacuum cleaners, motors, and welding equipment can cause noise in sensitive AV equipment, leading to poor performance and intermittent problems.
- Are there any existing power receptacles (outlets) available for this AV project, and if so, where are they located in the project's space? This information will help the designer decide whether to use the existing AC outlets or request that they be relocated.

Assessing Data System Capabilities

Similar to assessing the need for AC power, the site evaluation should consider the facility's IT technology to determine if the AV system's projected data needs can be met. Check for the following:

- Are data ports available for this AV project, and if so, where are they located in the project space? Determining the number of data ports required and the physical location of the equipment will help the IT staff ensure that the needs of the AV system can be met. Even though there may seem to be enough connection points available, the IT department might need to purchase additional equipment to make them active.

- Where are the data rooms that will provide connectivity? The location of the data rooms in relation to the space may affect the time line of the installation. In addition, these rooms are typically secured, so inquiring about access privileges will be important during the installation process. This work will most likely proceed in conjunction with the IT department.

- What is the current data capacity and does it need to be expanded? Expansion might be in the form of fiber-optic cabling, more data ports, or wireless access points. The IT department needs to know how the equipment you are installing will affect the network.

Identifying Cable Access Locations

Planning the cable routes for an AV system requires working with the electrician or a facilities manager. These professionals can indicate where cabling and conduit can be installed. Also, it is likely that you will need to discuss the location of additional conduit and floor boxes. In turn, the electrician and facilities manager will need your guidance on where AV racks, lecterns, and faceplates may be located.

Assessing HVAC Components

Air conditioning and ventilation ductwork can affect the location of AV equipment, such as projectors or loudspeakers that need to be installed in the ceiling. The noise from moving air can make it difficult for attendees to hear what the presenter is saying. Vibrations created by HVAC systems can cause images to appear shaky, making it more difficult to interpret what is on screen. Scheduling a walk-through with a mechanical engineer and architect can help you assess the impact of HVAC systems on AV systems.

The following are issues to consider when evaluating a site's HVAC systems:

- Where are the paths and locations of HVAC ductwork, diffusers, and equipment? This information is important for the placement of projectors and loudspeakers that will be installed in the ceiling. AV or HVAC designs may need to be adjusted due to the intended locations of ductwork.

- Are there any large mechanical equipment installations in the space? Documenting the locations of any HVAC equipment such as chillers and return vents can help ensure that there are no conflicts between the HVAC system and the AV systems. Try to assess whether large mechanical units are placed directly adjacent to the spaces that require a low-noise floor. Mechanical vibration and noise created by vibration are often major complaints in buildings. The low rumble of a chiller or large fan can resonate throughout a building if it is not designed properly and given the proper vibration isolation.

Chapter Review

A careful evaluation of the client's site environment can ensure that the AV system design and installation tasks produce AV systems that are compatible with the characteristics of the client's site. In this chapter, we reviewed the aspects of a site that the AV team should consider when evaluating the environment for an AV system. The team needs to assess the AV-enabled room, the overall site environment, and the site services.

Review Questions

The following review questions are not CTS exam questions, nor are they CTS practice exam questions. These questions may resemble questions that could appear on the CTS exam, but may also cover material the exam does not. They are included here to help reinforce what you've learned in this chapter. For an official CTS practice exam, see the accompanying CD.

1. What are the main concerns of the AV system designer regarding the audience area?

   A. Determine if the audience can see and hear the presentation, and determine whether movement within the seating area will be comfortable

   B. The architect addresses the audience area design, not the AV designer

   C. Determine if the proposed audience area HVAC and lighting are adequate

   D. Ensure that the audience seating is placed between 5 feet (1.5 meters) and 25 feet (7.6 meters) from the screen

2. What is the primary issue the AV designer should examine when evaluating the AV control or projection room area?

   A. If the room location is close enough to run cabling to the AV components

   B. That the room has a clear sightline to the presenter areas

   C. If the room meets the required size, power, and HVAC requirements, and provides other services needed for the AV system components

   D. If the proposed control room design and layout meet government construction standards and requirements

3. What is the typical primary concern of the AV designer when evaluating the HVAC systems at a client site?

   A. Whether the HVAC system will provide sufficient heating and cooling within the audience areas

   B. Whether the HVAC system will provide sufficient heating and cooling within the control room area

   C. Whether the HVAC system will interfere with AV system component placement or create excessive noise

   D. Whether the HVAC system will create electrical interference that impacts AV component operation

4. What is the primary issue that the AV designer should assess when reviewing the acoustic environment at a client site?

   A. Ambient noise and reverberation

   B. The required loudness level of the AV system

   C. The optimum locations for AV system loudspeakers

   D. The audio system needs for the presenter and presenter area within the room

5. A client needs a display system for a room that will be used for inspecting detailed drawings of computer system networks. What is the maximum distance that the viewers should be placed from a 4-foot (1.2-meter) high screen?

   A. 8 feet (2.4 meters)

   B. 16 feet (4.9 meters)

   C. 24 feet (7.3 meters)

   D. 32 feet (9.7 meters)

6. Based on your review of the client needs, the display should consist of an HD wide-screen flat-panel monitor that is 4 feet high. How wide is the image displayed on this monitor?

   A. 5 feet (1.52 meters)

   B. 6.41 feet (1.95 meters)

   C. 7.12 feet (2.17 meters)

   D. 8.1 feet (2.47 meters)

7. Your client is interested in installing a large projection screen in a front lobby area to display images of various projects for promotional purposes. The screen will be installed directly across from a street-level entrance with several windows and a revolving glass door. What should be the primary initial concern of the AV designer regarding the projection system?

   A. Ambient noise

   B. Audience sightlines

   C. Ambient light

   D. Projector placement

8. What is the primary factor an AV designer should examine when evaluating the potential to ensure a high contrast ratio for a front-projected image within a room?

   A. Proposed projector location and angle

   B. Proposed screen material

   C. Ambient light levels in the room

   D. Distance between the projector and the screen

9. The AV designer is evaluating the proposed AC power supply to an AV control room. What issue(s) should the AV designer consider?

   A. How much power is necessary to operate the AV system components

   B. What non-AV equipment is on the same panelboard as the AV equipment

   C. What existing outlets are available for the AV equipment

   D. All of the above

Answers

1. **A.** The AV designer should evaluate the overall room design to determine if the audience can see and hear the presentation, and that movement within the seating area will be comfortable.

2. **C.** The AV designer should determine if the room meets the size, power, and HVAC requirements and provides other services needed for the AV system components.

3. **C.** The primary concern of the AV designer regarding HVAC systems at a client site is whether the HVAC system will interfere with AV system component placement or create excessive noise.

4. **A.** The primary concern when reviewing the acoustical environment within a client site is the level of ambient noise and reverberation that is present within the room.

5. **B.** For inspecting detailed images, such as graphics and engineering diagrams, viewers should be placed no farther than four times the screen height from the screen. In this case, with a 4-foot (1.2-meter) high screen, the farthest viewer should be no more than 16 feet (4.9 meters) from the screen.

6. **C.** An HD display has an aspect ratio of 16:9, which is equal to multiplying the height by 1.78 to determine the width. In this case, the 4-foot (1.2-meter) high screen times 1.78 results in a screen width of 7.12 feet (2.17 meters).

7. **C.** The typical primary concern for AV designers regarding projectors and screens is the impact of ambient light on the quality of the image. In this instance, since there are several windows in this space, ambient light will likely be a substantial issue.

8. **C.** In a projection environment, the contrast ratio of the image is most affected by the ambient light levels within the room.

9. **D.** When evaluating the proposed AC power supply to an AV control room, the AV designer should consider how much power is necessary to operate the AV system components, what non-AV equipment is on the same panelboard as the AV equipment, and which existing outlets are available for the AV equipment.

CHAPTER 14

Maintaining Awareness of Changes to the Site Environment

In this chapter, you will learn about
- Recommending changes to address room environment issues
- Recommending changes to address issues related to the building environment

In Chapter 13, we examined some of the factors that can help you effectively evaluate an existing or planned site, including the acoustic characteristics of a room, lighting, sightlines, space utilization, and other room elements. In this chapter, we will look at how to assess whether changes should be made to the room environment to maximize the effectiveness of the AV systems. You can use this knowledge to make recommendations or advise a client about issues related to the building environment, such as services (power and HVAC), structure, and regulations that must be followed.

The knowledge and skills required to accomplish this task typically include the following:

- Knowledge of AV system and site design considerations
- An understanding of how to read scale drawings and other project documentation
- An ability to present recommendations to clients

Domain Check

Questions addressing the knowledge and skills related to maintaining awareness of changes to the site environment account for about 6% of your final score on the CTS exam (about six questions).

Addressing Room Environment Issues

As described in Chapter 13, many factors related to the room environment could affect the AV systems you plan to design and/or install. AV designers need to consider the following:

- Acoustics and how to manage noise levels
- Sightlines and how to improve the audience's view
- Lighting, including both ambient and artificial light
- Furniture and room finishes, such as how materials and colors affect the AV system

During your site evaluation, you will often need to suggest changes or adjustments to the environment so that the AV systems will meet the client's needs. Sometimes they are changes to the proposed or existing environment itself; sometimes they are elements of your planned design. Either way, at this stage, AV designers or integrators begin to deal with the room issues they identified while evaluating the site.

Acoustic Issues

In the previous chapter, you examined how to identify and assess elements of the acoustic environment at a client site. Based on the site assessment, the AV team should be able to identify potential acoustic problems in a room and report those findings.

The following are two common acoustic problems:

- Noises and vibrations within a room that may come from multiple sources
- Sound reflections and reverberation within a space

All surfaces reflect, absorb, or diffuse sound. These properties determine whether clear and intelligible audio can reach the listener. In addition, listeners positioned at different locations in a room may perceive direct or near-field sound, early reflections, echoes (late reflections), and/or reverberation as a result of sound-wave behavior.

When initially evaluating the acoustics of the site environment, the AV team considers how the room will be used. Is it mostly voice and video within a boardroom, or voice and music reproduction within a larger, versatile meeting space? The use of the space will help determine the level of noise reduction necessary to meet the performance criteria of the audio portion of the AV system. The acoustic needs of a courtroom or recording studio are much different than those of a boardroom or multipurpose divisible room.

The AV team should examine the site and/or site plans to identify possible sources of acoustic problems. As previously noted, a common source of acoustic problems is the HVAC system. HVAC noise can originate from any of the following:

- Fans
- Ducts

- Diffusers
- Mechanical room equipment (for example, pumps and chillers)
- Terminal equipment in ceilings (for example, fan/coil units and other distributed air-handling devices)
- Outdoor mechanical equipment (for example, cooling towers and ventilation fans)

These devices can create noise in the air or through structures. Airborne noise is generated by equipment or air movement. It may be controlled through mechanical system design, architectural treatment, or both. Structural noise is typically generated by equipment attached to the structure. It can radiate as airborne noise from walls and ceilings in the AV space. It can also cause AV equipment problems by transmitting vibration to projectors, loudspeakers, and microphones mounted to walls, ceilings, floors, and furniture.

The AV team should recommend appropriate duct design where practical. This may involve moving mechanical components out of the room, using larger ducts and diffusers, adding gradual bends in ducts, and other necessities. For existing sites, replacing HVAC components with low-velocity, dampening diffusers can help reduce ambient noise levels.

Managing Noise Within a Room

No room is acoustically perfect. All rooms and spaces will have undesired sounds and vibrations. This suggests that the sound characteristics of a room must be carefully managed. (But keep in mind that a room with too much sound absorption may be perceived by the listeners as odd or strange.) There are a number of reasonable steps an AV team can recommend to help manage noise in a room.

Room size and surface treatments determine the amount of reverberation in a space. Hard surfaces, such as untreated gypsum board, marble floors, and wood paneling, contribute to sound reverberation. You can minimize this reverberation by incorporating various soft surfaces in the room or treating existing surfaces with acoustically absorptive or sound-diffusing materials. Acoustical tile and fabric-wrapped fiberglass panels are two examples of absorptive materials that the AV team may suggest to a room designer. Figure 14-1 shows an example of the use of acoustical panels.

Figure 14-1
The main classroom at InfoComm University uses acoustical panels on its walls.

NOTE In some situations, more reverberation is better. You can increase reflective and reverberant sounds by using materials that are designed to intentionally reflect sound toward an audience. An orchestra shell is an application of this concept.

Another approach the AV team can recommend to control noise within a room is to use materials that diffuse sound. Diffusing surfaces are generally preferred over flat, reflecting surfaces, because diffusion tends to have a gentler effect on the sound source while still offering a sense of space. The ideal diffusing surface, such as an irregularly shaped or convex surface, sends sound energy equally in all directions. Diffusing materials are also often more suitable for a room than sound-absorption materials. Diffusion can reduce echoes in a room without the extreme effects of sound absorption.

Some additional questions to ask when making recommendations to reduce noise in a room include the following:

- What are the shapes and dimensions of walls, floors, and ceilings?
- Do the windows have treatments such as heavy drapes or curtains?
- Are the walls built from the floor and continue to the hard ceiling above?
- Does insulation already exist in the walls? If so, what is the rating of the insulation?

Quantifying Noise Levels

The following are two common methods for quantifying background noise levels:

- **Noise Criteria (NC)**   The NC method is commonly used by the building industry to specify or quantify the background noise levels from ventilation equipment or other sources in an occupied space. It is referred to as Noise Reference (NR) in countries outside the United States.
- **Room Criteria (RC)**   This metric is helpful in quantifying the tonal characteristics of ventilation noise, such as rumble or hiss. The RC method takes all of the components (walls, ceilings, diffusers, etc.) in a room as a whole to determine the criteria levels for room design.

NC and RC are single-number ratings that represent a complete spectrum of sound-pressure levels to meet a particular criterion. For example, to meet NC-35, the sound-pressure levels in the room cannot exceed the levels in any octave band on the NC-35 curve. Figure 14-2 shows some examples of NC curves.

Sightline Issues

In the previous chapter, we reviewed the issues associated with ensuring proper sightlines for viewing screens and displays. In many situations, the AV team may not be able to influence the space at the beginning of construction. However, a CTS-certified professional should work to make sure the client understands the sightline implications of a viewing environment, as well as the options available to address any issues.

Figure 14-2

Noise Criteria (NC) curves

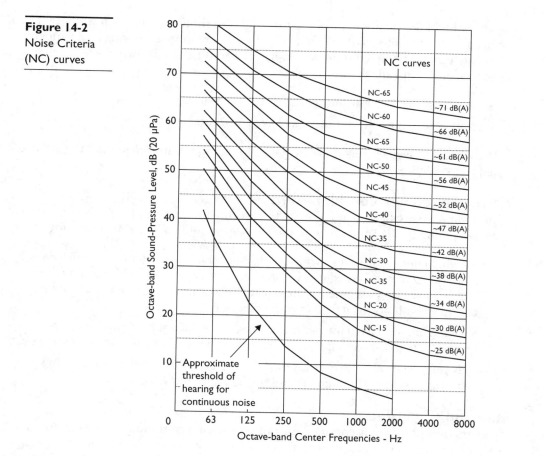

Sightline Studies

A *sightline study* helps the AV team determine the most appropriate seating layout to ensure the audience has a clear view of the screen. The sightline study will identify the lowest visible point on the display wall, appropriate lines of sight for the nearest and farthest viewers, distortion of the image from off-axis seat locations, and other ergonomic factors. The goals are to come up with a preferred field of vision, and identify and maximize viewing comfort.

In Figure 14-3, sightlines are drawn from the eye level of each seated row, passing just over the heads of the viewers in the next row. In an optimal viewing environment, the

Figure 14-3

Sightlines and head interference

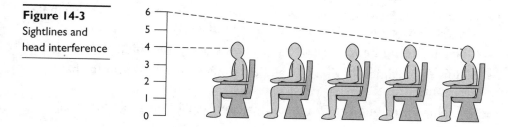

highest point of the sightlines drawn to the front wall should define the lowest point of the image. The example shown in Figure 14-3 is not an optimal viewing environment.

The sightline parameters that AV designers should consider when conducting a sightline analysis include the following:

- Average seated head height of 52 inches (1,300 mm)
- Average seated eye height of 48 inches (1,200 mm)

Therefore, the average seated eye height is 4 inches (100 mm) less than the average seated head height.

Viewer Placement Considerations

Correct viewer placement is also critical to ensuring that viewers see the best possible image. In planning viewer placement, the AV team should consider the following:

- **Viewing cone**   The best viewing area for the audience is referred to as the *viewing cone*. The term *cone* is used because the best viewing area has width, height, and depth, based on the center of the screen. This optimal positioning of viewers addresses both horizontal and vertical placement.
- **Viewing angle**   Generally, projected images should not be viewed at angles greater than 45 degrees off the projection axis. This is known as the *viewing angle*.
- **Viewing distance**   The maximum viewing distance is based on the viewing tasks (discussed in Chapter 13) and the height of the screen.

Improving Sightlines

If you discover problems accommodating the optimum positioning of viewers, you can make appropriate recommendations for enhancing the viewing environment.

Recommendations for changes to the viewing environment are typically based on the degree to which the client is willing and/or able to modify the space. If the client is able to fully accommodate the AV requirements, the team should propose the best solution.

One way to improve sightlines in a presentation space is to provide a sloped, raked, or stepped seating layout. Seating can also be staggered, as depicted in Figure 14-4, and the distance between seats can be varied. The AV system designer can experiment with alternative seating arrangements to create the best viewing environment for the audience.

In cases where the size, shape, and other characteristics of the space impose constraints on the viewing environment, recommended changes may include the following:

- Reducing the client's expectations of the size of audience the room can handle to meet viewing distance or viewing angle limitations
- Changing the position of the screen to allow better positioning of the audience
- Adding stepped tiers or risers for audience members seated near the back of the room
- Changing the position and/or mounting approach for the projector

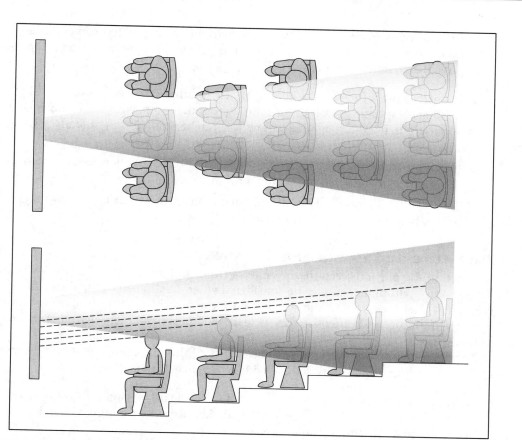

Figure 14-4 Staggered and stepped audience layout

As with the process for enhancing the acoustic environment, the AV designer should work with the client and the client's contractors to implement an optimal viewing environment that is compatible with the client's budget and needs.

Ambient and Artificial Lighting Issues

Creating an optimal viewing environment involves not only decisions about where to position the audience and AV equipment, but also decisions about the interior design of the space. These decisions relate to color, texture, lighting, and reflectivity of surfaces. The objective is to establish an environment that allows the viewers to perceive an image that comes as close to matching the original as practical. This means you must design the system to maximize brightness, contrast ratio, resolution, and visual intelligibility.

Ensuring appropriate ambient light levels in a space can be a difficult balancing act. For example, meeting rooms not only require an appropriate light level for projected images, but also sufficient task lighting for people in the room to take notes, read documents, find their way out of the room, and so on.

The AV team should evaluate the proposed lighting design and recommend changes as necessary to minimize the amount of ambient light that reaches the screen. Here are some possible recommendations:

- Instead of overhead fluorescent lights, use lighting systems that focus task lighting on specific work locations throughout the room. For videoconferencing applications, use direct lighting fixtures that can be pointed at the attendees, which will reduce the amount of light directly reaching the screen.

- Install lighting controls, such as dimmers with presets, to allow users to control the amount of ambient light within a room.

- Treat windows for both light infiltration and noise transmission. Heavy drapes or other covers are typical solutions for windows.

Furniture and Room Finish Issues

As noted in the previous section, the color, texture, and reflectivity of surfaces within the viewing environment should be considered when creating an optimal viewing environment. Light colors near the screen reduce the perceived level of image contrast. Reflective furniture and other surfaces increase the levels of ambient light that will fall onto displays.

The AV team should review the characteristics of the proposed interior, and make recommendations as necessary, to deal with issues such as the following:

- **Wall color** This applies to the color of the wall surfaces, both near the screen and throughout the room. Recommend darker surfaces near the screen to increase its perceived brightness and contrast.

- **Reflectivity levels** These apply to table and wall surfaces. Using surfaces that reduce reflections will reduce glare and the amount of ambient light reaching the projection screen and/or display.

If camera images will be captured by the AV system, as in a videoconferencing system, visual backgrounds may play a role. You need to consider how the room finish will impact the camera image. Bold or busy textures or geometric patterns visible on room surfaces may adversely impact the transmitted images. Also, choose colors that will prevent the subjects from blending in with the background.

Addressing Building Environment Issues

When it comes to the building environment in which the AV system will be installed, several issues may demand attention. These may involve building services, structure and substructure, and regulations.

Building Services Issues

The AV team should evaluate the building services systems (power, data, cable, etc.), and identify any areas that may impact the installation and performance of AV systems.

The following are primary concerns that AV designers should address regarding building systems:

- **Ensure adequate power**   The AV design team should request that power be supplied from a single-branch circuit, preferably with a dedicated panelboard. This will minimize potential problems with unbalanced electrical loads. The AV design team should work with the client and the electrical contractor to provide an appropriate power system.

- **Ensure proper access to AV program sources**   The client often specifies that an AV system should be able to support the display of content from non-AV and/or computer systems, such as information accessed via a corporate intranet or the Internet, or display broadcast television programming from MATV, CATV, satellite, or other sources. The AV team should evaluate any site-related issues that will impact the ability of the AV system to provide access to these content sources.

Building Structure and Substructure Issues

Where possible, the AV team should review the underlying structure of the room and identify any potential issues that may impact the installation of AV system components. The building structure includes components such as hanging points and wall supports. The building substructure refers to walls, ceiling height, and other elements.

For example, the ceiling may be used for hanging video monitors or projectors. Walls may be required to support loudspeakers and flat-screen displays. The underlying structure of these room features must be strong enough to support the weight of these devices. In some cases, additional structural support, such as blocking, may be needed to allow the AV team to safely install the desired system.

The AV team should identify any preliminary design considerations that may alter the building structure and communicate specific needs to the client and architect. In some cases, these issues may be addressed via modifications to the building design; in others, a work-around may be necessary to accommodate the intended AV functions.

Building Regulation Issues

Building regulation (code) considerations are a critical aspect of any building project, including AV system design and installation. While the specific codes or regulations that apply to a site vary according to the location and specific AHJ, the design and installation should comply with a range of generally accepted standards.

The AV team should evaluate the site to determine compliance with applicable codes, regulations, and standards, and make recommendations to the client if changes are necessary to meet those standards.

Fire Codes

Nearly every country has a set of written guidelines to govern the installation of wiring and equipment in commercial buildings and residential areas. These guidelines are designed to ensure safety from fire, smoke, and electrical hazards. Anyone involved in the installation process or in specifying cable must understand the local codes and regulations. When in doubt, contact the AHJ.

When considering fire codes, key issues for AV professionals to understand include design safety and cable requirements for breathable air spaces. In a fire situation, cables will burn and give off toxic smoke. This smoke can travel through breathable air spaces and end up vented throughout the office via air handlers. In cases where cables must be run in breathable air spaces, plenum-rated cable or low-smoke, zero-halogen cable must be used. When burned, these cables give off very little toxic smoke, so they are much safer than traditional cable.

Building regulations and codes may also specify the need for fire-suppression systems, generally consisting of an extinguishing agent, expellant gas storage tank(s), manual and/or automatic control hardware, piping or tubing to deliver the agent, and nozzles to effectively disperse the agent onto the fire area.

When installing AV equipment that can create heat, do not locate it near sprinkler or heat-sensor systems. Such incorrect placement can lead to false triggering of the system and potential damage to the facility.

Access for Disabled People

Many countries provide people with disabilities the right to public access and public accommodations. This covers a wide range of places, such as hotels, theaters, restaurants, and other public facilities. Each country will have specific regulations regarding disabilities. Some require the provision of auxiliary aids and services to ensure physical access to the facilities by disabled individuals.

NOTE In the United States, the Americans with Disabilities Act (ADA) requires that video walls and wall-mounted displays (and their mounts), among other systems, not extend farther than 4 inches from the wall in high-traffic areas. If a display solution needs to protrude farther, it must reach all the way to the floor, as in the case of a digital sign installed in a cabinet or similar housing.

One aid for people with hearing disabilities is an assistive listening system, which can use IR light, RF, or induction loop signals to transmit audio programming to hearing-impaired listeners. In the case of IR transmission, designers must allow for sufficient line of sight between the transmitted sound source and the listener's receiver.

For those with reduced hearing, the ability to view a presenter, performer, or program may be critical to their understanding of the material presented. Some codes also require other accommodations, such as flashing-light alarm systems, permanent signage, and adequate sound buffers.

Disabled individuals should also be able to access the AV system components that are operated by participants. This can mean ensuring that components are mounted in a position where people in wheelchairs can reach them. It also may require alternate presentation areas so that the lectern is not too tall for the presenter. Figure 14-5 illustrates the reach from a wheelchair.

Figure 14-5
Designers should consider access by people in wheelchairs.

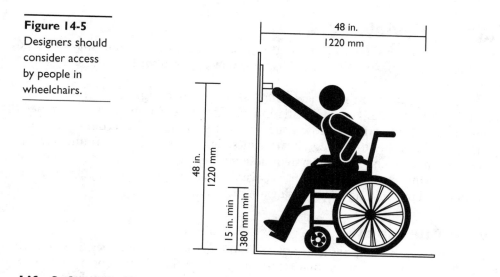

48 in.
1220 mm

48 in.
1220 mm

15 in. min
380 mm min

Life Safety Codes

Many building code requirements address issues surrounding the life and safety of building occupants. Although the client's architect and building contractor typically have primary responsibility for addressing these issues, the AV team should be aware of the implications of these codes and how they may impact the design of the rooms that will include AV systems.

For example, building regulations and codes usually define guidelines for ensuring that occupants can safely exit a building in the event of an emergency. These regulations typically include requirements for safe passage through rooms, doorways, hallways, corridors, and so on. These regulations specify standards for the size of door openings, location of building exits, emergency signage, lighting, and other safety measures.

A CTS-certified professional should understand the impact of these regulations. For example, issues such as emergency egress requirements may define the location, design, and use of a door. Regulations may also require exit signs in locations where they cast unwanted ambient light. The AV designer may not be able to change these elements of the room, and therefore must be able to accommodate for them in the overall design.

Room Occupancy

Building regulations also typically address requirements for room occupancy. These regulations may specify the following:

- Maximum number of persons allowed within a specific room
- Location and design of room exits
- Size and layout of aisles between seating

The AV designer must consider any applicable occupancy requirements when recommending site changes to the client.

Chapter Review

After the initial site evaluation (discussed in Chapter 13), the AV team may discover that changes are necessary. These modifications may apply to the space, to the AV system, or both.

The role of the AV team members is to assess the environment, identify relevant issues that impact the design and installation of the AV systems, and make recommendations to the client to address those issues, when appropriate. The recommendations may include approaches to address site acoustics, lighting, sightlines, furniture layout, and other room elements. The AV system designer may also be called upon to advise the client regarding changes to building services, as well as approaches to address limitations imposed by building code considerations.

Review Questions

The following review questions are not CTS exam questions, nor are they CTS practice exam questions. These questions may resemble questions that could appear on the CTS exam, but may also cover material the exam does not. They are included here to help reinforce what you've learned in this chapter. For an official CTS practice exam, see the accompanying CD.

1. What is the purpose of a sightline study?

   A. To determine the brightness of an image for a particular viewer

   B. To determine if a viewer can see the smallest items on a screen

   C. To identify the most appropriate location for a projector

   D. To determine if the audience has a clear view of the screen

2. Which of the following is *not* a recommended approach to minimize ambient light levels within a room?

   A. Use lighter-colored finishes on walls

   B. Use focused task lighting

   C. Install lighting controls such as dimmers

   D. Treat windows for light infiltration

3. What is the primary concern the AV designer has regarding the design of the building's HVAC system?

   A. Minimize noise levels within the room during HVAC operation

   B. Provide sufficient cooling and heating for audience comfort

   C. Ensure sufficient cooling to maintain appropriate temperature levels for AV component operation

   D. Locate HVAC system controls near the AV system control to allow for adjustment during a presentation

4. How do building regulations or codes impact the AV system design?

   A. AV systems are not affected by building codes.

   B. The layout and design of AV systems are usually strictly regulated via building regulations and codes.

   C. Building regulations and codes often specify the type of electrical wiring that must be used for specific AV installations.

   D. Building regulations and codes typically specify that a separate technical power system be installed for AV systems.

Answers

1. **D.** Sightline studies are used to determine which portion of the screen the audience will be able to see from each of the seating areas throughout a room.

2. **A.** Darker colors are better suited to reduce ambient light levels within a room.

3. **A.** The AV designer is typically mostly concerned with reducing the amount of noise generated by an operating HVAC system.

4. **C.** Building regulations or codes typically define the type of wiring used by the AV system, in addition to other factors, such as mounting requirements.

Developing a Functional AV Scope

In this chapter, you will learn about
- Compiling information for the functional AV system scope
- Documenting the functional AV scope with the program report
- Presenting your findings to the client

As recounted in previous chapters, during the initial stages of the AV systems design and coordination process, the AV team assembles the information it will use to identify the specific AV systems required to meet the client's needs. The result of collecting this information is an understanding of the required AV system functions, as well as the layout and installation needs of these system components.

Next, the AV team must review the information it has collected and develop a recommendation, or program report, for an AV system solution that meets what the AV designer or integrator has identified as the client's needs. This program report, which includes a functional AV system scope, should identify the overall AV system functions, the AV components required to provide these functions, the layout of components, the supporting systems, and other aspects of the AV system design and scope. The program report should also include the estimated cost of all AV equipment and installation services identified in the report, as well as requirements that are outside the scope of an AV contract but must be completed by other contractors, such as ensuring electrical power is provided to all AV racks.

The client uses the program report to evaluate the proposed AV system scope and determine whether its characteristics and costs meet its needs. After reviewing the program report, the client may request changes in the system scope or components. The client may also want to address other issues, such as cost, room layout implications, the installation schedule, and more.

> ### Domain Check
>
> Questions addressing the knowledge and skills related to developing a functional AV scope account for about 8% of your final score on the CTS exam (about eight questions).

Compiling Information for the Functional AV System Scope

AV designers use the information collected during initial site visits to identify client needs and evaluate issues at the site environment that will impact the AV system design, installation, and operation. "Developing a functional AV scope" refers to continuing the documentation process through what is sometimes called the *program phase*.

It is critical to include the end users in the program phase of the process. End users, administrators, and their managers are the people who will be using the system. Both the client/owner and the end users should provide input to the design team to ensure their needs are met, and that all parties are "invested" in the final product. Systems that are designed based solely on design team assumptions, without input from end users, are frequently off the mark.

This stage will also require priority decisions. Along with the client and end users, the AV designer needs to determine what is realistic to meet the client's needs, especially if there are budget constraints. Keep in mind that end users may be considering future requirements, which could have an impact on the current system design. For example, a university with a popular on-campus course may consider eventually offering off-campus presentations via videoconferencing. While this may not be a current requirement, the potential need may be included as an application when designing the AV system.

All the information gathered via these activities is interpreted and presented in a program report. Once this document is distributed, reviewed, and approved, it becomes the basis (sometimes part of the contractual basis) for the subsequent design phase. This review and approval process may take several rounds to complete and to obtain owner and/or end-user sign-off. Figure 15-1 illustrates the components of the program phase.

 NOTE Much of what a CTS candidate needs to know about the program phase is laid out in ANSI/INFOCOMM 2M-2010, Standard Guide for Audiovisual Systems Design and Coordination Processes. This standard provides a description of the methods, procedures, tasks, and deliverables applied by industry professionals in medium-sized to large-scale commercial AV system design and integration projects. Ultimately, the standard applies to all phases of an AV project, but it's particularly important during the program phase. You can find more information about standards at www.infocomm.org/standards.

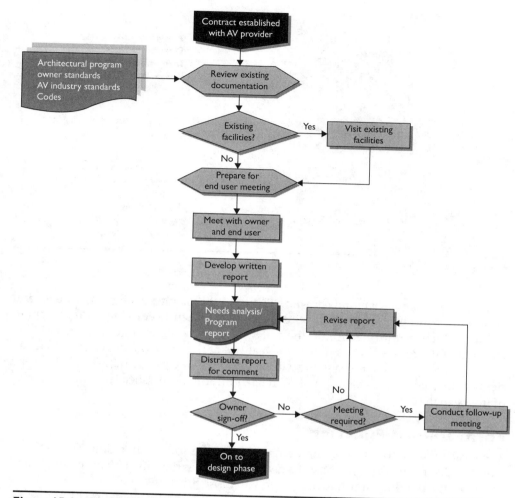

Figure 15-1 The program phase process

Needs Determine Applications

AV systems are intended to support the users' needs—the activities that the end users intend to perform. For example, a sales department may need to perform long-distance training and teaching activities to keep its sales staff informed and on target. A video-conferencing system is used to support this need. The AV designer must work to carefully identify the characteristics of the client needs before defining the features and functions of the AV system intended to support those needs.

AV technologies will often be used to enhance existing end-user applications, such as presentations, training, and conferencing. The tasks that make up these applications—such as displaying slides, making notes, and placing calls—are supported by the features and functions of an AV system installed in a room.

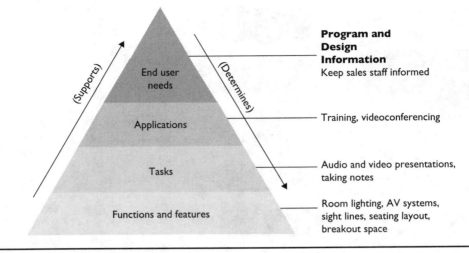

Figure 15-2 The needs analysis pyramid

The needs analysis pyramid shown in Figure 15-2 illustrates both the top-down and bottom-up concepts of examining end users' needs. Their needs determine the applications, which in turn determine the tasks, which then determine the functionality required of the AV systems, as well as the features of the space (architectural, electrical, and mechanical) where the systems will be installed. In Figure 15-2, the labels to the right of the triangle are examples of AV-related information associated with each level of needs analysis.

Form Follows Function

The basic premise for all building design is that form follows function. The designer works to determine the functions of a given building, as described by future users, and to design the building's form to accommodate those functions.

The same is true for professional AV systems. Once you have defined the specific system functions that will address the client's needs, you can then design the AV system itself and its supporting infrastructure.

The flowchart in Figure 15-3 provides an overview of how the end users' needs ultimately define the design team's ability to create the required AV system electronics, and how these have an impact on the architectural, electrical, and mechanical infrastructure design.

Gathering Information About Facilities

As discussed in previous chapters, you must spend the early part of the program phase collecting information about the physical, organizational, and technical characteristics of a project. Existing documentation may address a number of areas relevant to the program phase.

If the AV systems are to be installed in an existing facility, it is also important to tour those areas during the program phase. You will need to gather information about the

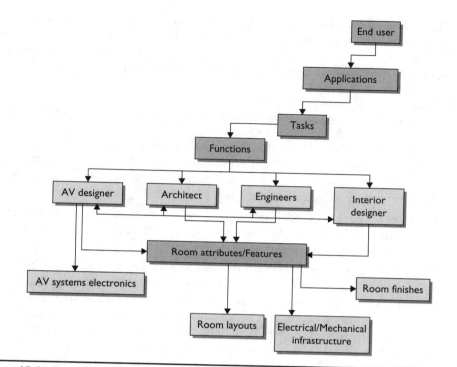

Figure 15-3 Translating the needs into a design

physical aspects of the facility's spaces and how they are currently used. Here, we'll review the types of documentation an AV designer needs to be able to access.

Architectural Documentation

Architectural documentation of the project may include the following:

- **Architectural and infrastructure requirement documents**   These contain information such as the number of proposed classrooms, conference rooms or auditoriums; the sizes of the spaces; and the number of people they will support. These documents may include a description of what the architectural team has in mind when it comes to AV systems, in terms of space allocation and/or budgeting. Architectural program documents also offer insight into the owner's operational needs and how those needs translate into space requirements.

- **Architectural drawings**   These may be "as-built" drawings for an existing building or drawings of a design in progress. These drawings may be available in paper and/or electronic form. They depict the size of the spaces, their relationships to one another, and planned finishes. The AV designer can gain an understanding of the layout and scale of the working spaces by examining floor plans, reflected ceiling plans, and section drawings of the spaces or buildings (these plans are described in Chapter 12).

- **Engineering drawings**   These depict the mechanical, electrical, structural, plumbing, and other engineered systems in the building. These drawings are used to help identify issues of concern to the AV designer, including power, conduit, and data infrastructure, as well as any obvious HVAC noise concerns.

Organizational Documentation

Organizational documentation that should be examined can include the following:

- **Project directories**   These are documents produced and maintained by the architect, owner, or construction/program manager. The project directory lists the major contracted entities on the design and construction team, their identified role, and the primary contact.

- **Contract scopes and roles**   This refers to the portion of the architect's contract that delineates the architect's role and may include terms that apply to the AV provider's role and responsibilities.

- **Owner and end-user information**   This is background and operational information that can be crucial to determining the AV project's purpose. This type of information is often contained in the end-user organization's resources, such as websites, procedural and operations manuals, organizational charts, and university catalogs. It is increasingly important that AV designers and integrators also obtain owner/end-user information relating to IT operations and personnel, as AV systems must interface with, or even operate on, IT networks and systems.

- **Industry standards**   Examples of standards include ANSI/INFOCOMM 2M-2010, Standard Guide for Audiovisual Systems Design and Coordination Processes, and ANSI/INFOCOMM 3M-2011, Projected Image System Contrast Ratio. Standards outline a consistent set of tasks, responsibilities, and deliverables required for professional AV system design and construction.

Technical Documentation

The following technical documentation may provide information relevant to the project:

- **Owner and design manuals**   These are documents that are not specific to a project, but include standards, criteria, and procedures that must be adhered to in any project involving the owner's organization.

- **Codes and ordinances**   The most common codes that must be addressed here are electrical, fire, and life safety. In addition, you must comply with accessibility requirements. Many U.S. federal regulations have international equivalents, and local authorities may impose additional restrictions.

- **Industry standards**   These include basic standards and recommended practices regarding audio, video, and control systems, such as the ANSI/INFOCOMM 1M-2009, Audio Coverage Uniformity in Enclosed Listener Areas standard. Additional research for specialty systems or certain vertical markets may need to be performed. For instance, interfacing with an audio and video broadcast system or health-care equipment may require adherence to audio, video, and digital standards that are not typical professional AV system standards. Other nonelectronic requirements, such as the ANSI S12.60, Classroom Acoustics standard, may be required by some owners within the education arena.

- **Network architecture**   This information may be easier to obtain in existing facilities where computer networks are already in use; however, in new buildings and/or spaces, you should coordinate with the client to obtain information about the network architecture that will support users' IT requirements. AV designers will want to understand network topology, security, bandwidth, and more. (Refer to Chapter 6 for information about networks and network architecture.)

Conducting Project Planning and Coordination Meetings

As noted in earlier chapters, throughout the program phase, the AV team should conduct several meetings. The purpose of a program meeting is to gather and exchange the information necessary to determine which functions are required to support end users' applications. This information should reveal what the end users currently do, what they need to do, and what they want to do.

Program Meeting Attendees

Program meetings should include representatives of both the design team and the owner, and all key stakeholders.

The owner and end users should be represented as follows:

- End users, including representatives from each organization or department that will use the new systems
- AV technology manager, representing the technical side of the owner's AV operational needs
- IT representative, to address issues in utilizing or accessing the IT system and how it relates to the users' activities
- Administrative representatives/stakeholders, including owner executives, facility managers, department heads, and funding organizations

If the facility will be used by people outside the owner/operator's organization (such as videoconferencing facilities for hire or a classroom building serving several departments), the perspectives of those groups should be included whenever possible. If there is an owner or end-user technical committee, it should also be represented in the meetings.

Program meeting attendees from the design team should include the following:

- AV consultant project manager or AV integrator project manager
- AV designer
- Architect, to provide input regarding space allocation and planning
- Program or construction manager, who may be onboard as a stakeholder in the overall project outcome

Program Meeting Agenda

The AV designer's main objective in a program meeting is to gain a clear understanding of the users' needs in order to develop a document that includes descriptions of the necessary systems and a preliminary budget. Depending on the scope of the project and the number of users to be interviewed, the meeting can take many forms. It may be a single meeting with a few individuals that lasts a few hours, or it may be a series of meetings with different groups over several days.

Regardless of the size and duration of the program meeting(s), the issues to be addressed are similar. The agenda for a meeting should typically include these items:

- **Technology trends**   Depending on the sophistication of the users, it's often helpful to initiate a discussion of trends in AV technology, particularly as they apply to the users' applications. Be careful as you lead the discussion to ensure it does not result in extravagant "wish lists."

- **End-user needs**   The meeting should focus on the end users' needs and applications, as determined via the needs analysis process described in Chapter 11.

- **AV tasks and parameters**   Once you have identified the needs and applications, you can define in more detail the AV systems and tasks that will support those applications. This determines the overall system configuration and proposed budget, which you will document in the program report.

- **Infrastructure issues**   Any issues related to the building design, layout, or facility that will impact the AV system design, installation, and operation should be identified. These can include potential impacts related to room design and layout, electrical systems within the building, HVAC, data/telecommunications, and so on.

- **IT issues**   Addressing IT issues allows the AV designer to get a better understanding of the client's IT networking systems and what's required to support the AV systems. Is the IT department willing to allow AV applications to access the same resources as IT applications, or must additional networks be built to support AV systems? If the IT department indicates a willingness to allow AV applications on its network (videoconferencing or digital signage, for instance), you will need to learn how bandwidth, security, and other IT issues may impact the AV design. Networks are discussed in Chapter 6, and you'll find information on designing and integrating networked systems in Chapters 16 and 19.

The outcome of the program meeting(s) should be an overall agreement on the needs and applications that will drive the final AV system design for each room and system.

Documenting the Functional AV Scope with the Program Report

At the conclusion of the program meeting(s), AV designers and integrators must capture all the information they have collected in a written report, which should include an interpretation of the users' needs with respect to the AV systems. The report should contain a conceptual/functional system description and, if applicable, information about its impact on spaces that have already been programmed, designed, or built.

Objectives of the Program Report

The program report is meant to provide those involved with the AV system project the information they need to proceed with the project. The following are the objectives of the program report:

- Communicate to the decision makers about the overall systems and budget.
- Communicate to the users the system configurations that can serve the needs identified during the program meetings.
- Communicate to the design team a general description of the AV systems and what impact they may have on the other trades working on the building.
- Communicate to the AV designer and integrator the scope and functionality of the AV systems to be designed and installed.

Contents of the Program Report

The program report typically consists of the following sections:

- Executive summary
- Space planning
- Systems descriptions
- Infrastructure considerations
- AV budgets
- Operational staff expertise level required
- Maintenance budget requirements and life-cycle expectations

Let's take a closer look at what each section contains.

Executive Summary

The executive summary provides an overview of the project, project programming process, systems, special issues, and overall budget. Keep in mind that the report may be read by a wide range of stakeholders: the CEO, CIO, architect, AV manager, IT manager, end users, electrical engineer, mechanical engineer, facilities department manager, and so on. It should briefly highlight the findings from the various interviews you conducted, as well as the systems you will design.

The executive summary should include the following:

- An overview of the project (what the facility is; how it serves the owner; and why it is being built, upgraded, or renovated)
- An overview of the programming process (who was interviewed at the departmental or organizational level and other individuals who were important during the information-gathering step)
- An overview of the systems (what and where the systems are, their type and quality, and which users they serve)

- An overview of any special issues
- A reference to the overall budget required for the systems

The client should be made aware of any special issues related to the proposed AV system design, installation, or operation. Examples include major project obstacles or limitations, project schedule issues, options for specific spaces, or overall system configuration options.

Space Planning

This section provides advice to the design team, if necessary, regarding any special requirements of the AV systems. This may include equipment closets, projection rooms, observation rooms, dressing rooms, head-end rooms, etc.

Systems Descriptions

The systems description documentation is a nontechnical description of the owner's desired functionality for each system. This section may include AV sketches, drawings, diagrams, photos, product data, and other graphics to exemplify the capabilities of the proposed systems.

Depending on the project size, the number of systems, and the different types of users, systems descriptions may be organized by space, user, system type, or a combination of the three. The nature of the project will determine how the system descriptions should best be presented. For example, some systems may be specific to a single room, or there may be systems that are in several similar rooms that span a number of departments. There may also be facility-wide systems that connect some or all of the AV systems together or that provide overall control, monitoring, interconnection, or help-desk functions to a building or campus.

Infrastructure Considerations

The infrastructure considerations section includes a description of the electrical, voice, data, mechanical, lighting, acoustic, structural, and architectural infrastructure needed to support the AV systems. The documentation identifies site environment issues that will impact the system design and installation and any recommended site modifications that may be necessary to facilitate optimum system operation.

AV systems often have a significant impact on architectural, mechanical, and electrical systems, so awareness of where the impact may occur is critical to a complete understanding of the project. The following areas should be addressed:

- Lighting
- Electrical
- Mechanical (both noise and heat)
- Acoustical
- Data/telecommunications
- Networking
- Structural

- Architectural (space plans, adjacencies, allocations, and other architectural issues)
- Interiors (finish requirements)
- Coordination of other trades with AV installation needs
- The budget impacts of these issues

Note that a general discussion of infrastructure budgets may be included in the infrastructure section. For example, if AV applications and systems will require bandwidth on the IT network, it may be necessary to add bandwidth capacity. But there may be more space- or trade-specific issues that require additional discussion or illumination. If you feel your infrastructure considerations are running long, you can address them in a section about special issues.

AV Budgets

This section outlines the probable costs to procure, install, and commission the proposed AV systems, as well as any additional costs, such as taxes, markups, service, support, and contingencies.

The AV program document should include an opinion, or estimate, of the final cost of the identified AV systems. This opinion establishes a more accurate cost estimate for the AV portion of the overall project budget.

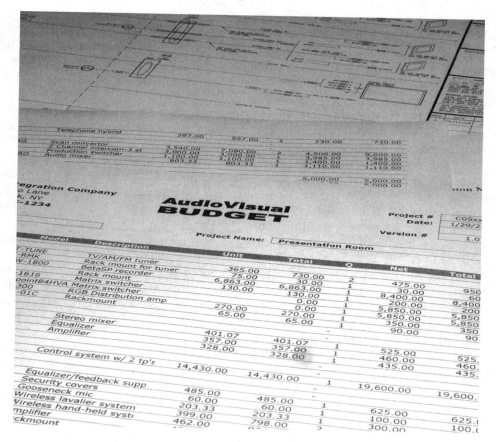

Keep in mind that AV professionals typically use standard terminology when discussing the potential costs of AV systems or other aspects of a project. The following are some common terms:

- **Opinion of probable cost** This term describes an early attempt to determine the cost of a system before there is enough detail in the design to produce a line-item estimate. The opinion of probable cost is an educated guess based on experience and may include some specific line-item costs for large equipment, such as projectors or large matrix switchers. Final costs cannot be applied until the system is designed and the actual equipment selected.

- **Estimate** This implies that there is a more objective basis for the cost provided. It is an approximate calculation that includes a line-item analysis for equipment and labor (perhaps including taxes and other ancillary costs), and would be more accurate than an opinion of probable cost.

- **Quote** This is a detailed and enforceable estimate. It should be provided and identified as a quote for an AV system only if the estimate is based on a program report and a detailed system design.

The term *budget* is often used to describe the preliminary estimate of costs, but by strict definition, budget refers only to the general amounts the owner or project team allocates for a particular system, trade, or whole facility. The general AV budget should be established based on an opinion of probable cost or an estimate. A quote is then submitted later by an AV provider based on an RFP, after the client has accepted the preliminary system specifications and budget. This quote will also include detailed system depictions that specifically define the system design, components, and installation approach.

Costs are typically organized in the same manner as system descriptions, unless the owner or architect requests a specific budget breakdown. You can report a basic breakdown by room or system (with no further cost detail) and a total cost.

You may further break down costs into the costs for components and the costs for installation of subsystems and major equipment, such as display systems, audio systems, video-processing systems, large switchers, control systems, data storage, and programming and signal-distribution systems between rooms. Any breakdown beyond this level of detail can be misleading to those unfamiliar with the design and construction process. It is also likely to be inaccurate later in the project—at least at the detail level—especially if final equipment choices are made a year or more later.

Small or short-term projects may allow for a more detailed estimate or quote to be developed in conjunction with the program report. Such an estimate or quote should include a caveat stating that revisions may be needed if there are changes in the scope or configuration of the AV systems after the owner review. You may also want to include a time limit on the validity of the quote.

Additional costs should also be identified and stated as separate line items, such as taxes, markups, and contingencies. For example, some owners are exempt from sales tax on all or part of equipment and labor purchases. This should be noted explicitly in text and shown in the cost breakdown, so that it is clear whether taxes are included and what percentage is being applied.

Depending on the anticipated contract arrangement, you may also need to include information about markups. These costs are usually associated with the subcontracting process. For example, if the AV systems and labor costs are contracted to a general contractor, the contractor may charge the owner an additional 5 to 10 percent or more over the AV system cost. This markup can be significant, depending on the cost of the AV system. A $2,000,000 AV system, for example, might have a markup of $100,000 to $200,000 applied to it.

While not yet a common practice in the AV industry, contingency budgets may be an important consideration in larger, longer-term projects (those costing more than $500,000 and lasting for one year or more). Such projects are more prone to potential site conflicts, equipment updates, and new technologies that require additional funding. A contingency budget item of 5 to 10 percent of the AV system's total cost can cover unanticipated overages. Sometimes, the building budget has a contingency that will indirectly cover overages in AV systems.

Presenting the Program Report to the Client

Once the system program report is completed, the AV designer or integrator should distribute it to all stakeholders for review and sign-off. The AV team should establish a period of time for recipients to review the document, and schedule a meeting to review stakeholder comments and make necessary revisions to the identified approach. The report will then be reissued to obtain formal approval from the owner or a representative.

Distributing the Program Report

The following stakeholders are generally included in this final stage and should receive a copy of the program report:

- **Owner**   The owner should review the suggested system design approaches, the end users' needs and requests, and the estimated pricing. Ultimately, the owner needs to agree and sign off on the proposed systems and on the cost for those systems. Keep in mind that the owner's organization may not include the actual end users.

- **End users and technology managers**   End users, technology managers, and/or AV/IT managers should review the document to determine whether their needs have been addressed and to understand the AV systems concepts, approach, and costs. It is especially important for the client's IT managers to review the report to understand how the proposed AV systems might impact the network, data security, and other areas. By reviewing and signing off on the program document, this group establishes its agreement with the scope and purpose of the AV systems.

- **Architect** The architect will be interested in the scope of the AV systems and the impact it may have on the facility design and the overall project budget. The architect does not usually sign off on the program—that approval is reserved for the owner and end users—but if the AV team is subcontracted to the architect, the architect often serves as the person responsible for distributing the report and collecting feedback. The architect would also distribute the report to the consulting mechanical, electrical and plumbing (MEP) engineer in high-end projects.

- **Project manager** The project, program, or construction manager, if there is one, will have a tremendous interest in the program report, because that manager must guide the AV system integration within the overall scope of the facility project. In some cases, the construction or program manager may even participate in the sign-off process, along with, or representing, the owner. Note that a project manager is a *required* position for construction projects in many locations outside the United States.

- **General contractor** A general contractor (also referred to as a main contractor) may be onboard during the design phase to assist with project costing or as a part of a fast-track project. In such cases, the architect may provide construction documents to the general contractor in stages, before completing the entire design package. If so, the general contractor might review the program report to become aware of activities during the AV program phase and any upcoming issues in the architectural design.

- **AV integrator**   If the AV integrator is preparing the AV program report as part of a design-build project, the internal team members should be included in the report distribution for their comments and education. The integration team should also be on the distribution list if the project is operating under a consultant-led design-build process. If the project is to be consultant-led, the program report (not including cost information) will be distributed to potential AV integrators during the bidding process.

- **Other professionals with special interests**   Program report recipients may include other professionals with special interests, such as members of a funding organization that is donating money for a facility, even if they are not directly involved in the project design and construction. The owner may specifically request distribution to these other individuals. Such additional distributions are acceptable, as long as the recipients are aware of the purpose of the program report, their responsibility for a response (if required), and the appropriate limitations on redistribution of the report.

Preparing for the Client Presentation

The review process will likely require several meetings in which the AV team presents the program report. The AV team should prepare for this presentation as follows:

- Identify the participants planning to attend the meeting. This will help focus your presentation to address the interests and priorities of the participants. For example, the owner or construction manager will be interested in different issues than end users.

- Create a meeting agenda. A detailed agenda document identifies the topics you plan to present, the planned sequence of topics, and the amount of time that you intend to spend on each topic. This will help you set audience expectations and maintain more control of the discussion.

- Once you identify the information you need to present, spend time gathering it for the presentation. Collect the appropriate number of documents you want to distribute; obtain samples or other items required to conduct a demonstration (if necessary); and locate illustrations, photographs, and drawings that depict system components, system design documents, room layouts, etc. In some cases, it may be appropriate to create a formal projected presentation using presentation software; however, for smaller meetings, this may not be necessary.

- Think about what types of questions and concerns might arise and what other information or backup documentation may be necessary to help you address them. Some examples are notes from interviews with end users or other stakeholders, information from benchmarking visits, and building/site plans.

Conducting the Client Presentation

Once you begin your presentation, try to keep the meeting on track and focused on the topics and objectives you previously identified. Address questions and concerns from the participants. If you don't know the answer to a specific question or concern, don't try to make one up. It is perfectly fine to say that you will look into an issue and provide the answer at a later time.

Because the objective of the meeting is to collect client feedback on your proposed approach, and because that feedback will be incorporated into an updated document or become part of the subsequent system design, it is usually a good idea to bring along a colleague to take notes. You may also wish to use an audio recorder, with the permission of all meeting participants. This will help ensure that you address client issues in a comprehensive and accurate manner.

Program Report Approval and Next Steps

Once the review process is completed (possibly requiring additional follow-up meetings), the program report should be formally approved by the owner and end users. Upon approval, this document can be used as a basis for the design of the AV systems and their infrastructure. The subsequent design document protects both the owner and AV system team by ensuring that everyone is defining the project in the same way before more money is spent.

If the project is being led by an independent AV consultant, the program report becomes the basis for the design of the AV systems. During the design process, the report guides the AV designer in preparing the bid package to be distributed to integrators later in the project.

If the end users' needs were not sufficiently defined before the program phase began, the contract for AV design services may not have been established or proposed. In this case, the program report also becomes the basis for the AV consultant's proposal for the continuation of the design and construction administration services.

Program Reports and RFPs

In some cases, the program report becomes the basis for publishing an RFP to obtain bids for the design and build processes. Two common scenarios may drive this decision:

- The owner has completed an internal program development process and wishes to pursue a design-build contract with an integrator.

- A fast-track project has been programmed by an independent AV consultant and there is insufficient time for a sequential design-bid-build process. The owner and/or consultant would then manage the contracting process and monitor the process once an integrator is engaged.

If an integrator had been contracted for a design-build project before completion of the AV program report, the contract may have been established based on an assumed scope and predetermined AV budget. In this case, the AV program document defines more fully what the system design and installation will be.

Modifications to the integrator's original contract may be required if the program results differ from what the integrator first proposed. If, as in the design-bid-build scenario, the project scope was not sufficiently well defined before programming, the integrator should develop a proposal for AV design and installation services based on the system characteristics identified in the program report.

Chapter Review

In this chapter, you reviewed the general approach for evaluating and documenting the AV system scope that will meet the client needs. This process includes the following:

- Compiling customer and site information to determine the required functional AV scope for a system

- Writing the program report recommendations in a manner that is detailed, clear, and accurate (requires skills in depicting the system layout and components, and calculating budget and costs for the system components and installation)

- Presenting findings to customers, either via documents or in person

This stage is a critical aspect of the AV design process, because it is intended to ensure that the needs of the end users and all other stakeholders are clearly identified and considered during the system design and specification process.

Review Questions

The following review questions are not CTS exam questions, nor are they CTS practice exam questions. These questions may resemble questions that could appear on the CTS exam, but may also cover material the exam does not. They are included here to help reinforce what you've learned in this chapter. For an official CTS practice exam, see the accompanying CD.

1. What is the objective of the program phase of AV system design?

   A. Create the computer software required to control an AV system

   B. Create detailed design documents depicting the AV system components and installation

   C. Describe the AV systems necessary to support the defined needs and the general cost of those systems

   D. Provide a detailed cost quote for the client to approve

2. What should the specific AV system capabilities be based on?

   A. Identified needs of the end users

   B. Results of the baseline visits

   C. Installation capabilities of the general contractor

   D. Recommendations from the client

3. What is the main objective of an AV program report?

   A. Communicate to the decision makers about the overall system's capabilities and budget

   B. Provide a detailed layout of AV components to the general contractor

   C. Provide a listing of specific components to be used within the AV system

   D. Describe the location and layout of AV components within a room

4. Which portion of the AV program report should describe the end users who were consulted to determine system requirements?

   A. Special issues

   B. Infrastructure considerations

   C. System descriptions

   D. Executive summary

5. Which of the following cost descriptions provides a general budget for the AV system for use within the AV program report?

   A. System quote

   B. System estimate

   C. Opinion of probable cost

   D. System "ballpark"

6. Once approved, what does the AV program report become the basis for?

   A. Identifying required AV system components

   B. Purchasing and installing the AV system components

   C. Developing more detailed AV system design documents and cost estimates

   D. Additional discussions of AV needs

Answers

1. C. The program phase should focus on defining the user needs, the general AV systems required to meet these needs, and the general costs of these systems.

2. A. The AV system capabilities should be defined based on what the end users need to accomplish. This will ensure that the system meets the needs of the users.

3. **A.** The program report provides an overall description of the system, its capabilities, and general cost. Once it is reviewed and approved, it will provide the basis for more detailed descriptions of the AV system.

4. **D.** The end users who were consulted during creation of the program report are typically identified within the executive summary.

5. **C.** The opinion of probable cost is a general estimate of the overall cost of the AV system that can be used by the client as a basis for creating a general budget for the AV system portion of a project.

6. **C.** Once the AV program report is reviewed and approved, it will provide the basis for creating detailed AV design documents and cost estimates.

PART IV

Designing and Building an AV Solution

You've done your prep work. You've learned about the clients, discovered their needs, and documented the environments where AV systems need to go. Part IV of the *CTS Exam Guide* deals with the next steps, including actually designing and building the AV solution. Along the way, you'll need to select vendors and purchase equipment, plus assign a price tag to your company's efforts and manage the project from start to finish.

Domain Check

Chapters 16 through 20 address tasks within Domain A (Creating AV Solutions) and Domain C (Conducting AV Management Activities). In this part of the *CTS Exam Guide*, we will focus on the Domain A tasks related to taking an AV solution from design to installation. The Domain C tasks covered in this part involve financial management, from purchasing to job costing, and project management. The tasks covered in this part account for about 30% of the CTS exam.

CHAPTER 16

Designing an AV Solution

In this chapter, you will learn about
- Designing the display system
- Designing an audio system
- Designing a control system
- Integrating an AV system design with IT networks
- Documenting system design

Earlier chapters focused on facility design considerations, such as accessibility, seating, and room layout—factors that help determine the AV technology required to meet a client's needs. This chapter focuses on the actual AV system design.

System design encompasses the electronics required to support the client's communication needs and how they're interconnected to create one or more systems. You'll need to select AV equipment based on, among other things, the performance required to successfully fulfill the finished system's functions. And you will need to document it. A significant responsibility of AV professionals is to provide system design documentation.

AV system design involves research, creativity, decisions, diligence, and rigor. AV designers start with a blank piece of paper or a bare computer screen and follow their own personal approach, mixed with industry best practices, to arrive at a system design. With so many different choices of equipment available, it is as much an art as it is a science. Therefore, designing an AV solution that meets the client's requirements, while being fully functional, can be challenging.

To complete the AV design task requires the following knowledge and skills:

- An understanding of the capabilities and applications of typical AV devices
- The ability to determine the required equipment performance to meet the client needs
- Knowledge to assess the issues that may affect the selection of AV devices
- Knowledge of how the devices should be integrated into an AV system design
- An understanding of how modern AV systems interface with IT networks and how that affects the AV design

- Basic technical writing skills for developing system specifications that include block diagrams and drawings that depict the layout and integration of system components
- The ability to create a bill of materials that identifies the proposed pricing of the elements of the AV system

Domain Check

Questions addressing the knowledge and skills related to designing AV solutions account for about 12% of your final score on the CTS exam (about 12 questions).

Designing the Display System

The visual display components of an AV system usually drive the room design. Designers must configure the overall room layout to ensure that the audience has a high-quality viewing experience, and then they must design the other elements of the room to fit that layout.

The *display system* (also referred to as the *video system*) consists of the display devices themselves (flat-panel display, projector and screen, etc.) and the components that support those devices (sources, switchers, etc.). Every aspect of the video system affects what the audience sees; therefore, it all requires careful consideration by the AV designer.

The designer must make sure the system can perform at the intended level. Doing so requires knowing the performance specifications of every part of the system, from video components to cabling. It is important to select components that can meet the required level of system capability and performance.

The designer also needs to understand the characteristics of the signals that will transport images from a source to a display. All the components along a signal pathway must provide adequate bandwidth, because bandwidth affects performance.

The AV system designer performs the following tasks when designing a video system for a client application:

- Determines how the system will be employed
- Identifies the required display device(s)
- Identifies the monitoring, feeds, and recording requirements
- Identifies the display signal sources
- Identifies the switching and distribution components
- Identifies display signal issues
- Adds any required processing and conversion

Let's take a closer look at each of these tasks.

AV technology is critical to the way people collaborate, communicate, and connect in the twenty-first century. AV systems increasingly incorporate information technology in order to simplify installation and operation, and to expand their capabilities. CTS-certified professionals understand not only the components of today's advanced AV systems, but also the important considerations of an AV design that meets the client's unique needs.

Increasingly, AV professionals are asked to design and build multipurpose rooms, which include flexible audio, video, acoustics, and lighting systems to support a variety of different uses. (See Chapter 11, the "Conducting a Needs Analysis" section.) Photo courtesy of RJC Designs.

Depending on what the client needs, such as a display for teaching a roomful of students using multiple content sources, you may determine that multiple displays integrated into one fits the bill. (See Chapter 16, the "Designing the Display System" section.) Photo courtesy of InfoComm International.

Cables in an AV system eventually lead to electronic equipment. An *equipment rack* is the skeletal framework where system components are housed, arranged, and interconnected. (See Chapter 19, the "Rack Building" section.) Photo courtesy of InfoComm International.

Producing a live event requires many of the skills that other AV professionals must learn. (See Chapter 19, the "Producing AV Events" section.) Photo courtesy of InfoComm International.

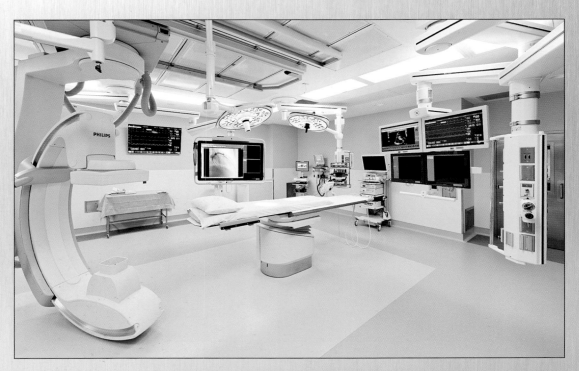

The health-care market is one of the fastest-growing in the AV industry. You must understand the market in order to address its AV needs. (See Chapter 25, the "Defining Your Market, Audience, and Value Proposition" section.) Photo courtesy of Human Circuit.

Ambient light from exterior windows will impact the location and type of a display. (See Chapter 13, the "Assessing the Visual Environment" section.) Photo courtesy of CMS AudioVisual.

Emergency operating centers, with multiple displays and display sources, rely heavily on AV technology and IT to share critical information. (See Chapter 16, the "Identifying Display Signal Sources" section.) Photo courtesy of AVI-SPL.

A properly planned and programmed control panel helps users manage AV systems without exposing them to the complexities of AV integration. (See Chapter 16, the "Designing Control Systems" section.) Photo courtesy of Westbury National Show Systems.

Confidence monitors are a common element of presentation systems, especially in large spaces. These secondary displays allow presenters to see the material that the audience sees without turning their backs to do so. (See Chapter 16, the "Identifying Monitoring, Feed, and Recording Requirements" section.) Photo courtesy of Westbury National Show Systems.

These days, technologically advanced AV systems, such as video walls, often require CTS-certified professionals to have an operational knowledge of computer systems in order to install and operate them. (See Chapter 16, the "Designing the Display System" section.) Photo courtesy of InfoComm International.

Digital signage—the practice of using video displays to advertise products and services or communicate messages—is a rapidly growing application of AV technology and requires integration with IT. (See Chapter 11, the "Obtaining Information About Client Needs" section.) Photo courtesy of InfoComm International.

A technician safely installs a mount for a small flat-panel display. The mounting location and approach are often defined in AV design documents, based on an earlier site survey. (See Chapter 19, the "Mounting AV Components" section.) Photo courtesy of InfoComm International.

Building an equipment rack and installing AV components involve care and planning. For example, you must take into account how much heat the AV systems will give off, as well as the types of signals the interconnected cables will carry. (See Chapter 19, the "Rack Building" section.) Photo courtesy of InfoComm International.

Terminating cable is a major part of what AV professionals do when installing AV systems. Termination is the method of creating connections at each end of a cable. (See Chapter 19, the "Cable Termination" section.) Photo courtesy of InfoComm International.

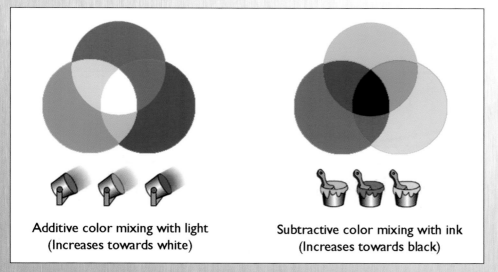

430 trillion Hz
(Lower critical frequency)

750 trillion Hz
(Upper critical frequency)

The visible light spectrum is just a small part of the entire electromagnetic spectrum. As frequency increases, wavelength decreases. (See Chapter 5, the "Light" section.)

Additive color mixing with light
(Increases towards white)

Subtractive color mixing with ink
(Increases towards black)

Mixing colors of light is unlike mixing colors of paint. With light, white is the perceived color when red, blue, and green light are added together. (See Chapter 5, the "Light" section.)

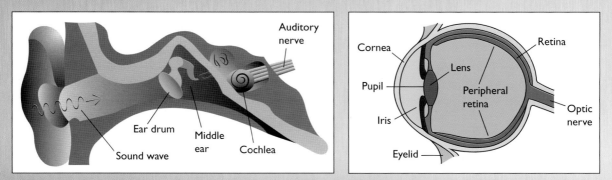

Auditory nerve

Ear drum

Middle ear

Cochlea

Sound wave

Cornea

Pupil

Iris

Eyelid

Lens

Peripheral retina

Retina

Optic nerve

To understand AV, you should understand how your body interprets light and sound. Your ears allow you to process sound by converting sound waves into mechanical energy and then into electrical signals. (See Chapter 4, the "Sound Waves" section.) Similarly, your eyes allow you to detect light by converting light waves into electrical signals. (See Chapter 5, the "Light" section.) Your brain then translates these signals into usable information.

Identify Your AV Cables

(See Chapter 19, the "Termination Methods and Types of Cables and Connectors" section.)

A five-conductor video cable

Two fiber-optic cables

A shielded twisted-pair cable

An HDMI cable

A coaxial cable

A Category 5 unshielded twisted-pair cable

An HDMI connector

A DVI-D connector

Analog Signal Bandwidth vs. Network Bandwidth

It is very important to understand the difference between analog signal bandwidth and network bandwidth. *Analog signal bandwidth* applies only to analog signals. It affects performance in that all devices in the system must have adequate frequency response characteristics to reproduce the signals in question. If a component can't accurately reproduce some of the frequencies in an analog AV system (the signal's frequencies are outside that component's band), then those parts of the signal will be distorted.

Network bandwidth (described in bits per second or bit rate) applies only to digital signals traveling over a TCP/IP network. It affects system performance differently depending on how the network is set up. In a nutshell, your signal will require a certain number of bits per second per stream. You network's "pipe" will be capable of handling a certain number of bits per second. If the bit rate of the signal exceeds the available network bandwidth, data could be lost or delayed, or arrive out of order.

Determining the Display System's Purpose

The first thing to address when designing a display system is how the client will use it. You must carefully and accurately determine the client's needs in order to understand what type of system is required. This information is critical to creating a system design that is cost-effective and provides sufficient image quality.

What and How the Audience Will Watch

The following questions will help you assess audience need:

- What will the audience watch? This information helps you determine the appropriate screen size and required resolution. For example, a videoconferencing application does not require the same screen size and resolution as a video system intended for the inspection of an engineering design.

- How many images are required? Are multiple video images required? Does the room/audience size require multiple screens to allow everyone to be able to view the images?

- Is it a group or an individual experience? In other words, under normal usage, will individuals view the display or will the audience be a group? And will individual displays be required for each viewer?

- Where will the displays be located? Does the room require one or two large displays at the front of the room? Are additional displays required for other sources? Are smaller displays required for each audience position?

- What else will the audience do within the room? Will the audience take notes, interact with each other and the presenter, or work on notebook or desktop PCs?

The answers to such questions will affect the room layout and audience space requirements.

Technical Features of the Video Display

You should also identify the required technical features of the display system. The following are some of the tasks involved:

- Determine which types of signal(s) will be used—such as live videoconferencing, videotapes or Blu-ray discs, PC-generated video (such as PowerPoint)—and how these images will be selected.

- Provide for conversions where necessary, such as PAL to NTSC, XGA to video, and so on.

- Consider audio "follow" issues, such as how to control audio when video cameras are switched among the presenters, audience, and program.

- Plan for controlling video components, such as who does the switching, the type of controller used for switching, and the location of the control system. (We will discuss the design of the control system later in this chapter.)

- Calculate the bandwidth necessary to provide sufficient quality throughout the video system.

- Allocate physical space for system components, either within the room or in a control room.

Identifying the Display Devices

Selecting the appropriate devices for a display system is critical to the AV design process. Displays are often among the most expensive components of an AV system, and they must support a wide range of specifications and requirements. Questions you must ask include the following:

- How large does the display need to be?
- How bright does the display need to be?
- What resolution is required to support the task(s)?
- What aspect ratio is required?
- What inputs does the display need to have in order to support output from the AV system?

In addition to meeting specifications for signal-quality requirements, a display should also support the proper types of connectivity and control so that you can integrate it with the complete AV system. In short, you should select display monitors that match the system's signal and connectivity requirements.

Keep in mind that not all displays can receive and process the same types of signals. For example, a flat-panel television does not always support the same connections as a professional video monitor, and a monitor may not always have the capability to display a computer-generated image properly.

Specific types of displays are designed to meet specific client needs. AV designers must be able to explain how each of these technologies operates and why specific types of monitor/projection technologies are appropriate for a specific application.

Identifying Monitoring, Feed, and Recording Requirements

An important consideration of the AV system designer is whether the video system must route a signal to additional destinations, depending on application, function, or task. The following issues need to be addressed:

- Are there additional system video requirements besides the presentation display?
- Is it necessary for a technician or presenter to monitor the program?
- Is technical monitoring of the signal necessary?
- Are there other locations the display signal needs to go?
- Is it necessary to record the video program?

Monitoring

Confidence monitors are a common element of presentation systems. These secondary displays allow presenters to see the material that the audience sees on the main display without turning their backs on the audience to do so.

There are three typical types of monitoring in a display system:

- Preview and program monitoring
- Multiple-source, simultaneous monitoring
- Technical monitoring

Preview and program monitors aid the presenter or an assistant in providing a smooth presentation. The presenter may view on the preview monitor the upcoming slide or source prior to sending it to the program monitor and the main display for viewing by the audience. For example, the preview monitor allows the presenter to find the correct camera, cuing location on a videotape, or file on a computer prior to displaying these sources to the audience. The program monitor displays the live event that is recorded, broadcast, or shown to the audience.

Preview and program monitoring may consist of a variety of sources, both video and computer-based. The monitors should match the source as closely as possible in aspect ratio, resolution, and scan rates. Monitoring systems may consist of dedicated monitors or displays built into touchpanels with video and graphic inputs.

Multiple-source, simultaneous monitoring is used when the presenter (or user, depending on the application) can choose from several sources. Sources could include overhead cameras, computers, video recorders, different cameras in a videoconference, and so on. This type of monitoring lets a presenter or technician see all the sources at once for convenience and cuing, as shown in Figure 16-1.

Figure 16-1
Security
monitoring
interface
(courtesy of
AMX)

NOTE If the display system you're designing requires multiple-source, simultaneous monitoring, keep in mind that it will also require switchers and controls to route the monitored sources to the monitoring display.

Technical monitoring requires special test equipment to help technicians adjust the signals for consistency and standardization. This is particularly important when the video program is to be broadcast. Technical signal monitoring is typically required for broadcast feeds, such as for CATV and press conferences. These types of feeds are common in public assembly areas, such as city council chambers, universities, sports venues, and even some houses of worship.

While these types of signal-monitoring systems are required for broadcast, monitoring signal quality is also important for other types of applications. The system operator should have a method to ensure that the video signal is within required technical parameters.

The following are some examples of technical monitoring instruments:

- Oscilloscope, which plots signal amplitude and timing
- Waveform monitor, which measures amplitude and timing of luminance and color signals
- Vectorscope, which measures amplitude, hue, and phase of video signals (color is represented by a vector's direction and amplitude)

These monitoring systems may be integrated into a computer that monitors and displays the signals, or they can operate as stand-alone components.

NOTE Technical monitoring is a function of the analog world. A digital signal is simply made up of ones and zeros, and it either arrives at its destination or it does not.

Feeds

Additional feed locations are places outside the main presentation room that must receive the video signal. Some examples of when you might need additional feeds are to accommodate closed-circuit video systems, feed video to overflow rooms, place a monitor in a lobby, or stream video over a network.

Issues you should consider when considering remote feed locations include the following:

- Where are the remote locations you need to send signal(s) to?
- How far away are the locations?
- Do you also need to distribute audio?
- Do you need to control the remote devices?
- Are the remote sites on the same technical power source as the main video site?
- What type of signal is required?

When sending video feeds to other locations, you may encounter various challenges. To anticipate these challenges, ask the following:

- Should the signal be converted from one form (analog) to another (digital) and carried via RF, fiber-optic, or Category (Cat 5, Cat 5e, Cat 6, etc.) cable?
- Does the distance to the remote location require special cabling and amplification?
- Do the systems require electrical isolation and special grounding?
- Are additional power panelboards necessary?
- Do the remote sites require control elements (such as camera control)?

Recording

Depending on the client and the video application, recording and archiving can be important features of an AV system design. Many types of venues—such as judicial, public assembly, and house-of-worship facilities—are either required to or prefer to record and store their proceedings.

A recording/archiving system can be simple or complex. Sometimes, the system may need to capture only video of a presenter. Other times, the system may need to record both the presenter and the content.

 NOTE Recording/archiving systems may introduce legal issues, such as privacy or intellectual property rights issues. They may also be used for legal purposes and require time-stamping or other content-management features. Understanding what, exactly, the system is likely to record should lead the AV designer to ask the client questions that might reveal legal issues and solutions.

When addressing the recording and archiving requirements of a video system, the questions you should address include the following:

- Is it even necessary to record presentations, meetings, classes, proceedings, etc.?
- Will the system record one copy of the presentation or will multiple copies be required?
- Will recordings be for archival purposes?
- Where is the recording device(s) located?
- What is the device's format?
- What type of signal is required for the recording device?
- Where will content be stored and how will people find it later? Will someone need to apply meta tags to digital content to make it more searchable?
- How available should archived digital content be? How quickly might users need to retrieve it?
- How long will the client need to keep archived content?

Identifying the Display Signal Sources

Once you've identified the back-end requirements of the video system (for example, the display components), you can move to the front end, which includes video sources and source materials. You need to understand the various media that the client intends to present, such as live and prerecorded video, computer presentations, or a combination of several sources. You must also understand what types of signals the sources will produce and how those signals will interface with the system's display devices.

Initial questions to address when identifying source needs for a specific AV system include the following:

- What media will presenters be using? What specific media sources should the AV system provide or support?
- What source components are required? The system design must include the appropriate source components, such as DVD, Blu-ray, or other video playback devices.
- Where will the sources be located? Will they be in a control room with an operator, or at the presenter location?
- Is there an audio component? The audio must be supported by the audio system design.
- Is there a control component? What elements of the video presentation will the presenter or operator need to control? Can you make the system easier to operate by including a control system?

Common Video Sources

The following are common types of video sources that an AV design may be required to support:

- Video sources, such as DVD and Blu-ray players, and media players
- Broadcast, cable, and satellite sources
- Cameras
- Document cameras
- Dedicated computers
- Computer interfaces
- Network appliances, such IP television (IPTV) appliances and streaming servers
- Videoconference codecs

Supporting a wide range of sources can present many challenges. For example, broadcast, cable, and satellite sources present a range of variations in signal and connectivity, making it difficult to determine the necessary components and how they should be connected to the system. Other issues can be technical, such as the following:

- Ensuring sufficient network bandwidth for signals traveling over IP, such as streaming video
- Ensuring sufficient analog signal bandwidth
- Matching the resolution of the source signal to the display, where possible
- Addressing the potential for unplanned sources
- Providing monitoring test points or device redundancy in the event of a failure
- Needing to license music for playback within a public area
- Difficulty distributing a digital signal that employs Digital Rights Management (DRM), such as High-bandwidth Digital Content Protection (HDCP), to multiple outputs

You'll need to thoroughly discuss this topic with the client in order to determine the full range of sources that the system will be required to support. Any components that come up later in the design process can have a big impact on your system. And keep in mind that each source will introduce additional corresponding requirements, such as computers that require network connectivity or television receivers that require a television network connection.

Simultaneous Video Sources

Simultaneous sources are another issue you must consider. You should determine which sources may be required to display in a simultaneous manner (such as a video-conferencing system). Here are some issues to consider:

- Will more than one source be viewed at one time?
- Will the sources be shown on one display or multiple displays?
- How will users view a source generated in one aspect ratio on a display device with a different aspect ratio?

Computer-Video Interfaces

Computer-video interfaces allow the client to connect a computer display output to the display system. They are analog devices that typically convert an analog computer display signal, such as VGA, to an analog, full-bandwidth five-wire video signal (RGBHV). In some cases, they convert a digital signal, such as DVI-D, to an analog video signal.

Computers are a major source component for AV presentations, due to the popularity of computer presentation software, such as Microsoft PowerPoint, and pervasive access to Internet-based applications and content.

A computer-video interface offers the following primary functions:

- To isolate and buffer the computer and display
- To provide a discrete signal for a local monitor
- To provide signal processing, such as peaking amplification to reduce the apparent effects of bandwidth loss in both the display and the interconnect equipment
- To give the user some measure of control over the displayed image, such as shift (which entails an image moving off-center)
- To provide buffered output, unity gain for all inputs, and support for a dedicated computer type

The universal computer-video interface is designed to accommodate a range of different computer types. In a fully digital system, however, they are not used. Modern computers and display devices can often be directly connected via HDMI.

Line drivers are examples of components used to interface computers with AV systems. Line drivers are used for gain and peaking in order to compensate for signal attenuation created by cable resistance in longer cable runs.

Wall Plates and Floor Boxes

When you address video sources, you must also consider devices such as wall plates and floor boxes. The table box/floor box example depicted in Figure 16-2 provides the client with a wide range of interface options, from inputs such as component video to digital and fiber-optic connections. It also provides AC power for notebook computers and network connection points. Supporting this level of connectivity for the presenter

Figure 16-2
A table box/floor box provides a range of interface options

requires careful planning, and consideration of issues such as cabling attenuation, cable isolation, and so on.

Identifying Switching and Distribution Components

Signal switching and routing can be a challenge for an AV designer. You must determine the most efficient route for the signals, minimize conversions, maintain the highest level of signal quality, and address control issues. The following are some questions to consider:

- Which components are necessary in order to connect sources to destinations?
- Where is processing or conversion required?
- Are any advanced capabilities required?

When designing the routing and distribution portions of the system, it is important to remember the system's purpose. Some components just switch or distribute video signals without any effect on the signal itself. Other components modify the signal or convert it to another type. Still other components maintain the same type of signal, but modify it to use a different transport standard, such as converting the signal transport from baseband video on copper coaxial cable to fiber-optic cable.

Here are some examples of connectivity components:

- Switching components, such as switchers, seamless switchers, and matrix routers
- Distribution components, such as distribution amplifiers, line drivers, isolation devices, fiber-optic cable, and Category cable
- Scan-conversion components, such as doublers, quadruplers, scalers, and scan converters
- Digital conversion components, such as for DVI, IP, and IEEE 1394 interfaces

Some applications may also require sophisticated video processors used for adding effects to presentations, either dramatic or practical.

Addressing Video Signal Issues

In order to determine the best signal type to use in a system, you need to understand the differences among them and how you can convert signals from one type to another. You also need to be aware of the issues presented by a display system's bandwidth requirements, because bandwidth is a primary criterion for selecting components. The following are some of the video signal issues that AV designers should understand:

- What are the different forms that display signals take?
- What is bandwidth, how do you calculate it, and how does it affect AV designs?
- What is bandwidth limiting?
- What is gain?
- What is peaking?
- How does cable affect bandwidth?
- What are some different digital video signals?

Adding Processing and Conversion

It is often necessary to process a video signal prior to displaying the image. For example, AV systems may require the following kinds of signal conversion and processing:

- Video signal format conversion
- International standards conversion
- Time-base correction
- Composite-video adjustment
- Scan-rate conversion (up or down)
- Resolution conversion (up or down)
- Analog/digital conversion

Signal-processing devices are used in a video system for the following purposes:

- To ensure technical integrity of a signal by adjusting the time or amplifying the signal for long cable runs
- To ensure continuous transitions by avoiding jumps and shifts between signal sources
- To correct deficiencies in the original signal
- To adjust the picture's color, brightness, and contrast
- To convert a signal from a incompatible video standard
- To change the format of a signal so that it may be transported to a destination in a standardized or more efficient manner

Best Practice for Determining System Bandwidth

Analog system bandwidth refers to the bandwidth of the entire system. This is limited by the bandwidth of the lowest bandwidth device within the system, including the method of interconnecting components. To determine the required bandwidth of a system, follow these steps:

1. Identify the highest resolution signal.
2. Calculate the required system bandwidth.
3. Determine the signal transport means.
4. Select components to exceed the bandwidth. Choose components with two to three times the highest signal bandwidth.

It is a good idea to recommend that the system be designed to exceed required bandwidth by a factor of three in order to give the customer room for future growth.

Designing the Audio System

The audio system is a critical component of any AV system. To design it correctly, you must be able to identify audio performance requirements and issues. Your goal is to create a system that ensures an audience can hear the presenter and/or the program materials via clear and undistorted audio.

In many cases, the audio system can be one of the most challenging aspects of an AV system design. Audio system design requires thoughtful consideration of room acoustics and careful analysis of system performance requirements. You'll need to apply standards and design best practices to achieve the required system performance.

Here, we will review the issues associated with designing an audio system to support a client's needs and requirements, including how to predict performance before an audio system is built and how to verify that the audio system design meets identified performance requirements.

The audio system design process is similar to the video system design process—you identify client needs and create a system design to meet those needs. The following are typical steps of the audio system design process:

- Identify the audio system's purpose.
- Address the three parameters for an audio system: intelligibility, loudness, and stability.
- Identify the audio sources.
- Identify the audio destinations.
- Consider audio mixing and amplification components.

- Add processing and conversion.
- Consider power amplifiers and loudspeaker wiring.
- Apply appropriate power to the loudspeaker.

Identifying the Audio System's Purpose

Audio systems generally perform one of two functions: live sound reinforcement or program playback. Live sound-reinforcement systems amplify the voice of a presenter or other sound source (such as a band). Applications for these types of systems include the following:

- Public address/paging
- Speech/performance reinforcement
- Audioconferencing and videoconferencing

Program playback systems reproduce a prerecorded audio program, such as a soundtrack. These systems are typically used for foreground and/or background music, as well as prerecorded program content.

You must work with the client to determine specific needs for voice amplification and program audio reproduction. This involves identifying the characteristics of the presentations and/or performances that will take place in the room in order to understand the overall audio system requirements.

Addressing the Three Parameters for an Audio System

The functionality of an audio system is often reduced to three easily understood concepts:

- Can the audio be understood? This is *intelligibility*.
- Is the audio loud enough to be heard? This is *loudness*.
- Does the audio system operate in a stable manner? This is *stability*.

The first parameter, intelligibility, determines whether the acoustic energy emanating from the audio system is understandable when the audience receives it. If there is intelligence in the audio signal, is it faithfully reproduced so that it may be decoded and heard? Two ways the intelligibility of airborne signals can be destroyed are by distortions introduced to the signal chain and due to a component's lack of frequency response.

The second parameter, loudness, is perhaps the easiest to grasp. Can the listener hear the intended audio? If the answer is no, you have a problem. Solving it may be as simple as addressing listener or loudspeaker locations, or just increasing the volume. But it may also be a sign of a greater issue, such as poor acoustics (acoustics are discussed in Chapter 13).

The third parameter, stability, has to do with system feedback. This applies only to rooms using microphones that are being reinforced in that same room. Stability compares the available gain or amplification of the system to the amount of gain or amplification required from the system to produce an adequate level, including the necessary headroom to address transient peaks.

Let's look at the audio issues you need to consider when addressing intelligibility, loudness, and stability.

Frequency Response

Each piece of equipment through which an audio signal passes has the potential of maintaining the original signal shape or distorting it. This is the *frequency response* of the equipment. The frequency response of a component tells us how the device reacts to the various frequencies that flow through it.

While you are most interested in the human-hearing range of audio systems, not all systems require the full frequency range of 20 Hz to 20 kHz. The target frequency response for audio should be based on the application. A recording studio requires a full-frequency response system, but a speech-only system does not. And although a performing arts center requires less frequency response than a recording studio, it certainly requires a more full-frequency response than a speech-only system.

Consider a boardroom system that's being used to preview video for an upcoming advertising campaign. The audio portion of the content is integral to getting the message across to viewers. Inadequate frequency response or poor fidelity won't accurately reflect the impact of the message. If, however, based on the needs analysis, the same boardroom requires only speech reinforcement, you can design a system that reproduces only the human-speech range of frequencies.

If you input an audio signal of consistent amplitude and continuously increasing frequency into an audio device, the output level for each frequency can be observed and documented, as shown in Figure 16-3.

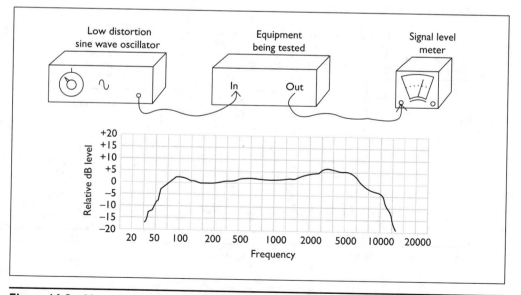

Figure 16-3 Measuring the frequency response (bandwidth) of a device

A typical frequency response reads like this:

40 Hz to 16 kHz, +/– 3 dB

One of the most important parts of the specification is the *tolerance*. In this case, the tolerance of the stated frequency response is +/– 3 decibels (dB). This tells us that from a normal or average level for any frequency within the stated range (40 Hz to 16 kHz), the maximum deviation is 3 dB. If this part of a frequency-response specification is omitted, the information is not very relevant, because the equipment could be insensitive to frequencies that are critical to your particular application. Be sure that the response of the equipment meets your design purposes.

Harmonic Distortion

Another way audio signals lose fidelity is through distortion. There are many types of distortion, but the one most common to professional AV is *harmonic distortion*. Harmonic distortion can occur in electronic components and loudspeakers. For our purposes, we'll focus on electronically generated distortion.

A *harmonic* is a multiple of a fundamental frequency. If the fundamental frequency is 800 Hz, the harmonics are 1.6 kHz, 2.4 kHz, and so on. If you inserted a pure tone of 800 Hz into an audio signal-processing device and found harmonics on the output, you would know that the harmonics are a distortion, because they do not exist in the original signal. Figure 16-4 shows an example of testing for distortion.

You can compare the level of any of the harmonics to the fundamental frequency level as a percentage, but that does not tell the whole story. A better method is to analyze *all* of the harmonics together and compare that composition with the fundamental level. This is called *total harmonic distortion* (THD). The lower the THD percentage, the better the quality of the audio signal.

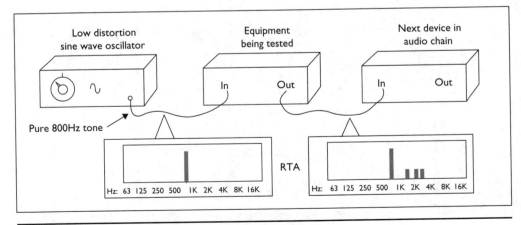

Figure 16-4 Measuring the harmonic distortion of a device

When an amplification stage is overloaded, the signal "squares," or changes its shape. The amount of squaring can be measured by totaling the odd harmonics (third, fifth, and so on). So, it's important to look at the total harmonics present in the electrical output of the system.

A typical equipment specification of frequency range and harmonic distortion reads something like this:

30 Hz to 20 kHz, +/– 3 dB with 0.1% THD

or

THD: 0.1% 30 Hz to 20 kHz

Sound Measurements

Loudness is not a scientific term. It is a subjective perception of sound level. That said, scientifically proven methods of measuring the actual movement of air can help you better quantify how much sound you have and need. The scientific measurement of sound is called the sound pressure level (SPL).

An SPL meter measures the intensity of an audio source. The meter consists of a calibrated microphone and the necessary circuitry to detect and display sound level. It converts sound pressure levels in the air into corresponding electrical signals. These signals are measured and processed through internal filters, and the results are displayed in decibels of sound pressure level (dB SPL).

TIP For good intelligibility, overall target sound pressure level should be at least 25 dB SPL above the ambient noise level. Put another way, the maximum background noise level should be 25 dB SPL less than the appropriate target SPL.

Measurements of sound are often *weighted* to take into account the manner in which humans perceive sound, as depicted in Figure 16-5. Weighting assigns variations in

Figure 16-5
Decibel weighting
curves

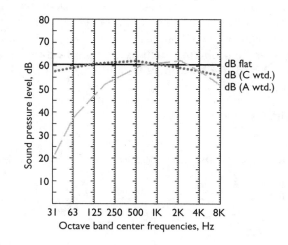

value according to typical human perception. A flat measurement of sound is unweighted and used as a reference. It doesn't reflect how humans perceive the different frequencies, but rather how instruments can be compared as they respond differently.

Loudness, Sound Level, and Uniformity

Whenever you're listening to a sound source, its sound level should be greater than any noise that may distract from detecting all of the details. If the background noise is too intense, it may mask the sound you're trying to hear.

At quieter sound levels, higher frequencies sound louder than lower frequencies at the same energy level. One type of sound-measurement standard—dB A-wtd—provides a more accurate representation of how humans perceive sound at these lower listening levels.

At higher sound energy levels (above 85 dB SPL), the human ear is less sensitive to variations in levels across frequencies. In these environments, engineers use another type of measurement standard—dB C-wt—which provides a better representation of how humans perceive sound at higher energy levels, such as at a concert.

A typical conference room requiring speech reinforcement may have a target sound level of 70 dB SPL A-wtd, not including headroom. Therefore, the maximum background SPL should be no more than 45 dB SPL A-wtd.

The *uniformity* of the SPL is another important consideration. A speech-reinforcement system should provide balanced coverage, so that it is not too loud at some locations in the audience and too soft at others. The system should provide the same sound level to all listeners within the space. If only one person has the benefit of having 70 dB SPL A-wtd, while others are straining to hear at 60 dB SPL A-wtd, the system is not performing its task very well. Uniformity can be achieved by proper loudspeaker component selection and placement. The ANSI/INFOCOMM 1M-2009: Audio Coverage Uniformity in Enclosed Listener Areas standard addresses this fundamental goal.

Audio System Stability

Audio system stability refers to the extent to which the reinforcement system can amplify sound from a microphone without feedback or distortion. One way to address stability is by controlling acoustic gain. Proper equipment selection and placement are large contributors to system stability.

Frequency-specific problems, such as strong reflections at specific frequencies that create distortion and feedback, may be addressed via a range of methods, including some combination of the following:

- Acoustic treatment
- Equalization
- Feedback suppression/extermination
- Mix-minus

Identifying the Audio Sources

Part of the design process is to consider all the audio sources. The following are some examples of transducers, or origins for sources:

- Microphones, for vocals (speech or singing) and any musical instruments
- Room or ambient microphones, to pick up sound in a portion of a room
- Wireless microphones (handheld, lavalier, and head-worn)
- Audioconferencing hybrids, which bridge telephone lines and separate far-end and near-end components into discrete signals
- Videoconferencing codecs, which have discrete signals for near-end and far-end components

Examples of program sources include the following:

- Playback units, such as CD, DVD, Blu-ray, MP3 players, and solid-state recorders
- Modulated sources, such as radio and satellite radio
- Computers
- Feeds from other systems

Program and live sources present a range of challenges and considerations. For example, speech systems should make sure that the sound seems to emanate from the front of the room, in order to maintain a connection with the presenter. This is less critical with some program sources, but for sources that are synchronized to visual images—speech where the speaker's face is shown, certain sound effects, musicians performing, etc.—it is important.

Systems that support speech sources must ensure intelligibility of the voice signal, while program sources generally require higher fidelity. You may also need to incorporate program source material intended for surround-sound effects within a room design.

Identifying the Audio Destinations

Loudspeakers—the most typical destinations for audio in an AV system—are available in a wide range of designs and configurations. The choice of a loudspeaker system is a critical aspect of audio system design.

Loudspeaker Configurations

Examples of common loudspeaker configurations include the following:

- **Two-way or three-way, with passive crossover**   These are usually cabinet units with multiple (typically two or three) loudspeaker types included. The individual loudspeakers respond to different portions of the audible frequency range, and the crossover acts as the dividing component that sends the appropriate frequency range energy to the appropriate loudspeaker. Crossovers may be passive (handling loudspeaker-level signals from a single amplifier) or active (processing line-level signals and feeding them to multiple power amplifiers, which is called *biamplification* or *triamplification*).

- **Self-amplified** Often called *powered loudspeakers*, these have an integrated amplifier within the loudspeaker cabinet. This design is very convenient because line-level signals can be routed directly to the units.

- **Constant-voltage (25V, 70V, and 100V)** Also called a *high-impedance* speaker system, this method of distributing sound over a large area through a network of speakers tends to minimize power loss.

- **Subwoofers** These special units are used when amplifying only the lowest frequencies. The subwoofer is a critical addition to a high-fidelity audio system.

- **Clusters** These groups of loudspeakers are suspended to cover a wide area. Each speaker component of the cluster is aimed at a portion of the audience.

- **Specialty** Specialty systems are designed to meet various situations, such as constant directivity, asymmetrical, and line. These are intended to deliver equal sound energy to the audience. Some do this by using irregular patterns or directing sound to a specific spot to avoid spill into surrounding areas. Some can even be steered electronically to respond to a changing situation.

Handling Monitoring, Feed, and Recording Requirements

When creating an audio system, the designer must determine whether audio signals need to be routed to other locations or components for the purposes of monitoring, reaching other rooms or devices, or recording the audio signal. Each of these routing needs may require additional design features and components.

Audio signal-monitoring elements include the following:

- Peak/clip indicators
- VU meters (analog or LED)
- Loudspeaker monitors
- Headphones
- Phase meters

 TIP Designers need to ensure that operators can see and hear everything the audience does.

Many recording options are available, each with its own set of issues to consider. Here are some examples:

- **Single-channel or multichannel recording** Do you need direct feeds (discretely captured input to a mixer) or mixes (captured outputs from a mixer)?

- **Analog or digital recording** Will you use tape, CD, hard disk, or solid-state storage? Will it be in compressed or uncompressed format?
- **Audio servers** These can record multiple channels digitally. How will the audio get into the server (what interface)? How much hard disk space should the server include? How will the disks be configured?

Audio feeds route an audio signal to other rooms or remote destinations. Here are some examples of common requirements for audio feeds:

- Nearby rooms (for example, overflow rooms and foyers)
- Stage monitors
- Press feeds
- Broadcast feeds
- Codecs
- Streaming servers
- Recording
- Monitoring and metering

Meeting Mixing and Amplification Requirements

Once the audio sources and destinations have been identified, you can begin to select the equipment necessary for all routing and processing between them. A couple common components in an audio system design are the mixer and the amplifier. These devices do most of the work to direct and amplify an audio signal for conversion to acoustic energy.

Audio mixers perform the vital function of combining different levels of audio signals so they can be routed to one or multiple destinations. For example, some mixers are intended for line-level mixing only, therefore one important job of the audio mixer is to take microphone-level signals and amplify them to line level for routing and processing within the mixer itself.

Be sure that you have the proper signal-level inputs available for your system. As you determine the number of inputs and outputs required of the audio mixer, don't forget to consider future growth and expansion.

TIP Consider specifying a mixer with phantom-power capability, even if you are only specifying dynamic microphones. This will allow you to easily add condenser microphones if you decide to do so in the future.

Adding Processing and Conversion

Many applications require processing and signal conversion. The AV designer must evaluate which components are necessary to maintain the levels and spectrum balance of audio signals, and whether additional components are needed to accommodate the design.

Audio-processing devices to consider for a given application include the following:

- Digital signal processors (DSPs), which typically combine one or more of the following items
- Equalizers
- Echo cancelers
- Gates
- Compressors
- Limiters and expanders with automatic gain control (AGC)

As examples, we'll look at the first three tools on this list.

DSPs

Digital signal processing and matrix mixing add a higher level of flexibility to an audio system. A DSP typically includes all the functions of discrete analog processors, such as equalizers, limiters, and compressors. A matrix mixer is actually a series of mixers that allow for multiple outputs, each discretely configurable depending on how much (if any) of an input is passed through. Figure 16-6 shows the configuration screen of the Lectrosonics AM16 matrix mixer software.

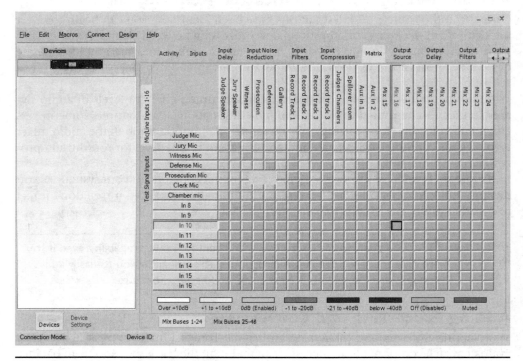

Figure 16-6 An example of DSP software

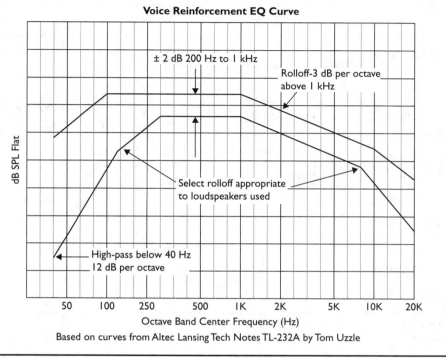

Figure 16-7 An equalization curve for a voice-reinforcement application

Equalizers

The ultimate goal of the sound system should be to deliver intelligible audio to the right places at the right level. An equalizer can be an effective tool to help achieve this goal. Equalizers allow you to emphasize those frequencies that are beneficial and deemphasize frequencies that are detrimental to the overall audio experience. As an example, a paging system does not require good low-frequency response. It would be best to optimize the loudspeaker response for the speech range.

It is common for the AV designer to provide instructions to the installer concerning the setup of equalization in order to achieve the designer's goal. This can be stated in words, but it is more commonly done by providing a *preferred equalization curve* diagram, as shown in Figure 16-7.

Note that equalizers also add a small amount of time-delay distortion, which can be perceived as slight phase shifts, ringing, and harmonics.

Echo Cancelers

In audioconferencing and videoconferencing, two distinct types of echo exist:

- *Electronic echo* can occur on a telephone line used for bidirectional communication. An electronic echo canceler, or hybrid, will attempt to discern which audio came from which direction and cancel it out to avoid feedback.

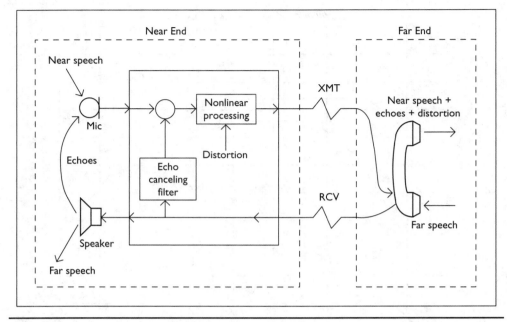

Figure 16-8 An echo-cancellation system

- *Acoustic echo* refers to the environmental echoes created by the far-site sound bouncing around walls and furniture and returning to the microphones. Acoustic echo cancelers are used in conferencing systems. Acoustic echo cancellation is often one function of a conferencing device or other audio component; it's not usually handled by a stand-alone "acoustic echo canceler." It may be integrated with other functions, such as microphone mixing or amplification.

The echo-canceling function is rated by the echoes' tail length, or the amount of reverberation memory the device can handle. Tail lengths may vary from 40 ms to as much as 270 ms.

Figure 16-8 depicts an echo-cancellation system configuration.

Considering Power Amplifiers and Loudspeaker Wiring

When it comes to power amplification for an AV system, the following are the main considerations:

- Number of available channels
- Expected load impedance
- Available power per channel

Let's take a closer look at each of these considerations and how they affect the AV system design.

Number of Power Amplifier Channels

A single-channel power amplifier is commonly used for voice support in conference rooms. Two-channel power amplifiers can be used for stereo playback. Multiple-channel power amplifiers are used for everything from combinations of speech reinforcement and program playback, to mix-minus systems, to multichannel (surround-sound) playback.

Whatever the design, make sure that you have the proper number of discrete channels available to distribute the audio properly.

Expected Load Impedance

Electrical circuits are wired together in series or in parallel; sometimes, a combination of both is necessary. In the case of audio system design, think of loudspeakers as the electrical circuits.

In series wiring, each loudspeaker is connected one after the other, in sequence. The power amplifier's positive output terminal connects to the positive terminal of loudspeaker 1. Loudspeaker 1's negative terminal connects to loudspeaker 2's positive. Loudspeaker 2's negative connects to loudspeaker 3's positive terminal, and so on, until the last one is wired. The last loudspeaker's negative completes the circuit back to the amplifier's negative terminal.

To calculate the total impedance, you add the impedance of each loudspeaker and compare the result to the power amplifier's rated impedance load.

Total Impedance $Z_T = Z_1 + Z_2 + Z_3...$

If you have four loudspeakers, each rated at 4 ohms, the impedance is additive:

$Z = 4 + 4 + 4 + 4$

$Z = 16$ ohms

Therefore, an amplifier rated for a 16-ohm load would be the best choice. An amplifier optimized for a 4-ohm load could not efficiently couple energy to this circuit of loudspeakers.

In parallel wiring, the positive output of the power amplifier connects to each loudspeaker's positive terminal, and each loudspeaker's negative terminal connects to the amp's negative terminal, as depicted in Figure 16-9.

Often, all of the loudspeakers in a parallel circuit are rated at the same impedance. In that case, the impedance of the circuit equals the impedance of one of the loudspeakers divided by the total number of loudspeakers wired in parallel:

Total Impedance $Z_T = Z_1/N$

where:

- Z_T equals the total impedance of the circuit.
- Z_1 equals the rated impedance of one of the loudspeakers.
- N equals the number of loudspeakers wired in parallel.

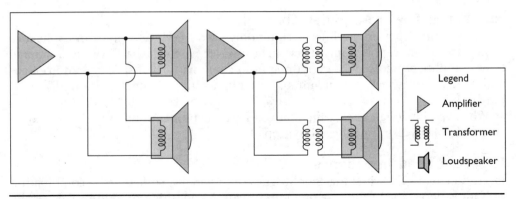

Figure 16-9   Loudspeakers connected in parallel

If you have four loudspeakers wired in parallel, each rated at 8 ohms, the circuit's impedance is 2 ohms.

$Z_T = 8/4$

If the loudspeakers are of varying impedance, you must use a different formula for finding impedance in a parallel circuit:

$$Z_T = \frac{1}{1/Z_1 + 1/Z_2 + 1/Z_{3\ldots} + 1/Z_N}$$

Sometimes, designers use a multiple series of loudspeakers, with each series wired in parallel, as depicted in Figure 16-10. You may choose to do this when you want a particular

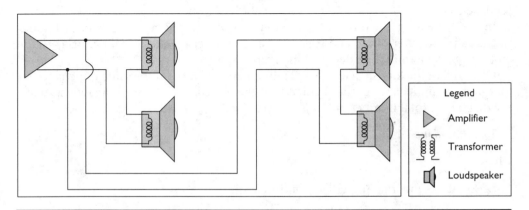

Figure 16-10   A series and parallel wiring configuration

amount of impedance to match the rating of a power amplifier, or when wiring in series or parallel results in too much or too little impedance based on the number of loudspeakers and their impedance ratings.

In general, series and/or parallel wiring are good solutions under certain circumstances, but they become impractical when more than several loudspeakers are involved. Installers must keep careful track of how they wire loudspeakers to ensure they match the amplifier impedance as planned.

Direct-Coupled vs. Constant-Voltage

Series and parallel wiring describe a system's physical connections. The following are two ways the system can be electrically connected:

- **Direct-coupled**   A *direct-coupled system* (also called a *low-impedance system*) is designed for low-impedance loudspeakers (16 ohms, 8 ohms, or 4 ohms) and works pretty much as its name implies. The system loudspeakers are connected directly to a power amplifier output. If multiple loudspeakers are installed on one power amplifier channel, they are wired in series, parallel, or a combination of series and parallel. This is done to make sure that the impedance of the loudspeaker circuit matches the rated impedance load of the power amplifier.

- **Constant-voltage**   When it becomes impractical to connect a large number of loudspeakers in a direct-coupled scheme, it's common to employ transformers (which are discussed in the next section). This setup is called a *distributed* or *constant-voltage system*. Constant-voltage systems are often identified by the maximum voltage allowable, such as 25V, 70V, or 100V.

Direct-coupled systems provide the cleanest sound, good low-frequency response, better quality, higher fidelity, and more power than constant-voltage systems. Direct-coupled systems are best for program audio and where the cable length from the amplifier is reasonable.

Direct-coupled systems have their limitations, however. Because of the low impedance of the loudspeakers, resistance in the wires can consume some of the power intended for the loudspeakers. To offset this loss, you may need a larger gauge of wire. Alternatively, you can position power amplifiers closer to the loudspeakers to minimize cable loss.

A constant-voltage system allows you to connect many loudspeakers to one amplifier, or to connect loudspeakers to the amplifier from a considerable distance. For example, an airport paging system's loudspeakers tend to be located a long way from the audio source, but each transformer receives the same strength signal. Even though the line is common to all loudspeaker points, the transformer wire (tap) determines how much energy is coupled to the loudspeaker itself. The AV designer must calculate the tap value for the transformer that will be used for the connection.

Types of Transformers

Transformers transfer energy from one circuit to another without physical connection. They use the principle of magnetic induction with two windings, or coils of wire. One winding is connected to the power source (primary winding) and another winding is connected to the load (secondary winding).

The following types of transformers are available (see Figure 16-11):

- **1:1 transformer** This type of transformer has an equal number of primary and secondary windings. It is is used for circuit isolation in order to solve problems such as audible hum, buzz, or rolling hum bars on a display.

- **Step-up transformer** This type has more windings on the secondary side than on the primary side. Voltage and impedance increase, while current decreases.

- **Step-down transformer** This type has fewer windings on the secondary side than on the primary side. Current increases, while voltage and impedance decrease.

Power will be equal on both sides of the transformer. (In reality, some power will be lost in the transfer from the primary to secondary side because the magnetic coupling is not perfect.) Using the formula $P=IV$, you can see that if voltage goes up, current (I) will go down because they are inversely proportional.

In a constant-voltage system, a step-up transformer (internal or external) is used at the power amplifier's output. The increase of impedance on the secondary side of the transformer allows multiple loudspeakers to be wired in parallel without risk of the impedance being too low in the circuit. A step-down transformer is then used at each loudspeaker in the circuit, as depicted in Figure 16-12.

Matching the amplifier's output impedance to the loudspeaker load helps maximize the energy transfer from amplifier to loudspeaker, which results in acoustic energy. If the circuit has insufficient impedance, the amplifier will produce more current than it's rated for and overload.

NOTE Transformers are not completely efficient. Some of the power going through them is converted to heat. The inefficiency is represented by a value called the *insertion loss*.

Transformer Connections for Constant-Voltage Systems

Each loudspeaker in a constant-voltage system has a transformer. The transformers allow you to select the desired wattage for each loudspeaker. Typically, the transformer

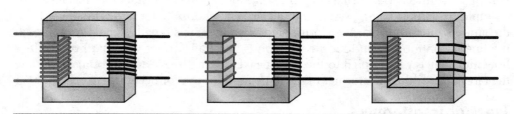

Figure 16-11 1:1, step-up, and step-down (from left to right) transformer windings

Figure 16-12
A loudspeaker
with a
transformer

has different colored wires on the secondary side, each color referring to a voltage level. The transformer manufacturer usually will provide a value chart for each of the wires (taps). The taps are rated based on the amount of power they deliver to the loudspeaker. Typical tap values for a constant-voltage system are 1/4 watt to 8 watts (although there are higher values).

To make things easier, many loudspeakers have a rotary switch with positions that correspond to the transformer taps. This way, the installer deals with just one set of wires. The tap power is indicated at each position on the switch. Figure 16-13 depicts how taps are displayed on a constant-voltage loudspeaker.

Loudspeaker Cables

Your choice of loudspeaker cables should balance cost against the amount of power lost in the line. The longer the cable run, the thicker the wire should be to compensate for loss. The lower the resistance in the cable, the more power will be available to the

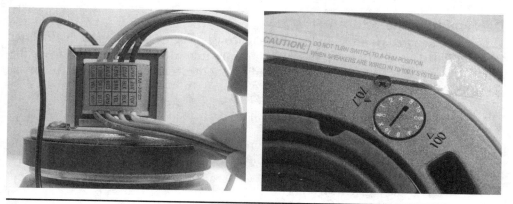

Figure 16-13 Transformer taps

loudspeaker. The percentage of loss in a given cable is available from its manufacturer and is expressed as a loss per distance (ohms per 100 feet or 100 meters).

The following are a few other issues related to loudspeaker cables:

- When it comes to specifying loudspeaker cabling, you must observe all appropriate electrical codes and regulations.

- Conduit or plenum-rated cable may be required for a distributed-sound system.

- When determining system specifications, keep in mind that local building codes may impose limits on loudspeaker cables that are *not* run through conduit.

Calculating Power Required at the Loudspeaker

Before you can select a power amplifier, you need to calculate the amount of power required for the loudspeaker circuit. In order to perform that calculation, you need to know the sensitivity specifications of the loudspeaker.

Loudspeaker Sensitivity

Loudspeaker sensitivity defines the way in which the loudspeaker converts electrical energy to acoustic energy, relative to the efficiency of the loudspeaker. It's the response you get when you input a known amount of electrical energy.

Loudspeaker sensitivity is expressed as an SPL at a specific power input and distance. For example, here's a common measurement for a loudspeaker:

90 dB SPL, 1 watt @ 1 meter

Designers must determine an appropriate SPL for the listener position in a sound application, plus any necessary headroom, and then calculate the power required at the loudspeaker to achieve that SPL.

A loudspeaker sensitivity rating of 90 dB SPL, 1 watt @ 1 meter, offers a reference point from which you can determine how much acoustic energy would be produced at a different distance. Use the inverse square law to make these determinations.

Determining the SPL

The inverse square law for sound states that acoustic energy will drop to one-quarter of the original amount of energy if you measure it at a distance equal to twice the original. This is caused by the energy spreading out over a broader area. The following is the actual formula for this calculation:

Difference in dB = 20 log ($distance_1$ / $distance_2$)

If you double the distance from the sound source (from 1 meter to 2 meters), the formula looks like this:

dB = 20 log (1 meter / 2 meters)
dB = 20 log (0.5)

dB = 20 (–0.301)
dB = –6.02

In this case, because you have doubled the distance from the source, you find that you lose 6 dB.

Now let's determine the SPL at a listener position. Suppose that the distance from a loudspeaker to the farthest listener is 8 meters, and we have a loudspeaker with a sensitivity of 92 dB SPL, 1 watt @ 1 meter. The formula looks like this:

dB = 20 log (1 meter / 8 meters)
db = 20 log (0.125)
dB = 20 (–0.903)
dB = –18 (rounded off)

This tells us that the SPL we can expect 8 meters from the loudspeaker (the position of the farthest listener) is 18 dB less than the SPL 1 meter from the loudspeaker (the reference distance). Therefore, if you input 1 watt of energy into the loudspeaker, the sound level you can expect 8 meters away is 74 dB SPL (92 dB – 18 dB = 74 dB).

NOTE Even after doing the math, there may not be as much decibel loss as you calculated. Don't forget about the acoustical environment. Depending on the environment, acoustic energy that normally expands in all directions under free-field conditions and follows the inverse square law may be affected by boundaries, such as the walls, ceiling, and floor.

Adding Power to the Loudspeaker

In the previous example, we determined that the farthest listener would receive a level of 74 dB SPL if 1 watt were applied to the system. But let's say, for a speech-reinforcement application, that the appropriate SPL at the listener position is 70 dB SPL, plus 10 dB of headroom. This means we need to achieve a total of 80 dB SPL at the listener position. We're currently 6 dB below that target level (80 – 74 = 6 dB).

What we need is more than 1 watt of power to attain the required 80 dB SPL. How much power do we add to raise the level by 6 dB? The formula looks like this:

6 dB = 10 log ($Power_2$ / $Power_1$)
6 dB = 10 log (x / 1)
0.6 = log x
3.98 = x

where:

- 6 represents the additional decibels required from the loudspeaker.
- x represents the power required at the loudspeaker.
- 1 represents the 1 watt reference from the loudspeaker specification.

Designing Control Systems

One significant design goal of any AV communication system is to keep the technology out of the way so it doesn't interfere with the client's task of actually communicating. This objective can be achieved through an easy-to-use control system.

Rather than presenting users with a confusing mix of controls and switches, or leaving disparate pieces of equipment out in the open so operators can access them, a control system combines multiple functions into a single interface. Figure 16-14 shows an example of a touchscreen control panel for an AV system.

A control system allows for the integration of various manufacturers' equipment and provides the ability to control it all through a central processing device. It is a hub for user input and controlled devices.

Control systems incorporate software and microprocessors that are configured for each installation. In a control system, a command goes into the processor, and then the processor sends that command to the appropriate port, which communicates that signal to the intended device.

The devices within the system must be capable of remote control via a specific control format. Generally, anything electronic can be controlled.

When designing a system, the AV designer should examine the devices that need to be controlled, identify the different control methods and signals they use, and understand the possible applications for each control type. There are advantages and disadvantages to each control type.

Defining the Control System

In order to determine control system requirements, designers must define the control system's purpose within the parameters of the room. The designer's goal is to make sure the controllable items in the room perform the functions that the user asks the control system to perform. The good news is, thanks to the substantial capabilities of modern control systems, AV designers have a chance to create an AV system that exceeds client expectations, at little additional cost.

Figure 16-14
A touchscreen control panel (courtesy of Crestron Electronics, Inc.)

The following are the issues to consider when defining a control system:

- **Functionality** The first step is to define the functionality of the control system. Begin with a blank sheet of paper. On the left side, list the equipment that requires control. On the right side, list which control functions are needed from each piece of equipment. Table 16-1 shows an example of how designers can organize their thoughts on which components and functions should be under control. Creating this type of control and function list is the first step toward communicating the control system requirements to the control system programmer.

- **Interoperability** The next step is to determine the interconnected nature of the control system. The control system design needs to include a control path layout. The control path layout defines what initiates the control, the control protocol, the wiring, and the hardware to be controlled. The layout determines what happens when a control option is selected, and what occurs when the control signal is received. The designer should keep in mind that a control signal can be used as much to stop an action as to start one.

- **Interconnectivity** To create the control system program, a programmer uses the interoperability information, along with information about component locations and other aspects of the physical system layout and design (such as floor box connectivity, distances of cable runs to components, interfacing, and infrastructure).

- **Capabilities** The AV designer considers a wide range of issues when defining the control system requirements. For example, each component has a defined series of controllable functions. These functions are accessed via programming instructions that are transmitted using a compatible remote-control protocol, through an interface, and back to the component. The communication between the component and the control system is two-way, with devices providing feedback that communicates device status back to the controller.

Equipment to Be Controlled	Functions
Satellite TV	Play, pause, stop, rewind, fast-forward, power
Blu-ray	Play, pause, stop, reverse, fast-forward, power, next disc, search
Video projector	On, off, inputs, lamp life
Screen	Up, down, stop
Lights	Preset 1–4, off
Recorder	Play, pause, stop, rewind, fast-forward, power, record
Computer	Keyboard, mouse
Switcher	Inputs
Conferencing unit	Dial, mute, camera controls

Table 16-1 Defining the Functions of a Control System

NOTE AV systems are increasingly complex and interconnected with other building systems, such as lighting, HVAC, building automation, and IT networks. Increasingly, AV professionals need to understand how these systems communicate in order to establish control.

Integrating with IT Networks

The discussion of converging AV and IT systems is over; convergence has already happened. These days, many modern AV systems have network interfaces (such as Ethernet ports) and can communicate using standard network protocols. This means that the audio, video, and control data that networked AV systems process and transmit from source to endpoint can run over the same type of IT network that handles your e-mail.

In many situations, it makes sense to integrate your AV system with an IT network. Many clients already have IT networks in place. Also, networks can be more economical to build and maintain than other types of AV communications. In some situations, an IT network is almost a requirement. Today's videoconferencing, digital signage, paging, and mass notification systems are examples of AV solutions that use IP-based networks.

This means that the AV design team must coordinate with the client's IT staff and/ or its IT integrator (if it's a new facility) so that the IT team understands, among other things, how you plan to interface with their network, what data you plan to transmit, and how the networked AV may or may not impact other data on the network.

In many situations, the IT department is averse to having AV systems on its network. In other cases, it doesn't make practical sense to share a network, from both a technological and security standpoint. So, sometimes the best option for designing an AV system that requires an IT network may be to build a separate network, dedicated to AV traffic. This can be done physically (an actual network of wires, switches, and routers) or virtually (basically, a network within a network).

Network Considerations for AV Systems

Any AV professional working today should be versed in networking principles and terminology. Chapter 6 provides foundational knowledge of network types, topologies, layers, protocols, and more.

AV professionals also must understand what IT professionals think about when they work with networks, and reflect that thinking in their AV designs. For example, IT gauges much of its success by a *service-level agreement* (SLA). Just as AV designers and integrators must discover and document what clients or end users expect from their AV systems, IT professionals go through the same process with their users. Often, the result is an SLA, which lays out how the network is expected to perform, how the IT staff and the end user will measure the network's performance, who is responsible for maintaining the performance, and other design aspects.

To the extent that networked AV systems impact an SLA (or even become part of the SLA), AV designers must address a variety of issues, including network bandwidth, bottlenecks, quality of service, latency, and packet loss.

Bandwidth

AV professionals are accustomed to thinking of bandwidth as it pertains to analog signal bandwidth, measured in hertz. Whenever the term is used in the context of networking, however, it refers to data throughput—the amount of information that can pass through a network at one time. If any part of your AV system design is to traverse the client's network, you must, in consultation with the IT department, ensure the network has enough bandwidth (measured in bits per second) to carry AV data and network data without adversely affecting the performance of either. AV streaming usually requires at least 100 Mbps.

LAN-to-WAN Bottlenecks

On large projects, chances are your AV system design will span offices. For example, a videoconferencing system may necessarily connect an office in one city to an office in another city. The networks inside each office are local-area networks (LANs); the network that connects the two offices is a wide-area network (WAN).

In general, LANs operate faster than WANs. Therefore, it's important to plan your AV applications so that they don't slow down unexpectedly when they traverse the WAN. You can expect a WAN to have about one-tenth the bandwidth of a LAN, but you should confirm the WAN's speed during design. As much as possible, optimize your system design to keep high bit-rate traffic within the LAN and minimize the number of high-bandwidth streams that must be sent over the WAN.

Quality of Service

Quality of service (QoS) refers to technology designed into a network that ensures certain data and applications get to their destination faster than others. For example, for an interactive videoconference to be successful, the data from each end must reach the other end as quickly as possible. However, no one will notice if an e-mail message arrives a few seconds later than normal. Video-on-demand (VOD) programs may also be lower priority than other network-based AV applications, such as streaming media, because with VOD, the video data can travel the network piecemeal and be collected, or buffered, at the user's location for playback.

QoS was developed with networked-based multimedia in mind, particularly voice over IP (VoIP) and telephony. It prioritizes network traffic to ensure high-performance applications get the bandwidth they need. However, QoS must be built into various stages of the network, so coordination with the IT staff is critical.

Latency

Latency describes a network's response time. It is expressed as the amount of time, in milliseconds, between a data packet's transmission from the source application and its presentation to the destination application. For network-based AV systems, too much latency can cripple the application, resulting in packet loss and jitter.

You must understand the factors in your client's network that could introduce latency, such as network security measures, and coordinate with the IT department to meet your latency requirements. Conferencing applications require low latency, but other

AV systems are not so cut-and-dried. IPTV can tolerate more latency than emergency notification, for example.

Packet Loss

Packet loss occurs when some of the data traveling over a network fails to reach its destination, which could be due to lack of bandwidth, signal degradation, faulty hardware, incorrect configuration, or some other factor. No one wants to lose packets, especially in an AV system, but customers are rarely willing to commit the infrastructure required to provide guaranteed delivery of all AV traffic.

In designing a network-based AV system, you must determine how many packets your application can be expected to drop given the network environment, and set client expectations accordingly. To do so, do the following:

- Calculate your AV application's peak and average bandwidth consumption.
- Compare that consumption to the network's average available throughput.

Popular Protocols for Networked AV Systems

Depending on your AV system and how it will utilize IT networks, you may choose to incorporate one of several IT protocols that have been developed over the years to transmit AV data over Ethernet, including the following:

- **CobraNet** CobraNet transports real-time audio with a maximum sampling rate of 96 kHz/24 bits, plus embedded control and monitoring, using Ethernet frames. It does not require dedicated bandwidth on switched networks and uses any Ethernet topology.

- **EtherSound** EtherSound transports up to 512 channels of audio, with a maximum sampling rate of 96 kHz/24 bits, plus embedded control and monitoring, using Ethernet frames. It requires a separate network with dedicated bandwidth and may run within a virtual LAN.

- **Audio Video Bridging (AVB)** Ethernet AVB is an IEEE standard that transports uncompressed video and up to 200 channels of 48 kHz/24 bit audio in real time, plus embedded control and monitoring, using Ethernet frames. It requires AVB-enabled switches and network components, but does not require separate network infrastructure or dedicated bandwidth.

- **Dante** Over gigabit Ethernet (1 Gbps), Dante transports up to 1,024 48 kHz/24 bit channels of audio, or half that number at 96 kHz. Over 100 Mbps Ethernet, it can transport up to 96 48 kHz/24 bit channels of audio, or half that number at 96 kHz. Dante is fully routable over IP networks using standard Ethernet switches, routers, and other components. It requires no separate infrastructure.

- Assess the impact of dropped packets on AV quality.

- Coordinate with the IT department to reserve the bandwidth your application needs, or make adjustments to the application to require less bandwidth.

 TIP For more advanced information about networked AV systems, IT networks, and working with IT professionals, InfoComm offers a new course titled NET212 Networked Audiovisual Systems. As an intermediate to advanced level course, it requires a CompTIA Network + certification–level understanding of networking technologies and design principles.

Planning for Networked AV Systems

Having assessed the network environment in which your AV systems will operate, you should coordinate with the IT staff to ensure your design delivers the performance the client needs while also meeting the IT department's requirements.

Networking Technology Issues

Here are some of the networking technology issues you may need to go over with IT staff during system design:

- **IP addressing**   This refers to how your AV devices will be assigned IP addresses on the network. There are various ways of handling IP addressing. Often, IT staff members will tell you how you'll need to work with them to obtain IP addresses via the client's existing system.

- **Subnetting**   This is the process of subdividing a single physical network into several logical networks, in this case, to support AV traffic.

- **Virtual LANs (VLANs)**   These can further isolate AV information to address security and performance needs. If an AV designer, in consultation with the IT staff, determines that a VLAN is appropriate for a certain AV application, the designer will need to request it from the IT department.

- **Virtual private networks (VPNs)**   These can be especially important for AV applications, such as videoconferencing, that may traverse the public Internet to link offices. A VPN is a secure way of extending the security of a private network across public networks.

- **Directory integration**   This may be necessary so that certain AV systems may be configured as part of an organization's scheduling, e-mail, and other systems. Directory integration is discussed in the next section.

When designing your AV system, make sure you discuss all of these issues with the client's IT staff. There will likely be other networking issues to address, such as security provisions, which the IT staff will bring to your attention. We discuss IP addressing, subnetting, VLANs, and VPNs in more detail in Chapter 19, in the "Configuring Networked AV Systems" section.

Planning Ahead for Directory Integration

One of the principal benefits of converging AV systems with the network is the ability to integrate the client's directories into the AV system. Directories can contain user contact information, calendar and scheduling data, device configuration data, and so on. Tapping into these systems can increase the functionality and ease of use of networked AV systems.

Directory servers are typically accessed and queried using the Lightweight Directory Access Protocol (LDAP). Using LDAP, you can search for and retrieve directory entries; add, delete, or modify entries; and perform other tasks. In the context of networked AV systems, this allows you to do the following:

- Integrate the enterprise's directory of contacts information into a conferencing system
- Automatically generate schedule information on digital signage outside a conference room
- Remotely view and maintain device profiles

Microsoft's Active Directory and IBM's Lotus platform are probably the most common LDAP integrations for AV applications. For example, a videoconferencing codec may use Active Directory as a user directory, automatically pulling in contact information for scheduling and initiating video calls.

Assessing Network Readiness for AV Systems

To determine whether an IT network can support the AV devices your design requires, you should inventory those devices and their network characteristics. This inventory, in the form of a detailed spreadsheet, will go a long way toward coordinating networked AV systems with a client's IT staff.

For starters, create an entry for every AV device that will be connected to the network. You are not expected to discover all relevant networking at once; much of it will be determined throughout the needs assessment, conceptual design, product selection, and installation processes. The goal is to provide as much information as you can, as soon as you can provide it. During the conceptual design phase, list the devices you anticipate will sit on the network and explain the rationale for connecting each device. For instance, what information will it send and receive? What client need will it meet?

From that point, start collecting and detailing the network characteristics of each device. You may organize the information in whatever manner you see fit, but remember to be thorough. Some details you'll learn through collaboration with IT. Listed in the following sections is a sampling of the information that should be in your networked AV device inventory, organized into categories.

Inter-device Communication

During the design phase, determine the following:

- What transport layer protocols the system will use (TCP or UDP)
- Quality-of-service (QoS) tags associated with each traffic stream

- Whether the device will send data using multicast, unicast, or both
- Anticipated incoming and outgoing peak bits per second (bps)
- Anticipated incoming and outgoing average bps

Routing and Addressing

As you select components, collect the information needed to address each device and route data to and from the address, including the following:

- The physical location of each device (e.g., in the videoconference room, at the ISP site, etc.)
- If known, the default gateway (router) IP address of the location
- The scheme used to assign an address to each device
- A static IP address, if applicable
- Subnet mask
- Host name and DNS server(s) if applicable
- Required ports
- Required protocols

Device Properties

As you select components, record each device's relevant manufacturer specifications.

Optional/Application-Based Fields

For Simple Network Management Protocol (SNMP)–managed devices, document the SNMP server, SNMP community name, admin username, and admin password. For conferencing devices, document the gatekeeper address, system name, and E.164 address.

Where applicable, and if requested by the customer, document the device username and password.

Providing System Design Documentation

An AV designer should clearly document the AV system in a manner that provides the information necessary to understand and implement the AV design and project requirements. The design documentation must address the following areas:

- **Fabrication and installation** The main purpose of the system documentation is to convey the intent of the designer with respect to the selection and procurement of equipment, cable, and other materials, in addition to the fabrication, installation, and software programming tasks.
- **Functions** The documentation should describe the intended operation and functions of the system.

- **Quality** The documentation should communicate the desired level of quality intended for each aspect of the system—from the equipment to be procured, to the actual installation (such as wiring, terminations, rack assembly, and signal quality), to the optimal functionality for end users.

The process of competitive bidding for the design and installation of an AV system typically requires detailed written specifications of the AV system equipment and installation approach. Because these AV specifications usually will be incorporated within the overall project construction documents, these documents should conform to the accepted construction documentation standards or requirements for the jurisdiction.

System Drawings

Drawings that depict the AV system should reflect graphically the equipment configurations, interconnections, details, plate layouts, and other elements of the system installation. AV system documentation should complement related architectural, engineering, and construction documentation for the relevant areas of a building, as well as address any issues that require coordination with other trades.

System drawings depict the audio, video, control-signal switching and routing, and other details, in order to convey the complete AV system design.

Functional diagrams, commonly known as *block, flow,* or *schematic diagrams*, offer an overview of the system's signal flow and illustrate the interconnection of the audio, video, and control subsystems. Functional diagrams also include references to related subsystems and help clarify the boundaries of the AV scope of work. The diagrams include contractor-provided and owner-furnished equipment, as well as equipment supplied by others.

System drawings include compiled or separate drawings for the following:

- Audio subsystems
- Antenna distribution systems
- Digital signal processing
- Video subsystems, including any special video subsystems with interconnections to the main video subsystem, such as matrix routing, closed-circuit television (CCTV), camera automation, and editing
- Control equipment
- Voice/data system interface components
- Other diagrams and notes provided for other AV subsystems, such as lighting, drapes, shades, HVAC, and security interfaces

A signal-flow diagram logically arranges equipment, as depicted in Figure 16-15. Sources are depicted on the left side of the diagram, processors in the middle, and outputs on the right. This type of depiction allows you to consider the signal types, processing, conversion, and routing required for the system.

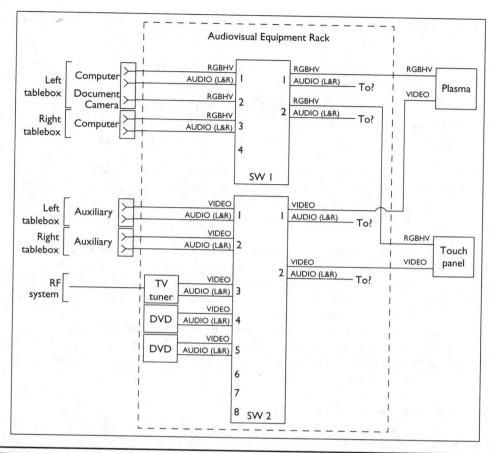

Figure 16-15 A display-system signal-flow diagram

Signal-Flow Diagram Considerations

When creating signal-flow diagrams, the AV designer should consider the following:

- How to represent the system graphically so that it can be built (include equipment labeling, physical location of equipment, cable requirements, line impedances, expected dB SPL readings, etc.)
- How to represent the interconnectivity of the subsystems
- How to represent the connectivity with IT systems and other external systems
- How to define the functionality and performance of the system

Selecting specific devices for the AV system presents a number of challenges. Including them in your diagram can help address issues such as these:

- What are the main building blocks of this design? Are there equivalent components from multiple manufacturers?

- Are there advantages to using proprietary components or multiple components from the same manufacturer?

- Will these devices satisfy video, audio, and control requirements?

This is the point where you move from generic building blocks to identifying specific manufactured components. This stage may present opportunities to combine functionality within a single component, rather than using separate components, keeping in mind that relying on one component instead of several could introduce a single point of failure.

You should strive to understand the functionality and limitations of each piece of equipment to ensure that it fits in with the AV system design. The signal-flow diagram helps follow every signal through the system to ensure it will be accepted by the inputs of the devices to which it flows. If signal conversion is necessary, these components should be indicated within the design.

Make sure you also provide for connectivity to and from the room, with reference to the room architecture, in your system-flow diagram. You should also reference the physical location of components such as equipment racks, credenzas, lecterns, closets, etc., along with the corresponding input/output connectivity devices.

Architectural Connectivity

The system design should also depict architectural connectivity (connector panels, floor boxes, and other elements). Including these elements in your diagrams helps identify relevant room parameters and determines which infrastructure elements must be prepared, how input/output plates should be designed, and which connectors are required. Cable lengths from external locations should also be evaluated in order to determine if additional compensation is required.

The architectural elements can affect the design and installation of a system, so be careful to ask the following questions:

- Where are the infrastructure connection points in the room(s)/facility?

- How are the connections made?

- What connectors are necessary?

- What cables are necessary from the connection point to the AV device?

The architectural elements of the design should provide sufficient detail so that installers know how to connect the components in the field. You need to carefully evaluate these elements—such as cabling and conduit, labeling, wall plates, and IT—and document them in your drawings.

Examples of System Drawings

Figures 16-16 through 16-19 provide examples of system drawings created by an AV designer to communicate the intended design and functionality of the AV system.

Supplying Other Drawings

Along with the system drawings discussed so far, you may include the following information with your AV system design documentation:

- **Equipment rack elevations**   These diagrams depict AV and other components within a rack. Equipment rack elevations confirm proper space allocation, ergonomics, accessibility requirements, thermal management, weight distribution, and cable management. For bidding and engineering purposes, equipment rack elevations provide a means to estimate the quantities of patch panels, bulkheads, interconnect cables, patch cables, equipment rack accessories, and other hardware necessary for a professional installation.

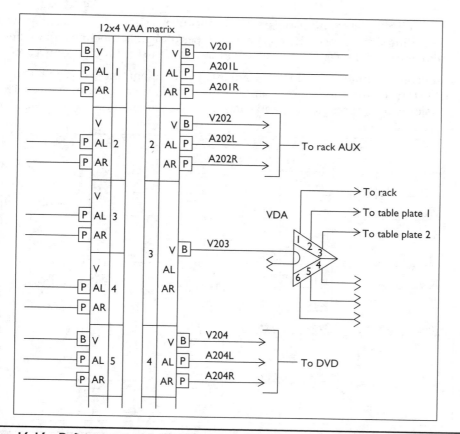

Figure 16-16   Defining inputs and outputs, creating wire labels, and showing terminations on unused video lines

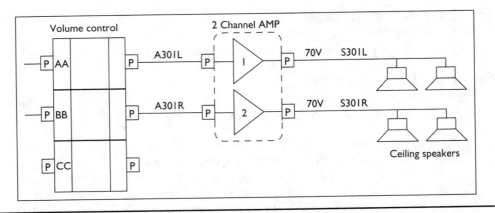

Figure 16-17   A diagram depicting speaker connections to an amplifier

- **Specialized plates and panels**   These documents show all plates and panels, including size, connector arrangements, types of signals, labeling requirements, and any special integration or installation requirements for that connector.

- **Specialty connector and miscellaneous wiring diagrams**   These diagrams illustrate connectors, pinouts, and wiring used in special circumstances. The majority of connectors, pins, and wires in an AV system follow industry standards

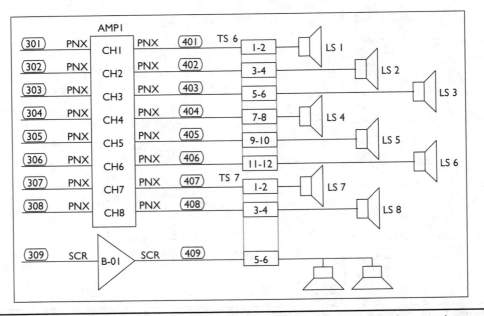

Figure 16-18   A diagram depicting the connections from the amplifier to terminal screws, then out to speakers

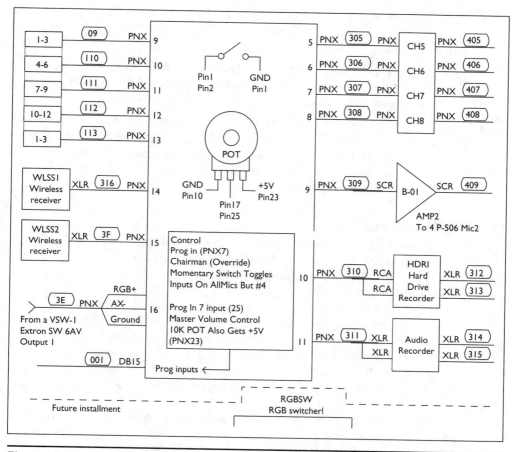

Figure 16-19 A system interface with details to assist installers in fabricating the system

when it comes to termination. Where a connector configuration is used atypically, details for the terminating connector are provided in the drawing set. The details include an illustration of the connector and its pins, as well as a numbering schedule or other labeling of the pins, and specific coordination of the pins to individual wires.

- **Custom configurations and modifications** These documents depict elements of the AV system design that may require internal modification or special device configurations to achieve the intended function in the system. They may include the setting of DIP switches, installation of internal jumpers, and other information. In all of these cases, details of the special requirements and settings are provided in the drawings to ensure that people who read them understand the requirements and their impact on functionality.

- **Patch-panel layouts and labeling**   These clarify the number and types of panels (by signal), as well as the number of connections per panel. The patch-panel layout should be consistent and labeled with the signal type.
- **Custom application documentation**   This documentation conveys information necessary to illustrate additional requirements for configuring software, hardware, and other programmable equipment and subsystems.
- **User interface descriptions**   These are visual or written descriptions of the logical sequence of the user interface. They may include a GUI layout and/or button-by-button descriptions. The documentation may also outline the requirements and best practices necessary to ensure that programmers have a firm understanding of the AV control system design and scope requirements.

Bill of Materials

The AV design team or design/build integrator should also create a bill of materials (BOM) or other document that identifies the required equipment and materials necessary for the system installation, including proposed pricing. The BOM should include an itemization and description of each piece of equipment to be provided as part of the project, as well as the quantities of each component.

The BOM is intended to provide an overall summary of the equipment costs necessary for creating the AV system. Creating the equipment list is usually left until the end of the design process, once the system designer knows which components are required.

When creating the BOM, components should be listed in categories according to component types—such as sources, processors, and destinations—along with any required accessories. This will allow for easier review by others. Keep in mind that some equipment may be considered part video, part audio, and part control, so the designer needs to be flexible in deciding how to classify specific components within the equipment list.

Listing the equipment associated with the audio system infrastructure is especially important. The list should include often-overlooked items, such as the following:

- Loudspeaker backboxes
- Interface plates
- Display mounts
- Operator and console tables
- Microphone mounts

The specification of these infrastructure components may also require the designer to include more detailed descriptions of the options, such as the style and appearance.

Other Design Considerations

As you've learned in this chapter, the AV system designer must consider many issues. A few more include future provisions, value engineering, and design and measurement tools.

Future Provisions

The term *future provisions* refers to designing extra capability into the system to support future upgrades or upcoming advancements in AV technology. The following are the main issues to consider to ensure expandability:

- Are there components that are likely to be integrated in the future that can be accommodated for now?
- Is there future expansion capability built in?

Anticipating future expansion of the system can save the client trouble and money later. For example, if you are aware of future needs, such as additional rooms and new projects, it is good practice to discuss these potential needs with the client and determine cost-effective methods to build extra capability into the system to support those needs.

Future provisions often include expansion features such as the following:

- Extra inputs and outputs on matrix switchers
- Extra conduit runs
- Extra equipment rack space
- Different projection screens (for example, 16:9 instead of 4:3)
- Extra inputs on audio mixers
- Larger control system mainframe

Value Engineering

Value engineering refers to a process of reducing cost by streamlining a design while still retaining functionality. System designers evaluating a design from a value-engineering perspective ask questions such as these:

- Is the conceptual draft design efficient?
- Are there components that may combine functions for cost-effectiveness?
- Are there components that may be separated for cost-effectiveness?

Clients inevitably assume there is room to cut costs in the system design. Designers, therefore, should be prepared to explain the functionality and efficiency of their design, or else have alternatives in mind to present to the client.

Keep in mind that value engineering is not simply cost-cutting or selecting cheaper components. Value engineering strives to retain the necessary performance—or anticipate long-term efficiency value (such as designing to support future expansion)—without sacrificing quality. For example, if instead of incorporating an expensive routing switcher, you specify a switcher and a distribution amplifier with higher bandwidth at less cost, that's value engineering.

Design and Measurement Tools

The AV system designer has a wide range of tools and resources available to support the design task, including specification template software, drawing software, measurement and diagnostic tools, and modeling software for acoustics or lighting.

For example, acoustic modeling tools enable the designer to do the following:

- Enter room geometry, room finishes, and seating positions
- Enter absorption and reflection systems
- Enter diffusion characteristics
- Vary the audio system design and placement in order to evaluate the proposed system design within the simulated environment

Incorporating the right tools into your process can help you deliver the right AV design for the client the first time.

Chapter Review

In this chapter, you looked at the issues an AV designer should consider when designing a solution to meet the client's needs. System design encompasses the electronics required to support the communication needs, as well as how devices are interconnected to create a system. It requires an in-depth understanding of the capabilities of AV components, the ability to assess how to ensure the system will perform properly within a certain environment (the room), and how to meet the client's identified needs and objectives.

To properly perform this task, the AV system designer must understand the capabilities and applications of typical AV devices, determine the required equipment performance to meet the client's needs, and assess the issues that may affect the selection of AV devices and their integration into the larger AV system.

Review Questions

The following review questions are not CTS exam questions, nor are they CTS practice exam questions. These questions may resemble questions that could appear on the CTS exam, but may also cover material the exam does not. They are included here to help reinforce what you've learned in this chapter. For an official CTS practice exam, see the accompanying CD.

1. When designing a video display system, what should be the first consideration of the AV system designer?

    A. How the client is going to use the video display system

    B. The resolution of the video display system

    C. The size of video displays

    D. The location of screens

2. The AV designer is creating a system for a client who also needs to route video and audio signals to various other locations at the site. At what level should the AV designer consider this need?

   A. The AV designer is concerned with only the display systems within the room.

   B. The AV designer should ensure that a standard video and audio signal output is available in the control room to allow other users to connect external systems.

   C. The AV designer should determine the needed monitoring, feeds, and recording requirements, and ensure that they are supported by the system design.

   D. Accommodating monitoring and feed needs is the responsibility of the AV installation team.

3. How should the AV design ensure that the AV system addresses the necessary display signal sources?

   A. By including a signal switcher in the system design

   B. By including all potential signal sources in the AV system control room

   C. By establishing a clear, up-front understanding of the media that the client needs to present

   D. By providing appropriate system inputs for each type of signal source

4. How should an AV system designer address system bandwidth?

   A. Minimize the bandwidth of signals transmitted within the system

   B. Maximize the bandwidth of signals transmitted within the system

   C. Select components that minimize bandwidth

   D. Select components that maximize the bandwidth of the entire AV system

5. What are the three main performance parameters that an audio system should be designed to achieve?

   A. Directivity, intelligibility, and consistency

   B. Intelligibility, frequency response, and headroom

   C. Loudness, headroom, and frequency response

   D. Loudness, intelligibility, and stability

6. How should an AV system designer determine the required frequency response of an audio system?

   A. The frequency response should be based on the type of applications the audio is intended to support.

   B. The designer should work to achieve a frequency response as wide as possible within the project budget.

C. The audio system output frequency response should match the frequency response of the audio source components.

D. The audio system frequency response should meet the industry standard of 20 Hz to 20 kHz.

7. Which of the following audio processors would a system design specify to enhance the intelligibility of a videoconferencing system?

A. DSP matrix mixer

B. Echo canceler

C. Crossover

D. Compressor

8. What is the total impedance of a system consisting of three loudspeakers, each with 4 ohms of impedance, connected in series?

A. 2 ohms

B. 4 ohms

C. 12 ohms

D. 32 ohms

9. In audio systems where the loudspeakers are located far from the amplifier, what type of loudspeaker system is typically used?

A. Direct-coupled system

B. Constant-voltage system that uses transformers

C. Series/parallel wired loudspeaker system

D. Low-impedance loudspeaker system

10. As a listener moves away from a sound source, such as a loudspeaker, the sound energy drops. According to the inverse square law, which formula would you use to determine the drop in acoustic energy if the user moved from 6 to 12 meters away from a sound source?

A. $dB = 20 \log (6 \text{ meters}/12 \text{ meters})$

B. $dB = 20 \log (12 \text{ meters}/6 \text{ meters})$

C. $dB = 20 \log (6 \text{ meters} \times 12 \text{ meters})$

D. $dB = 20 \log (6 \text{ meters} + 12 \text{ meters})$

Answers

1. **A.** The primary initial consideration of the AV system designer when designing a video display system component of an AV system is determining how the client will use the display system. This will drive the subsequent decisions about resolutions, screen size, screen location, and so on.

2. **C.** The AV designer should determine the needed monitoring, feeds, and recording requirements, and ensure that they are supported by the system design.

3. **C.** During the design stage of the AV system, the designer must have a clear understanding of the client needs, such as the media that the client must be able to present, and use this information to design a system that accommodates the necessary sources.

4. **D.** Since the system bandwidth is limited by the lowest bandwidth device within the system, the AV system designer should select components that maximize the bandwidth of the entire system.

5. **D.** The three main performance parameters that an audio system should be designed to achieve are loudness, intelligibility, and stability.

6. **A.** As with all AV system design decisions, the frequency response of an audio system should be based on the client needs and intended applications. For example, an audio system that will transmit only videoconferences does not need to achieve the same wide frequency response as a system intended to amplify music performances.

7. **B.** An echo canceler can reduce the echoes that sometimes impact the intelligibility of the audio portion of a videoconference.

8. **C.** When wiring loudspeakers in a series, the impedance is additive, meaning that three loudspeakers that each have 4 ohms of impedance will result in a system of 12 ohms of impedance.

9. **B.** Audio systems where the loudspeakers are located far from the amplifier typically use a constant-voltage loudspeaker system that includes transformers for each loudspeaker.

10. **A.** The formula you would you use to determine the drop in acoustic energy if the user moves from 6 to 12 meters away from a sound source is dB = 20 log (6 meters/12 meters).

CHAPTER 17

Conducting a Vendor-Selection Process

In this chapter, you will learn about
- Identifying potential equipment vendors
- Reviewing and evaluating vendor proposals
- Communicating with vendors to clarify proposals
- Awarding contracts

Installing an AV system at a client site often requires that an AV company work with other vendors to obtain the equipment and services necessary to meet its client's needs. For example, in many cases, the AV company is responsible for procuring the AV equipment identified in the system design. This may require the AV company to obtain proposals from various equipment suppliers or manufacturers.

Moreover, a single AV company may not have sufficient staff to perform all the required installation tasks. Therefore, the company may need to hire subcontractors to perform specific portions of the system installation or configuration, such as control-system programming. In these cases, the AV company may be required to go through a subcontractor-selection process.

The goal of the vendor-selection process is to identify companies that are able to supply the required equipment or services in a reliable, professional manner, at an acceptable price. Depending on the complexity and scope of the project, this process may range from reviewing the offerings from various equipment suppliers to conducting a complex, formal procedure. You may need to identify several qualified vendors, create an RFP document that describes the products or services you need to procure, review proposals from several vendors, and select the vendor that offers the best value.

This chapter reviews some of the issues that you may encounter when identifying potential vendors, reviewing vendor proposals, and selecting vendors to assist you with your AV system installation. The knowledge and skills required to accomplish this task typically include the following:

- Knowledge of AV design and system installation issues
- An ability to create RFP documents and review vendor proposals
- Professional communication skills

Identifying Potential Equipment and Service Vendors

An AV system proposal typically involves the procurement of equipment and services. The AV company awarded the project is usually responsible for obtaining the equipment specified within its proposal and performing the identified installation tasks.

Small projects may involve simply ordering equipment from suppliers and using the existing company staff to install the equipment at the client site. Large projects may require the AV company to obtain proposals from equipment suppliers. In addition, there may be tasks that the AV company can't address with in-house staff. In that case, the company will need to get bids from other companies (AV or allied trades) to act as subcontractors.

Once you determine the equipment and services you need, you must identify the vendors capable of providing the specific equipment, services, or both.

Equipment and Service Considerations

The AV design or system proposal that the client accepted will typically include a list of specific AV equipment. Alternatively, it may identify the required technical specifications for components in a manner that allows the AV company to select the specific components that meet the identified specifications.

Equipment can include the following:

- AV components, such as projectors, displays, screens and mountings, amplifiers, loudspeakers, microphones, mixers, and switchers

- Network components, such as routers and switches, gateways, firewalls, and servers

- Installation supplies, such as cabling and mounting brackets

- Special-order or long lead-time items, such as furniture designed to fit within a specific space, custom wall or equipment plates, custom-sized screens, and large displays

 NOTE Since special-order items may require additional time to fabricate, ship from the supplier, or assemble, they should be ordered as early as possible to ensure that they are available when the installation task commences.

For smaller projects, the AV company may be able to order the components and supplies from a manufacturer or equipment vendor, using a supply catalog or an online system. In these cases, the project team should be familiar with the capabilities of the individual suppliers and manufacturers, including the products offered by each supplier, the prices charged by each supplier for specific types of equipment and materials, terms of the sale, delivery costs and options, reliability of the supplier, return policies, etc.

For larger projects, it may often be advantageous for the AV company to provide equipment suppliers or manufacturers with a list of required equipment and supplies, and ask the supplier to submit a bid for fulfilling the requirements as a single purchase, as illustrated in Figure 17-1. In these cases, the AV company typically expects to receive a bid that provides better terms than ordering from a catalog, such as a lower overall price.

In all cases, it is important to ensure that the supplier is capable of providing the requested equipment in a timely manner.

Figure 17-1
An example of
an equipment
list provided to a
vendor for a bid

Audiovisual Systems Project Documentation Sample

PROJECT #1234   NEW CLASSROOM
OCTOBER 18, 2012   NEW INFOCOMM ACADEMY

Audio Equipment:

Digital Wireless Microphone System
RF sensitivity −98 dBm at 10^{-5} BER
THD −12 dBFS input, system gain @ +10
Latency <2.9 ms
RF transmitter power output 1/10/20 mW
Transmitter frequency range 470 MHz to 534 MHz
Handheld (ULXD2) and body pack (ULXD1)
Shure ULXD4 receiver or approved equivalent

Digital Sound Processor Matrix
16 inputs, 16 outputs
Echo and noise cancellation
Frequency response: 20–22,000 Hz, + 0.1 /− 0.3 dB
Analog input gain: −20 to 64 dB on all inputs in 0.5 dB steps
Output gain: −100 to 20 dBu in 1 dB steps
Polycom SoundStructure C16 or approved equivalent

Assistive Listening RF System
SNR of 80 dB or greater
Frequency response: 50 Hz to 10 kHz ± 3 dB
Simultaneous transmitters: 3
Tx range: 3000 feet (914 meters)
Listen Technologies LT-800 or approved equivalent

Audio Amplifier
Frequency response: 20 Hz to 20 kHz, ±1 dB
400 watts RMS output power (2x100 watts @ 8 ohms)
Energy Star qualification
100 dB signal-to-noise ratio with 0.1% THD+N
Extron XPA 2002 or approved equivalent

INFOCOMM   APPLIED VISUAL COMMUNICATIONS
                                                          27 41 16.51 - 20

©2012, InfoComm International®

In situations where the AV company needs to subcontract parts of the system installation, it will need to procure those services. These are examples of tasks that another party may perform:

- Specialized fabrication and installation of furniture
- Decorative frames around display screens
- Programming of system controllers
- Additional electrical services
- Professional engineering services

Criteria for Selecting Potential Vendors

Once you identify the equipment and services necessary to meet the client needs, you will need to select vendors that are capable of providing them. The objective at this point is to assemble a list of qualified vendors that you will invite to submit bids for providing equipment and/or services.

Vendor selection can be as simple as reviewing equipment vendor catalogs and choosing vendors that provide the best combination of price, quality, and service. However, for some projects, it may be a more complex effort that requires you to evaluate the ability of vendors to meet the identified needs and requirements.

Typical criteria for selecting potential vendors include the following:

- **Vendor capabilities**   If you're looking for equipment, does the vendor carry the items you need for the project? If you need services, is the vendor capable of performing the specialized fabrication, installation, engineering, programming, or other tasks you require?

- **Value**   Price is usually a primary consideration. However, the price should be balanced with issues such as the reliability of the vendor, professionalism, and quality of the product or service. You should also consider vendor support, including warranties, replacement, and installer training.

- **Experience with similar projects**   Has the vendor proved to be capable of providing the equipment or services that you require? A history of performance on similar projects can help you decide if a vendor can meet your needs on the current project.

- **Location**   In cases where the vendor will work on-site, you will need to identify vendors that provide services at that location.

- **Terms**   Does the vendor promise delivery within a specific period, allow returns, and provide acceptable warranty terms?

You can obtain this type of information about potential vendors from a variety of sources, such as experience with past projects, assessments from other industry professionals, and company information from website listings and advertisements. Through a quick review of candidates, you can usually determine which potential vendors to eliminate and which warrant closer examination.

After you have identified qualified vendors, the next step is to send them a description of the project elements on which you want them to bid.

Creating an RFP for Equipment and Services

The work description that you send to potential vendors should define the equipment, labor, or service that you want the vendor to deliver. This work description and specification is often called a request for proposal (RFP) or request for quotation (RFQ). This RFP will likely be similar to what you received from your client when the client asked you to bid on the project. However, in this case, you are the client, requesting a bid from the other vendors.

Creating a detailed RFP can be a complicated task, depending on the scope of the project. However, as an AV vendor, you've probably received RFPs from potential clients that can help guide you in creating RFPs for your company. Your experience responding to RFPs should also help you write an RFP for your project. Consider what information you need to bid on a project, and then ensure you provide this information in your own RFP. You may even be able to use relevant portions of your client's RFP for your project description.

Equipment Vendor RFPs

For equipment vendors, an RFP may include the following:

- An equipment list that shows either the required technical or performance criteria, or specific makes and models of equipment required
- Delivery methods, locations, and dates
- Special requirements, such as installer training and extended warranties

Labor and Service RFPs

For labor and service providers, an RFP may include:

- Detailed descriptions of the specific services that the vendor will provide
- AV drawings depicting component configuration/installation
- Experience you require of staff working on your project, including any certifications they should hold, such as CTS, and security clearances where necessary, such as for access to government buildings
- Relevant project site information
- Work deadlines or proposed schedule
- Insurance requirements
- A request for descriptions of recent similar projects completed by the vendor

Other RFP Information

In addition to the work description, all RFPs should include the following information:

- Contact information for the project manager
- The deadline date for vendor response
- Required format (such as hard copy, PDF, or editable spreadsheet) and required number of copies of the bid document

Other required information may include items such as these:

- Payment terms
- Warranty information

- Technical support after sale
- Insurance requirements
- Information about the company that may indicate its ability to complete the work

Contacting Potential Vendors About Your RFP

Once you've identified the vendors that you want to bid on portions of your project and created the RFP(s), your next step is to contact the vendors to describe the project to them and determine which vendors will be placing bids. In some cases, prior to sending vendors a copy of the RFP, your client may require that the vendors sign a confidentiality agreement that prevents them from revealing proprietary information about the project.

Potential vendors may have questions about the project, your approach to addressing the client needs, or the terms of the proposal. It is usually a good idea to collect all of the vendors' questions in writing and organize the questions and answers in one document that you can deliver to all of the interested vendors. This ensures that all vendors responding to the RFP will work from the same assumptions when providing their bid.

Reviewing Vendor Proposals

Vendors will respond to your RFP with their proposals. You should evaluate their proposals and determine which vendor provides the best value.

The criteria for selecting a vendor will vary for each project, but typically includes the following:

- **Price**   The proposed price is usually the primary issue that drives the selection of a vendor. However, there are a number of other issues to consider when evaluating vendor bids. Some companies make it known that they exclude the highest and the lowest bid from consideration to discourage underquoting or overquoting.

- **Approach**   For some aspects of the project, each vendor may propose a different method of achieving the desired results. For example, they may specify different materials or a different technical approach. You should evaluate the proposed approach to determine if it meets the client's requirements and provides the best value for the project.

- **Quality**   In some instances, vendor proposals may offer a trade-off between quality and cost. Low-cost proposals may achieve their reduced cost by using lower-quality materials or fabrication than competing vendors. In some cases, this may be acceptable (if the proposed quality level still meets what you've promised to your client). In other cases, this may disqualify the bid.

- **Ability to fulfill the requirements**   You must determine whether the vendor will be able to follow through on the work. For example, does the vendor have staff with the required skills and ability? Is the vendor financially solvent? Does the vendor have a history of meeting commitments? This can sometimes be difficult to determine, but it is an important aspect of evaluating potential team members.

Communicating with Vendors

When reviewing proposals, you may need to contact vendors to clarify issues in their proposals. For example, you may need to ask questions about the vendor's approach or terms. In some cases, it may be appropriate to ask the vendor to make specific changes to its approach or terms and resubmit the proposal.

Always communicate with the vendors in a professional manner, suitable for companies engaging in a business transaction. This is important for establishing a good working relationship with a vendor that may eventually be awarded the project. Be sure to use the same professional communication skills that you practice when interacting with your clients.

The Contract-Award Process

Once you have selected the vendor that you determine provides the best value for the project, the next step is to award the contract.

The contract-award process will vary according to the company policies and practices. The following are some paths the process may take:

- You may simply use the vendor's purchase contracts or processes, such as when you are procuring equipment and supplies from a manufacturer or supplier.

- In some straightforward cases (a simple contract for supplying equipment), a purchase order may suffice.

- The RFP and the vendor proposal may be sufficiently detailed to function as either all or a portion of the contract. However, it is usually necessary to supplement the proposal with other contract clauses that define issues such as payment terms and conditions, nonperformance remedies, and so on.

- You may need to work with the vendor to create an acceptable contract agreement. For example, when working with another AV vendor as a subcontractor, it may be necessary to negotiate some of the terms of the contract agreement in order to create a work contract that is acceptable to both parties.

Chapter Review

In this chapter, you reviewed the process of selecting vendors and suppliers to assist your company in fulfilling the client's needs. This process can range from reviewing catalogs from various equipment suppliers to conducting a complex, formal process of receiving bids from various vendors. You may need to identify vendors, create a detailed RFP document, review and evaluate proposals from vendors, and finally, award the project to the vendor(s) offering the best value.

Review Questions

The following review questions are not CTS exam questions, nor are they CTS practice exam questions. These questions may resemble questions that could appear on the CTS exam, but may also cover material the exam does not. They are included here to help reinforce what you've learned in this chapter. For an official CTS practice exam, see the accompanying CD.

1. An AV company that is planning a client system installation typically works with vendors to obtain which of the following?

   A. AV equipment, such as projectors, displays, audio, and control systems

   B. Installation services and assistance, when necessary

   C. Fabrication of custom components, such as furniture, defined within the proposal

   D. All of the above

2. When working with a vendor or subcontractor for AV services, what does the abbreviation RFP stand for?

   A. Request for pricing

   B. Request for proposal

   C. References for projects

   D. Request for products

3. What is the typical best practice for selecting potential vendors to bid on providing product and services for your AV project?

   A. Select local vendors that are based in your city or geographic area

   B. Select vendors that you have worked with previously on other projects

   C. Select vendors that advertise only name-brand products

   D. Select vendors based on information about their capabilities, experience, terms, and value

4. When creating an RFP, what is the best approach for describing the work that you want the vendors to address within their proposal?

   A. Describe the specific project work that you want the vendor to address in as much detail as possible, and ask the vendor to propose a technical approach and cost

   B. Describe the project needs in general terms that will allow the vendor to be creative when proposing a technical solution and cost

   C. Send the vendor the relevant portion of your original proposal to your client, and ask the vendor to propose a price for providing that portion of the project

   D. Send the vendor the relevant portion of your original proposal to your client, along with your proposed pricing, and ask the vendor to agree to complete the work on that portion of the project for the identified price

5. Which of the following best describes the approach for evaluating proposals for products and services from potential vendors?

   A. Select the proposal that provides the lowest cost.

   B. Select the proposal that provides the best combination of cost, value, and terms.

   C. Select the proposal that most closely matches the specifications within the RFP

   D. Select the proposal that has the best technical approach/specifications

Answers

1. **D.** An AV company may work with vendors to provide any or all of the noted products and services, depending on the scope and complexity of the project.

2. **B.** RFP stands for request for proposal. This is a document that defines which products and services the company is asking the vendor to provide. This document may also be called a request for quotation (RFQ), or other similar designation. Regardless of the name, it refers to the same specification of the project needs.

3. **D.** While issues such as location, past experience, and use of high-quality equipment and materials are important, they should be part of a number of criteria that you consider when selecting a potential vendor to bid on the project. You should also obtain information about the vendors' capabilities, their experience with similar projects, the standard terms that guide their purchase agreements, and any other information that helps you assess the overall value of their offerings. This will help you to qualify the vendors before asking them to provide a proposal in response to your RFP.

4. **A.** Your RFP should describe the products or services that you want the vendor to provide in as much detail as possible, and ask the vendor to submit a proposal that describes how the vendor will meet this need, along with the vendor's proposed pricing.

5. **B.** When evaluating proposals, it is usually appropriate to look for the proposal that offers the best combination of price, value, and terms. Each proposal may offer trade-offs on cost, technical approach, and terms of purchase and performance, making the evaluation process difficult at times.

CHAPTER 18

Performing AV Finance and Job-Costing Activities

In this chapter, you will learn about

- Estimating the costs for equipment and materials
- Estimating labor costs
- Purchasing AV supplies and equipment
- Obtaining pricing and delivery schedules
- Producing interim invoices
- Analyzing staff utilization rates
- Creating profit-and-loss statements

As a result of InfoComm's 2012 JTA (see Chapter 2 for details) several previously separate CTS exam tasks were combined into one. The new task, perform AV finance and job-costing activities, combines knowledge and skills that previously fell under three tasks: conduct estimating activities, conduct purchasing activities, and conduct job-costing activities.

AV professionals are often called on to provide estimates for the labor, equipment, material, and other costs associated with designing, installing, and repairing AV systems. Work estimates must be carefully considered. An estimate that is too low may result in the company losing money on a job. An estimate that is too high may cause the client to accept a lower bid for the work.

AV companies purchase AV supplies, materials, and equipment for internal use, for use as rentals to clients, or for sale to clients as part of an installed system. The primary objective of a properly managed purchase process is to obtain the best value for the prices paid for items procured by the company. To achieve this objective, AV professionals should understand how to work with supplier companies and follow some simple procedures to minimize cost.

AV professionals must also determine what it actually cost the AV company to complete a project. This assessment reveals in which areas the estimate of the project was accurate and in which areas the project incurred extra, unanticipated costs. Gaining a better understanding of the actual costs of a project can help increase the accuracy of future estimates, increase the effectiveness of project-management activities, and help you ensure that the business continues to be profitable.

Creating Estimates for AV Activities

The purpose of an estimate is to quantify the equipment, materials, labor, and ancillaries required to deliver a specific product—an AV system or live-event equipment and support—and provide an estimated cost for delivering the product. Before looking at the steps for creating project estimates, we'll cover some general considerations and the estimate markup.

Estimate Considerations

Typically, the process for estimating the costs for an AV project includes the following steps:

- Ensure that you understand the client needs, project scope, and system design (if the design has been produced by a different firm, such as a consultancy). Your estimate is based on the nature and scope of the work that the client is expecting to be accomplished. Make sure that you ask the right questions and confirm that you accurately understand the work the client is asking you to perform.

- Ensure that you understand the nature and scope of the work required. In order to accurately estimate the cost of an AV project, you must have a full understanding of the actual work needed to complete the project. This knowledge is usually gained from years of experience with similar projects, encountering common (but not always obvious) issues and problems related to this type of work. Therefore, it is important that a highly experienced AV professional be involved with developing the estimate.

- Create a breakdown of the required labor. Break down the project into separate elements and determine which specific tasks you must perform to accomplish each element of the proposed work, along with an estimate of the amount of time you'll need to perform each task. This will allow you to better assess all the small tasks and issues you need to address to complete the entire project. And because each type of employee working on the project may have a specific labor cost, you should also identify the staff member or contractor who will perform each task.

- Create a list of the equipment, materials, and supplies necessary to support and install that equipment for each task.

- Identify any other costs or issues that will impact the project. For example, remote projects may require travel and accommodation costs. Other projects may require costs for accommodating special regulations or historical considerations, removing construction debris, proper recycling of used electronic equipment, construction permits, equipment and/or truck rentals, business license fees, insurance, bonds, and so on.

- Identify any potential problems or constraints that will affect the cost. For example, does the project need to be accomplished in a short time frame? Is it necessary to perform work during holidays or business off-hours? These requirements can increase overall labor costs.

Estimate Markup

Your calculations of the goods, labor, and costs for the project will form the basis for the estimate you provide to the client. Typically, the costs in your estimate will be marked up (increased) by a predetermined percentage to pay additional costs in providing the work (overhead, insurance, etc.), plus a standard amount for profit. Some estimates may also include an additional amount to cover unforeseen risks or costs, called a *contingency* or *management reserve*.

Most companies use customized spreadsheets or other software that will automatically calculate the markup and create a final itemized estimate to be sent to the client for review and approval. This software solution will allow the estimator to enter the estimated amounts for standard items, such as labor hours in each labor category, project supplies, and the actual procurement costs for individual items of AV equipment and other materials.

Creating a Project Estimate

You create a project estimate based on each of the main project elements: equipment and material costs, project labor costs, and other project-related costs. When you've completed the estimate, you will need to present it to the client.

Estimating Equipment and Material Costs

The AV equipment required to complete a project will be based on the AV system design and specifications. In some cases, particular AV devices are specified by manufacturer and model number. In other cases, the estimator may select devices based on general specifications. For example, the AV system specifications may simply call for a 200-watt audio power amplifier, freeing the AV team to select the exact model.

Materials and supplies typically include items such as cables, connectors, paint, solder, cable ties, nuts and bolts, and hardware. The estimator should review the system drawings and determine the amounts of each type of materials and supplies necessary to install the system.

The estimator can then create a list of the required AV devices, materials, and supplies—called a bill of materials (BOM), as described in Chapter 16. The equipment section of the BOM should include specific manufacturers, models, and accessories for major components, noting the quantity of each, any shipping charges, and the price the client will be charged for each item. As noted earlier, the cost for each item is marked up by a percentage to cover the cost to the AV company of procuring the items and making a profit.

Estimating Project Labor Costs

The first step in scoping the labor required for the project is creating a work breakdown structure (WBS) listing all the elements of the project. The WBS does not take into account how long tasks will take or the labor needed to accomplish them; it considers only the general work areas.

Next, create a project plan that depicts the timeline of the work to be accomplished, the sequence required to complete the tasks, and milestones that are used to measure progress. This timeline should include an estimate of the labor and duration of work required to complete the tasks. The estimator should also add some time to provide room in the schedule to accommodate any unexpected delays. If the project depends on other vendors' completion of their tasks, that must be factored into the timeline. A number of software tools are available to create these types of project timelines/schedules, which we will discuss in Chapter 20. See Figure 20-4 in that chapter for an example of a project plan timeline, which also includes project milestones, in the form of a Gantt chart.

Estimating Labor Costs

The cost of providing labor for the project is based on the amount of labor hours or days that each task will require to complete. The charge-out rate for the individual worker designated to perform each task should also be specified. Some tasks require less skilled, general labor that junior staff can perform. Other tasks require more skill and experience, or they require specialists, such as control-system programmers.

The time and cost required to travel to the client site, and any meals or lodging necessary when working at a remote site, should be included in labor costs, if not itemized separately. And if the work schedule includes any weekends or holidays, you should consider rates for overtime or weekend time.

In short, the amount a client is charged for labor is not just the hourly rate for each staff member; it should represent the full cost of that employee. This typically includes salary, taxes, benefits, vacation, other nonbillable time, a percentage for company profit, etc.

Estimating Other Costs

The estimator should also consider any other costs that may be needed to complete the project. These can include fees for construction permits, bonds required by the client, financing costs incurred when borrowing to purchase equipment, insurance, travel costs (if not included in labor), delivery or transportation, parking, printing, etc. As with labor and equipment estimates, the estimates for these costs should also include a markup.

Providing Estimates to Clients

The estimates that you provide to the client should be as complete and accurate as possible, and presented in a neat format that is easy to follow. Each company may have different policies regarding the amount of detail that the client receives within a project estimate. Some companies provide a detailed breakdown of equipment, materials, and labor for each task. Others provide "rolled-up" (total) costs for each major task/category.

Purchasing AV Supplies and Equipment

In Chapter 17, we discussed acquiring products and services on a per-project basis. But AV companies must purchase AV supplies and equipment at other times, too.

Obviously, the first step in the purchasing process is to identify which items your company needs to buy. In some cases, those items are for your company's internal use, such as computers, furniture, vehicles, tools, test equipment, uniforms, etc. Smaller items may be purchased throughout the year as needs arise. More expensive items, such as vehicles, may be purchased according to a company's acquisition schedule or plan. In other cases, you will purchase items listed on the BOM for a project.

Some AV companies keep a warehouse of common materials—such as connectors, cabling, and hardware—and equip their AV system installation projects from that stock. This allows the company to obtain price discounts for larger quantity purchases, and to avoid paying expedited shipping fees when a project has an immediate need for those items. Therefore, the company will need to purchase items as necessary to replenish the warehouse stock.

The objective at this stage is to collect information to create a list of items your company must purchase. The items on the list will be sent to appropriate vendors for bids or used to purchase items directly from AV catalogs or online ordering systems.

Identifying General Categories of Purchase Items

When creating a list of items for purchase, it is a good idea to sort the items into general categories. This keeps things more organized and is especially helpful when you need to order items from suppliers that specialize in specific areas. The following are some examples of product categories:

- AV equipment and supplies
- Computer equipment
- Materials used for creating AV systems, such as cabling and connectors
- Services or labor, such as specialized installation or fabrication labor or software programming
- Tools needed for specific installation tasks
- Programming services

Creating Purchase Lists

When creating your purchase list, identify the specific quantity of each item needed, as well as specific products when required. For example, in many cases, the system designer or client will identify a specific product to meet the need and provide the vendor, item model, or part number. In other cases, the items will be identified by general specification, such as 20 pairs of male/female XLR cable terminations, which means that you can obtain this general type of item from any manufacturer.

Instances where products come from an internal company inventory should also be identified on the purchase list. The company will need to update the company inventory to ensure that it does not run out of these items.

Here are some other points to keep in mind when creating a purchase list:

- Expensive or specialized items may require that the AV company receive bids from several vendors.
- Smaller items may be priced and purchased via selection from a catalog or an online source.
- Items with long lead times that require specialized fabrication, such as AV furniture, should be ordered as early as possible in the purchasing process.
- Specialized tools may be rented or purchased for a job, depending on the potential for an ongoing need to use the tool.

Identifying Vendors and Suppliers

An AV company depends on its suppliers. If a supplier fails to meet delivery deadlines or provide the requested equipment and materials at the needed quality levels and prices, the AV company will not meet the client's expectations. It is in the interest of an AV company to identify reliable and cost-effective sources of equipment and supplies, and work to develop a long-term partnership relationship with these vendors.

We reviewed some of the issues associated with selecting vendors in Chapter 17. The following are some typical considerations when identifying vendors to fulfill the needs of a specific project:

- **Location**   Is the vendor located close by, or is the vendor a regional/national presence? Each may have advantages and/or disadvantages for your particular project. For example, dealing with local vendors that allow you to pick up your order in person reduces shipping costs, especially for larger size items.
- **Company size**   Does the specific project require a small company that may provide more personalized service, or a large company that may be able to offer better pricing and selection?
- **Reliability and quality**   The selected vendor must be able to meet your specific project needs in a reliable manner, providing the promised products at the required quality levels. Ask the vendor about past performance for similar types of projects, or search for other information that can give you an idea of the vendor's reputation, such as industry review websites.

- **Pricing**   Price is only one factor in the value proposition offered by a vendor, but it is often the most important. Obtain some price information for items needed for your project so that you can compare vendors.

- **Specialized products**   In some cases, you need to obtain items from a specialized vendor for specific project requirements, or to obtain products from specific manufacturers. Identify which vendors are able to meet these specific project needs, if necessary.

Tips for Working with Vendors

Here are some ways to improve your experience with vendors:

- If you use one vendor for a large portion of your order, it is usually appropriate to request a package price for the entire order that provides a discount on the standard pricing.

- It is often a good idea to combine orders where possible to achieve quantity discounts and to take advantage of any vendor-specified minimum purchase amounts to qualify for free shipping.

- Once you have established a relationship with a vendor, you can often negotiate "end-column" or "off-sheet" discount pricing that is not based on quantity discounts.

- Some vendors may provide discounts for prompt payment.

- Pay attention to delivery schedules to determine if there is any potential to coordinate deliveries to reduce shipping costs.

- If a specific product is no longer available, or a vendor identifies a product that offers a better value for the customer in terms of price, performance, or technology, it may be necessary to discuss advantages or issues with the client. You should obtain a client sign-off to use the substitute product or approach before purchasing it for the project.

- When obtaining pricing for labor or services, define the specific labor needed as accurately and comprehensively as possible, so that the vendor will clearly understand your needs. This allows the vendor to respond with an accurate bid for labor to support the project.

- Be sure to review your orders for correctness and consistency to ensure that you did not forget any items. This way, you avoid needing to pay additional fees for expedited shipping.

- Even if you have an ongoing vendor relationship, periodically review the offerings of other vendors to determine if they are able to provide better pricing or service.

The same types of general considerations apply when identifying vendors or subcontractors to provide services or specialized labor to support specific project needs. When selecting subcontractor vendors, be sure to follow a careful due-diligence process to ensure that the subcontractor has the capabilities to provide the needed labor or services at the required level of quality.

Your review of vendor capabilities should allow you to identify several vendors that appear to be capable of meeting your needs. These are the vendors that you will contact either to obtain bids on your purchase needs or to make your initial purchases.

Obtaining Pricing and Delivery Schedules

Once you have your purchase list and the list of vendors to contact for quotes, contact the vendors. Find out their process for obtaining a quote for purchasing and delivering the equipment, materials, or supplies, and then follow that process. Some vendors may use an online ordering system; others may request that you send a list of the items and quantities.

Performing Job-Costing Activities

Earlier, we addressed the process for estimating labor and other costs required for AV projects. These estimates are the basis for the price you propose to perform the work. They are based on labor rates for your staff, including overhead, insurance, and other costs associated with running a business.

Your initial estimate is based on your assessment of how much labor, equipment, and materials you think will be required to complete a specific series of tasks, such as installing an AV system at a client site or producing a live event. An estimate is based on your experience in completing similar projects, so it follows that you can create better and more accurate estimates if you track how close your estimate was to the actual amount of labor and materials required to perform the work. You can use that knowledge to guide your approach to future estimates.

In addition to tracking the final costs of a project, in order to verify the accuracy of an estimate, you must keep track of the progress and cost for ongoing projects to determine if they are on track to match the estimate. If not—if it is taking more labor and materials to complete than estimated—you need to know that, too. In cases where the ongoing project is exceeding the estimates, it may be possible to adjust the work approach in a manner that will allow you to bring the work back in line with the estimate.

You can evaluate the progress of a project by comparing a *job status* or *field report* against the project documents and estimate. A job status or field report is a current evaluation of the amount of the work that has been completed, and the tasks and materials remaining to complete the project.

Other project documents useful for assessing the amount of work and cost incurred to date include labor time sheets and invoices from subcontractors. Project managers can assess the actual amount of supplies and materials that have been consumed by the project and their costs. They can use this information to compare how much has been spent on the project to date with the estimated costs for each of the completed tasks.

Analyzing project expenses in this manner allows the project manager to identify the following:

- Where the project may require a change in installation approach
- Where more experienced staff may be needed to complete tasks
- Where project expenses and planned labor can be shifted to cover overages in expended labor or materials

This type of information is critical to creating new estimates for similar projects because it provides the estimator with feedback on what is actually required to complete the work.

When the actual work exceeds the estimates, the estimator can either increase the estimates to match the amount of work required for the task or change how the task is conducted. For example, if labor estimates for assembling racks at the client site are too low, the estimator may increase the estimate. However, the estimator may determine that in the past, assembling the racks at the company site took less labor, due to the direct access to tools and supplies, so it may be more cost-effective to assemble the racks at the company site and transport them to the work site.

This analysis also helps the project manager identify items that may have been missed in the original estimate or uncover unforeseen issues that impacted costs. Unforeseen issues that might affect a project estimate include the need to rent specialized equipment or shipping delays for equipment and materials, both of which may slow installation tasks. You can factor these types of unexpected costs into future estimates, or use them to alter the project-management approach in order to reduce risk. You may decide to order equipment and supplies earlier or use different vendors with more reliable shipping for future projects.

Producing Interim Invoices

Most project proposals define a series of partial payments based on completing specific portions of the work. These are called *interim payments*. They typically cover the labor, equipment, and materials necessary for completing a specific portion of the project. The client agrees to provide a partial payment at specifically defined intervals throughout the project.

Using an interim payment schedule ensures that the AV company continues to receive revenue throughout the project. The AV company does not receive the entire payment until the project is complete.

Because interim payments are based on work that has been completed, job progress or job status reports become an important part of the invoicing process. The interim invoices are based on the descriptions of work completed provided by the job progress reports, as discussed in Chapter 20. These reports must be accurate and complete to prevent any problems (including embarrassment) that might occur if the client is invoiced for work that has not been done yet.

Analyzing Utilization Rates

For staff members who typically work on client projects, as opposed to those who work on general administrative tasks, the project manager should track the amount of billable time they spend on the project. The total percentage of the staff member's time that is billable to a client's project tells the manager how efficiently that staff member is being utilized within the company. A fully utilized staff member is maximizing profits for the company, while less than full utilization may mean the company is overstaffed, given the amount of work currently available.

You can easily determine each staff member's utilization rate by comparing the amount of hours spent on individual client projects with the amount of time spent on other activities. For example, if a work week is 40 hours, a staff member may spend 16 hours on one project, 16 hours on another project, and 8 hours on internal, nonbillable work, such as receiving and warehousing AV materials and supplies from a vendor. In this case, the staff member utilization would be identified as 80 percent on client work and 20 percent for internal administrative work.

Companies may have a different target for each staff member or staff labor category. For example, personnel who staff live-event projects may have fewer hours per week that can be directly mapped to a specific client project than AV installation staff. In these cases, as in the previous example, the company may assign these staff members other internal work, such as maintenance of rental equipment, to ensure that their time is fully utilized.

Creating Project Profit-and-Loss Documents

The manager and accounting staff at a company generate a *profit-and-loss* (P&L) statement, also known as an *income statement*. This document adds up the gross (total) revenue coming into the company from clients and other sources, and subtracts the expenses and costs (such as burdened labor, materials, and overhead), to determine the net profit—the amount of profit left over after paying all of the expenses. The P&L statement covers a specific period of time, and it may be issued quarterly or yearly. Figure 18-1 shows an example of a P&L statement.

P&L statements can be hard to create. They must factor in a large range of issues, including depreciation, cost of funds (interest on loans), and so on. They are typically created by an accountant. However, it is important for staff members to understand what issues impact profit and how the profit is determined so that they can better contribute to making the company a success.

Assessing the profit and loss for specific projects is also important. This consists of tracking the costs of equipment, materials, and labor expended on each task of a project, and determining if these amounts exceed the budgeted amounts for each task. Some areas of a project task may exceed estimates, such as labor, while others may end up costing the company less, such as using fewer cables than originally estimated. When actual costs are greater than the estimates, the project is suffering a loss in those areas.

The project manager can use this information to identify areas in which the project activities should be adjusted or project resources shifted to reduce the opportunity for loss within a specific area of the project. This type of information is a valuable tool for project managers because it allows them to adjust the way the project tasks are being performed to reduce losses and maximize profits. For example, labor may be exceeding

Company				
Profit & Loss [With Year to Date]				
February 2012				
	Selected Period	% of Sales	Year to Date	% of YTD Sales
Income				
Dues - Branch/Hotel				
Dues - Branch/Hotel - Aust.	$0.00	0.00%	$249.10	0.60%
Total Dues - Branch/Hotel	$0.00	0.00%	$249.10	0.60%
Dues - Commercial Affiliate				
Dues - Comm. Affiliate - Aust.	$0.00	0.00%	$2,480.00	5.60%
Total Dues - Commercial Affiliate	$0.00	0.00%	$2,480.00	5.60%
Dues - Dealer				
Dues - Dealer - Aust.	$3,389.10	47.00%	$12,687.75	28.80%
Total Dues - Dealer	$3,389.10	47.00%	$12,687.75	28.80%
Dues - Manufacturer				
Dues - Manufacturer - Aust.	$1,240.91	17.20%	$1,240.91	2.80%
Total Dues - Manufacturer	$1,240.91	17.20%	$1,240.91	2.80%
Total Income	$7,209.56	100.00%	$44,042.31	100.00%
Cost of Sales				
Freight	$0.00	0.00%	$36.16	0.10%
Education Products				
Edu - Accommodation	$3,078.87	42.70%	$3,078.87	7.00%
Edu - Venue Charges	$8,451.81	117.20%	$8,451.81	19.20%
Total Cost of Sales	$17,364.57	240.90%	$17,400.73	39.50%
Gross Profit	($10,155.01)	−140.90%	$26,641.58	60.50%
Expenses				
Empl. Compensation (Wages)	$4,357.70	60.40%	$10,894.25	24.70%
(Empl.) Training & Development	$348.62	4.80%	$871.54	2.00%
Employee Travel				
Travelling - Local	$77.73	1.10%	$77.73	0.20%
Total Employee Travel	$77.73	1.10%	$77.73	0.20%
Insurance				
Workers Compensation Ins.	$357.55	5.00%	$357.55	0.80%
Insurance	$0.00	0.00%	$427.48	1.00%
Total Insurance	$357.55	5.00%	$785.03	1.80%
Total Expenses	$18,311.85	254.00%	$28,253.02	64.10%
Operating Profit	($28,466.86)	−394.80%	($1,611.44)	−3.70%
Other Income				
Other Expenses				
Edubucks	$2,579.55	35.80%	$15,730.00	35.70%
Earlybird Discounts	$0.00	0.00%	$349.50	0.80%
Total Other Expenses	$2,579.55	35.80%	$16,079.50	36.50%
Net Profit/ (Loss)	($31,046.41)	−430.60%	($17,690.94)	−40.20%

Figure 18-1 An example of a P&L statement

budgeted costs because the company staffed a task with less-experienced people who are taking significantly longer to perform the task than estimated. In these cases, the project manager may decide to bring in more experienced staff members to perform these tasks or use a subcontractor who can perform the tasks at a fixed cost.

Chapter Review

In this chapter, we examined the process for estimating AV project costs, including estimating the costs for equipment and materials, labor, and other project necessities. Creating an accurate project cost estimate requires that the estimator have a thorough understanding of the nature of the work required, the type of and amount of labor necessary to complete the work, the supplies and equipment required for the project, and any other potential costs for the project.

We also discussed a basic process for obtaining the best value for items purchased by your company. This applies whether you're purchasing AV supplies, materials, and equipment for internal use, for use as rentals to clients, or for sale to clients as part of an installed system.

Finally, this chapter focused on how to determine what a project actually cost the AV company to complete. Understanding the actual costs of a project can tell a project manager where a project estimate was accurate and uncover the sources of unanticipated costs. This type of information can help increase the accuracy of future estimates, aid in managing the resources within an ongoing project, help change the project approach for future projects, and help ensure that the AV business continues to be profitable.

Review Questions

The following review questions are not CTS exam questions, nor are they CTS practice exam questions. These questions may resemble questions that could appear on the CTS exam, but may also cover material the exam does not. They are included here to help reinforce what you've learned in this chapter. For an official CTS practice exam, see the accompanying CD.

1. On what is a project labor estimate based?
    A. The charge-out rate for labor in each labor category needed to complete all of the necessary project tasks
    B. The weekly wage of the average paid staff member in the AV company
    C. The standard, published labor cost for installing an AV system
    D. A standard company cost for installing each AV component
2. What is a BOM?
    A. The invoice for parts and supplies that the AV company sends the client
    B. The invoice for parts and supplies that the supplier sends the AV company
    C. A list of equipment, supplies, and materials required for the project
    D. A list of the equipment, supplies, and materials required for the project, along with the quantity and cost for each item

3. What is included in the charge-out rate for labor in a project estimate?

   A. The staff member hourly or weekly wages

   B. The staff member wages, plus vacation time and taxes

   C. The staff member wages, plus vacation time, taxes, and other company overhead costs applicable to that person

   D. The actual total time that the staff member spends working on the project

4. You are interested in getting a bid from a vendor for equipment and supplies for a client AV system installation project. What information do you give the vendor to provide the basis for the bid?

   A. A copy of the accepted and approved project proposal

   B. A description of the components of the project, so the vendor will propose products to meet your needs

   C. A detailed list of all the items you want to purchase for the project, broken out by categories

   D. A list of the specific items that you want the vendor to provide

5. Your company has several projects that require a large number of standard items, such as connectors, cabling, conduit, etc. What is likely the best approach to obtain items for these projects at the lowest total cost?

   A. Purchase a large amount of these items at a lower cost per individual item, and keep them in stock in a warehouse until needed for projects

   B. Purchase only the amount needed for each project, since the client is billed the cost for purchases

   C. Negotiate an up-front price for the products with a vendor, and purchase only the amount needed for each project

   D. Combine several individual orders to save on shipping costs

6. What is the main objective of evaluating the cost of completing an AV project?

   A. To determine if the client received a good value for its funds

   B. To determine if the AV company made money on the project

   C. To determine what to charge the client once the project is complete

   D. To assess the cost of materials and supplies

7. How do project managers use information obtained from evaluating the cost of completing an AV project?

   A. To improve project estimates and the project management approach

   B. To determine if the staff members are working hard enough

   C. To assess if the project used the proper quality of materials

   D. To evaluate the work of the AV designer

8. What is the purpose of evaluating staff utilization rates?

A. To determine if some staff are working harder than others

B. To determine what type of work each staff member is performing

C. To determine how staff should be scheduled

D. To evaluate if each staff member's time is being used as effectively as possible

9. How does an AV manager use a P&L statement to assess the company's performance?

A. It tells the manager which projects were profitable and which ones were not.

B. It provides general information about the financial health of the overall company.

C. It is provided to the client to show how the project funds were used.

D. It is used by the manager to assess the performance of individual AV installers based on the profit made on their projects.

Answers

1. **A.** The project labor estimate is based on the charge-out rate for labor in each labor category needed to complete all of the necessary project tasks.

2. **D.** A BOM is a list of the equipment, supplies, and materials required for the project, along with the quantity and cost for each item.

3. **C.** The charge-out rate for labor in a project estimate includes the staff member wages, plus vacation time, taxes, and other company overhead costs applicable to that person.

4. **D.** You should give the vendor a list of the specific items that you want that vendor to provide.

5. **A.** If a company has several projects that require many standard items, usually the best approach is to purchase a large amount of these items at a lower cost per individual item, and keep them in stock in a warehouse.

6. **B.** Once you determine how much the project cost to complete, you can compare that amount with the price that you charged the client. This gives you an idea of whether the project will end up being profitable.

7. **A.** Evaluating the cost of the completed AV project helps the project manager determine if the estimates were accurate and if the project was managed properly.

8. **D.** Evaluating staff utilization rates tells you how efficiently each staff member's work time is being used by the company. Managers can use this information to adjust the assignments of staff to maximize the amount of revenue produced by each type of staff member.

9. **B.** The AV manager uses a P&L statement to assess the overall financial health of the company.

CHAPTER 19

Building AV Solutions

In this chapter, you will learn about
- Installing AV equipment and systems
- Setting up and configuring AV equipment
- Configuring systems for network connectivity
- Producing live AV events
- Providing support, documentation, and training for end users

A key part of the AV professional's skill set is the ability to build and install AV equipment and systems at a client site. This requires the AV professional to be able to read and understand AV system design documentation, know how to physically install AV components, and know how to configure the components within the overall AV system. This expertise is typically built up over a period of time, both from work experience and study.

This chapter reviews some of the basic tasks and procedures that you will perform when building an AV solution for a client.

The knowledge and skills required to provide AV solutions include the following:

- Installing AV equipment and systems in a technically correct and professional manner requires an understanding of the AV design specifications created for the client system.

- Installing AV cabling and racks requires basic skills in construction, such as mounting, pulling cables, and working safely at construction sites.

- Setting up AV systems requires knowledge of how to properly configure AV components and interconnect them via a control system or over a computer network. This can include designing a control interface and/or configuring AV systems to operate within the security, bandwidth, and other requirements of an IT network, if applicable.

- Producing live AV events requires knowledge of equipment setup and configuration in a temporary setting in order to support a presentation or performance.

Domain Check
Questions addressing the knowledge and skills related to building and providing an AV solution account for about 8% of your final score on the CTS exam (about eight questions).

Installing AV Equipment and Systems

In this section, we will review some of the issues, tasks, and procedures associated with installing typical AV components and systems at client sites. We'll cover the following topics:

- How to prepare for the installation
- Safety concerns when working at a job site
- Cable termination
- Rack construction
- Cable handling and pulling
- Mounting AV components

Preparing for the Installation

Preparation is central to a successful and efficient system installation, both permanent and temporary. CTS-certified professionals are responsible for understanding how their work impacts the company's overall strategy.

Before installing an AV system, you should prepare as follows:

- Know the priority of each project.
- Review the project documentation with the sales representative, designer, and/or project manager.
- Understand the scope of your responsibilities and how successful completion of your work is measured.
- Know which job site(s) you are going to work at each day.
- Determine a plan for when you get to the job site.
- Examine the relevant facility and system drawings and specifications prior to beginning work.
- Verify you have the necessary tools, equipment, consumables, and manuals for the installation tasks.
- Open and inspect each piece of equipment upon arriving at the job site. Check for any damage, and power on each device to verify that it is in good physical condition.

TIP Each installation, whether permanent or temporary, will require different equipment. Make a checklist of the equipment needed for each type of job and keep it in the job file. When picking up the equipment and assembling it for installation, a quick glance at the checklist will confirm that everything is ready to go. If the installation is temporary, use the checklist when you disassemble it to account for all the equipment.

Job Site Safety Concerns

Safety should always be your primary concern when working at a job site. Job sites may present numerous hazards and safety issues, ranging from falling and electric shock hazards to exposure to sharp surfaces and head injury.

As you learned in Chapter 12, AV professionals working at a job site must comply with all applicable safety regulations, as well as local, regional, and national safety codes. Depending on conditions at the work site, anyone working there may be required to use PPE (such as hard hats, steel-toed footwear, work gloves, hearing protection, and safety glasses). Workers should also use the appropriate tools and equipment necessary for safely performing the work, such as approved ladders and fall protection when necessary.

Cable Termination

Cable termination is a significant part of what AV professionals do when installing AV systems. It refers to the method of establishing connections at each end of a cable. In this context, *cable* refers to any number of insulated wires—with or without a shield—protected in a jacket. Cable termination requires selecting the appropriate cable and connector for a specific need, and using the proper method and tools to install the connector on the cable.

Poor cable terminations, which cause disruption in the flow of electricity, are the single most prevalent problem in electronic systems. Poor terminations are expensive to track down and fix after the system is built. Termination-related problems are particularly hard to troubleshoot because the problem can be intermittent and difficult to locate. You can easily avoid these problems by using the appropriate tools and connectors, as well as taking the time to properly perform the termination. A CTS-certified professional must be an expert in terminating cable.

NOTE Poor cable terminations are the number one cause of problems in AV systems, yet they can be difficult to track down.

Termination Preparation

Cable stripping is the process of preparing the cable for the connector by removing the jacket from the end of a cable (and sometimes the shield and/or insulation from individual wires) to reveal the conductor. Precise stripping assures a good physical

Figure 19-1
A cable stripper
tool

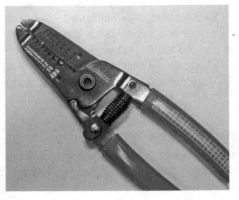

match with the connector. Stripping cable improperly can distort the shape of the insulation, thereby altering the impedance properties of a cable.

The technician should use a stripping tool designed for the specific type of cable and connector. The stripper depicted in Figure 19-1 is often used for stripping solid or stranded cables, other than coaxial cable. Special multibladed strippers are used for coaxial cable.

The outer jacket of the cable is the first thing to be removed. When doing so, you need to be careful not to disturb the wire(s) underneath the cable jacket. For some termination types, such as insulation displacement, you remove only the jacket. For most connectors, you will need to continue by removing the individual insulation from each conductor and cutting each conductor to the correct relative length.

Termination Methods and Types of Cables and Connectors

The type of termination depends on the cable and what you will connect it to. Termination methods include the following:

- **Direct-connection termination**   Also known as *screw-down termination*, this is a method in which the wire is *directly* connected to a piece of equipment, as depicted in Figures 19-2 and 19-3. This connection type is used for permanent connections. The most basic direct-connection terminations are the barrier

Figure 19-2
Direct
connections
using solid
conductors

Figure 19-3
Direct
connections
using stranded
conductors

strip, utility block, and screw terminals. These connectors are often used for loudspeakers, remote-control relay switches, and points of demarcation. In such cases, a solid conductor is wrapped around the screw, or a spade is crimped to a stranded conductor, and the screw is tightened to secure the wire or lug to the connection.

- **Compression termination**   In the compression termination method, a screw applies pressure to a flat plate to squeeze the wire and hold it in place. It uses a captive screw connector, as depicted in Figure 19-4.

- **Soldering termination**   This is a method of joining metal by employing heat and a metallic alloy. Here are some connectors that usually require soldering:

  - The XLR connector, shown in Figure 19-5, is used as a standard low-impedance connection for balanced audio circuits and other applications. These connectors can have three to seven pins.

Figure 19-4
A captive screw
compression
connection

Figure 19-5
An XLR
connection

- The phono or RCA connector, shown in Figure 19-6, is used for interconnecting video and unbalanced audio component, as well as a variety of other applications.

- The phone or 1/4-inch connector, shown in Figure 19-7, is available in either a tip, sleeve (TS) or tip, ring, sleeve (TRS) configuration. It is used primarily for line-level audio signals.

  Different connectors may require slightly different soldering techniques to create a proper termination. Figure 19-8 depicts the process of soldering a wire to part of a phone connector, using a pencil-type soldering iron and solder.

- **Crimping termination**   Applying pressure with a hand tool enables a connector to be properly attached to a cable. Crimping is a permanent action.

Figure 19-6
A phono or RCA
style connection

Figure 19-7
A phone or
1/4-inch
connection

Figure 19-8
Soldering a wire
to a phone or
1/4-inch
connector

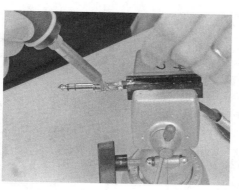

- **Linear-compression termination**   The cable is inserted into a special linear-compression connector and hand tool. A BNC connector, such as the one in Figure 19-9, may be used for interconnecting video and RF equipment. You can secure the AV signal cable in a BNC connection using linear compression, as depicted in Figure 19-10. Linear compression is a permanent action.

- **Insulation-displacement termination**   This type of termination requires a special hand tool with a slot for an insulated wire. As the wire is pushed through the slot, the insulation is cut, and a metal strip makes contact with the conductors. The pressure displaces the insulation and tightly grabs the conductor. You can use this method with solid or stranded conductors, employing any of a number of hand tools, each specific to its application. The ubiquitous, eight-pin, eight-conductor modular connector, the 8P8C or RJ-45 (shown in Figure 19-11), is terminated using insulation displacement. Such connections are created using a compression tool that displaces the cable insulation to create the termination, as depicted in Figure 19-12.

Figure 19-9
A BNC
connector

Figure 19-10
A BNC
connector
terminated to
coaxial cable
by linear
compression

Figure 19-11
An 8P8C or
RJ-45 connector
terminated by
insulation
displacement

Figure 19-12
An 8P8C and
6P6C insulation
displacement tool

Evaluating and Testing Terminations

For quality control, you must evaluate and test all terminations. It is more cost-effective to test the quality of a termination after it has been completed than after the entire system has been built.

You should visually inspect terminations for the following:

- Stray wires that may produce short circuits
- Use of the correct wires at the appropriate contact points or pins
- Excessive or insufficient solder
- Sharp protrusions, extensions, or evidence of foreign matter

CTS-certified professionals should use the appropriate tool to test terminations for continuity or shorts. A multimeter is a common testing tool for many types of terminations (shown in Figure 19-13). The 8P8C modular connector has a specific device to test pin-to-pin continuity (shown in Figure 19-14).

Proper termination, regardless of the type or method, requires the AV professional to understand and adhere to specific procedures, use the correct connectors and tools for the wire and connection type, and follow appropriate safety procedures. Termination is a key element of the AV technician's skill set and requires practice to master.

Figure 19-13
A multimeter

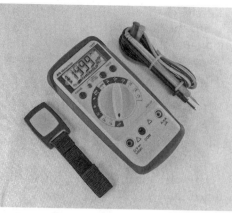

Figure 19-14
An 8P8C
continuity tester

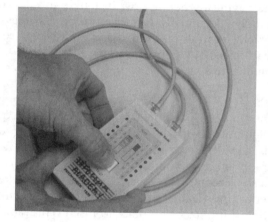

Rack Building

All the cables in an AV system eventually lead to electronic equipment. The equipment is often located in a rack. An *equipment rack* is the skeletal framework where system components are housed, arranged, and interconnected. As with cable termination, AV installation technicians spend a great deal of time building racks, dressing cable, and troubleshooting.

Rack layout and installation is a process of evaluating options that meet the client's needs. You should not mount equipment haphazardly in a rack. Here are some considerations for rack layout:

- How much heat a piece of equipment emits
- How the heat will be managed in the rack
- The types of signals present on interconnecting cables
- Whether the equipment installed in the rack has interface components that a user must be able to access
- The weight distribution of components in the rack

The AV system designer usually lays out a rack configuration for the AV technician to assemble, including the number of rack units (RUs) necessary for each piece of gear. However, it is important for the AV technician to understand the basic design concepts and techniques associated with rack assembly.

Rack building consists of three distinct tasks:

- **Building the rack itself**   Assembling the frame, shelves, panels, cable management hardware, power distribution, etc.

- **Installing the equipment**   Loading and securing the equipment in the correct place.

- **Wiring the hardware together**   Creating the interconnections that enable the system to function.

AV designers typically define how installers should arrange equipment in a rack using system design documents, such as elevation drawings or layouts and system wiring diagrams. A rack-elevation diagram is a graphical representation of the rack and the location of each piece of equipment in the rack. Figure 19-15 shows an example of a rack-elevation drawing.

Ergonomics

AV equipment should be positioned in a rack based on a typical user's need to access the equipment. Equipment with user-interface components (such as optical media players) should be placed within easy reach. Devices that do not require frequent access can be placed at the top or bottom of the rack.

You may also be required to locate equipment in a rack so that disabled persons can access its interface. Any feature of the equipment that normally requires user interaction (such as a control interface or a place to insert removable media) should be between 15 to 48 inches (380 to 1220 mm) above the floor. This limits the space available for accessible equipment, depending on the system design, to about 33 inches (840 mm) per rack. In some cases, additional racks may be required.

Weight Distribution

You should place equipment in a rack in a way that takes into consideration the weight of the equipment. Basically, you want to avoid a top-heavy configuration, which may result in an unstable rack. Unless the equipment rack is fastened in place, the best place for heavy components is at the bottom of the rack. If you place them near the top of a free-standing rack, the rack could tip over.

Most equipment is supported entirely by the rack screws in the front-mounting rail. Heavier equipment typically occupies two or more RUs, providing more support. However, some equipment may need the additional support of rear mounting rails. This is especially true with heavy pieces that occupy only one or two rack spaces, because the weight of the equipment can cause considerable torque on the front panel.

Figure 19-15
A rack-elevation drawing

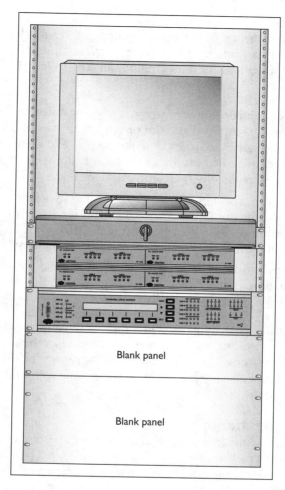

Blank panel

Blank panel

RF and IR Reception

Many racks contain transmitters and receivers for communicating with remote devices using RF or IR signals. For example, it is common for an AV system design to include a rack-mounted RF wireless microphone receiver. The antennas for the receiver must be located so they can receive the RF signals from the transmitter. If an antenna is located in the back of a metal rack, the rack itself acts as a shield, greatly reducing the RF energy before it can reach the antenna. Some manufacturers design rack-mounted receivers with antennas that face front. Alternatively, you may be able to locate the antennas away from the rack, closer to where the transmitters will be used.

IR receivers must be within line-of-sight of transmitters or else be equipped with a dedicated relay emitter.

If a rack has a metal door and wireless receivers are rack-mounted, it will interfere with RF and IR signal reception. Generally speaking, a transparent acrylic door will not interfere with these signals.

Heat Loads

The amount of heat that electronic equipment generates is related to its power consumption. Racks contain heat-generating AV equipment mounted close together, so they require a method for removing hot air. The following are the three main ways of cooling a rack:

- **Convection cooling** This type of cooling relies on the fact that hot air rises to move heat upward and out of vents at the top of the rack. The vents must be large enough to accommodate the output, and there must be a means for cool air to get in at the bottom. Convection cooling is suitable only for racks that produce a relatively small amount of heat.

- **Evacuation cooling** This type of cooling uses fans to draw hot air out of the rack, usually through top vents. Cool air is passively drawn into the rack from the bottom vents to replace the hot air. Also, some electronic equipment has vents in the chassis so hot air can escape. It is important not to cover such vents.

- **Pressurization** This type of cooling uses an approach that's the opposite of evacuation. It uses fans to suck cooler air into the rack, and vents at the top of the rack to provide an exhaust for the hot air.

You should also manage heat in a rack by separating equipment that produce a lot of heat and/or placing the equipment that produces the most heat near the top of the rack. Consider using vented panels placed at various positions in the rack to increase airflow. You may also use blank panels to separate components and allow for sufficient airflow inside the rack.

Signal Separation

As shown in Figure 19-16, *signal separation* is the process of grouping equipment and cables within a rack according to function and signal type in order to reduce crosstalk among cables resulting from electromagnetic interference (EMI). For example, a typical rack may include cables carrying low-voltage microphone-level signals (such as 0.001V) and cables carrying loudspeaker level signals (such as 30V). Maintaining physical separation between cables carrying different signal levels can help to maintain the integrity of the signals.

Rack Assembly

Rack assembly is often best performed in the AV installer's workshop, where the technician has access to components, supplies, tools, and other experienced personnel. The basic process of rack assembly includes the following steps:

- In some cases, the first step in building an equipment rack is actually assembling the rack itself. It may come in many pieces that must be put together prior to placing electronic equipment inside.

- Install all rack mechanical accessories. These can include power and grounding strips, horizontal and vertical cable management systems, casters, and fans.

Figure 19-16
Signal separation
in an equipment
rack

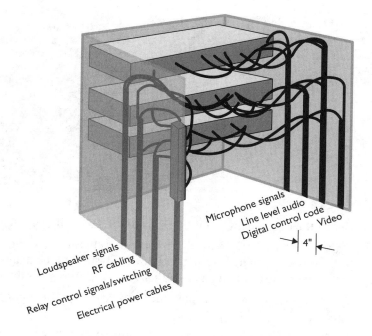

Microphone signals
Line level audio
Digital control code
Video

4"

Loudspeaker signals
RF cabling
Relay control signals/switching
Electrical power cables

- Assemble the equipment to be mounted in the rack.

- Document any information, such as equipment serial numbers, to include in manuals you will provide to the client. Some serial numbers or product labels can be difficult to locate after the equipment is in the rack. Therefore, it's important to gather this information before loading the rack.

- Install mounting ears on equipment where necessary. Mounting ears are metal brackets you attach to an AV component so that you can mount it in a rack. They could be optional accessories purchased from the equipment manufacturer, or they could be included with the equipment. Some AV equipment has built-in mounting ears.

- Identify the intended locations for specific components and install them in the rack appropriately. Keep in mind that attempting to hold individual pieces of equipment in place while installing the mounting screws can be hard work. A common best practice is to lay the rack on its back on top of a protective sheet so that the rack is not scratched or damaged. You can then insert the equipment from above, as shown in Figure 19-17.

TIP If you must insert equipment in an upright rack, have a colleague help, because it is difficult to hold equipment steady while securing the rack. Fill all rack spaces with equipment, blanks, vents, or power strips, so that it matches your rack-elevation drawing.

Figure 19-17
Installing
components
in a rack

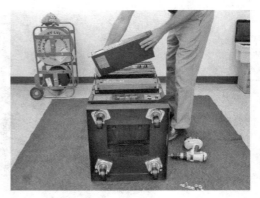

- Fasten the equipment mounting ears to the rack rail using screws.
- Make all internal rack-wiring connections between equipment inputs and outputs and organize the cabling. Use the AV system block diagrams to determine the signal path through a system. This step requires several additional tasks:
  - Select the appropriate cables and connectors for the equipment and signal type.
  - Carefully consider placing cabling runs so that they rest gently, without any tight bends, and are grouped in a manner that will maximize signal separation.
  - Ensure proper cable lengths. You may want to provide extra cable length so that the user (or your company's support staff) can move components for maintenance, but try to keep cable lengths close to the minimum necessary to reduce the potential for signal degradation.
  - Terminate cables using the correct technique for the termination method.
  - Label cables clearly.
  - Tie off cabling, initially using temporary strips, and then using permanent cable ties.
  - Test cable continuity.
  - Attach cables to the equipment.
- Install the AC power components defined in the rack design documents and applicable codes and standards. This step may include configuring the system to access more than one AC power circuit, proper placement of transformers (to allow for service access and minimize electrical interference), and ensuring proper grounding both for safety and to minimize the potential for hum.

When it's finished, a properly configured AV rack should appear neat and professional, both inside and out. All interior cabling should be routed in a neat and orderly fashion, and properly secured to the rack's cable management infrastructure.

Cable Handling and Pulling

In addition to racks, AV systems typically include components located throughout a room, or several rooms in a building, and most of them will be connected with cabling. CTS holders must be able to install cabling in a manner that meets applicable codes and regulations, and ensures proper system operation. They should also ensure the cabling remains as unobtrusive as possible. AV installers must understand the requirements for properly handling and installing cabling according to the approved system design.

Technicians install cable along a path from source to destination, usually through building conduit, junction and pull boxes, or in cable trays. Typically, electricians install conduit, trays, junction boxes, and pull boxes where they are needed in a building, while AV technicians use the infrastructure to install cables.

As Figure 19-18 illustrates, conduit connects to junction and pull boxes at various locations. Junction boxes are installed at the ends of conduit runs. Pull boxes are installed at intermediate positions along the conduit run so that technicians have access to pull cable through the conduit. Pull boxes should be specified for every 100 feet (30 meters) of conduit, or after 180 degrees of bend in the conduit.

At its ends, a cable is terminated to a panel in a wall, to a floor box, or directly to AV equipment. If the termination is at a wall or floor box, it may actually be to a socket for connecting another cable (such as for a microphone) or to a switch or control panel.

Conduit and Other Cable Support Systems

Conduit can refer to any pathway, but in the AV and electrical industries, conduit is a circular tube that houses cable. Polyvinyl chloride (PVC) conduit offers no protection against EMI, so it should not be used for AV cabling. Metal conduit, shown in Figure 19-19, offers a level of protection from EMI. Additionally, the thicker the steel tubing, the more protection it offers from external noise. In jurisdictions where using conduit is not required, it is still best practice to use it because of the physical and signal protection it affords.

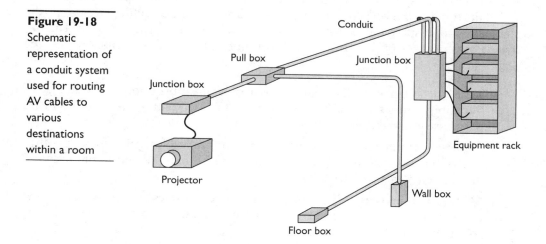

Figure 19-18

Schematic representation of a conduit system used for routing AV cables to various destinations within a room

Figure 19-19
Various types of
conduit

AV systems can use a range of metal conduit, including the following:

- Electrical metallic tubing (EMT) is probably the most popular type of conduit. It is called *thinwall* because the metal is not very thick. It offers minimal protection from EMI and minimal physical protection because any heavy pressure will crush it. You should not use EMT in an environment where the cabling will take a lot of abuse, such as a gymnasium or theater stage.

- Intermediate metal conduit (IMC) is heavier than EMT, and has a threaded metal end to join pieces together with couplers. It offers much better EMI and physical protection than EMT.

- Flexible metal conduit, called "flex" for short, is typically connected to a movable piece of equipment. For example, flex may be used to house cable for a document camera on a cart or trolley that may be pulled away from a wall.

- Rigid-metal conduit (RMC), also called "rigid," is the heaviest conduit and offers the best physical and EMI protection, making it the best choice for electronically noisy and/or harsh physical environments. It comes with prethreaded ends. Because RMC cannot be bent on-site, bends require the use of appropriately angled elbow couplers.

A conduit's internal diameter determines how much cable will fit inside. The maximum amount of space (cross-sectional area) that cables can occupy in a conduit is called the *permissible area*. Permissible area is typically described in fill percentages (the share of conduit space the cables may take up) and will vary by jurisdiction. The following are examples of permissible areas within conduit for one to three cables (see Figure 19-20):

- **1 cable = 53 percent** If you are running only one cable through a conduit, that cable's cross-sectional area may take up to 53 percent of the conduit's internal cross-sectional area.

- **2 cables = 31 percent** If you are going to run two cables inside a conduit, the sum of the two cables' cross-sectional area may be up to 31 percent of the conduit's internal cross-sectional area.

- **3 cables = 40 percent** If you are going to run three cables through a conduit, the sum of the three cables' cross-sectional area may be up to 40 percent of the conduit's internal cross-sectional area.

Figure 19-20 Examples of conduit capacity

When pulling cables through a conduit, the technician needs to determine the amount of space in the conduit that the cables will occupy. To do this, the technician determines the cross-sectional area that each individual cable will require and adds them together. The technician then compares the total area with the permissible area of the conduit to see if the planned cable quantity will fit.

Other cable support systems include ducts, cable trays, and troughs. Plastic wiring ducts may have snap-in slots to hold or help comb cable. A cable tray is an assembly of units made of metal or other noncombustible material to provide rigid, continuous support for cables. *Troughs*, also called *wireways*, are sheet-metal wells with hinged or removable covers for housing cables.

Preparing Cables for Pulling

Cable pulling, like rack assembly, requires preparation. Because the installer needs to pull all the cables through conduit at the same time, it is industry best practice for the installer to carefully plan which cables go into which conduit before pulling the cables. Some AV designs include a conduit riser drawing, which depicts the interconnection of all the AV devices in a system. The CTS-certified professional should check the required cabling against the system drawings before proceeding, because making a mistake at this point can be very costly.

Upon arrival at the site to pull cabling, the installer should review the drawings and look around the room for boxes in the floor, ceiling, and walls. Boxes located in the ceiling are often hidden from view, so the installer may need to move ceiling tiles to find them. Wall boxes may be mounted at various heights, as appropriate for the devices that will be connected. Inputs for microphones and computers will probably be near the same height as general electrical and data outlets. Boxes for plasma displays may be approximately 70 inches (1800 mm) above the floor. Loudspeakers are typically mounted higher.

Installers should be aware that empty boxes can look similar. It can be easy to mistake the fire-alarm box for your loudspeaker location, or a networking box for a microphone connection point. So be careful when identifying boxes. Also, follow health and safety regulations and site-specific safety requirements, including the use of PPE.

In most cases, installers must group together and pull multiple cables. A technician may combine and hold together multiple cables using a constricting metal grip, or create a bundle with electrical tape and a pull string, as shown in Figure 19-21. This is commonly known as a *pull-through bundle* (sometimes called a *cable snout*).

When combining multiple cables, ensure the appropriate level of signal separation by providing physical distance between cables that carry different signal levels. This helps to protect weaker signals from crosstalk. And because of signal-separation requirements, it's important that the installer, when grouping cables together in a bundle, ensure all cables in that bundle are for only one signal type. Generally, group cables according to these categories:

- Microphone
- Line-level audio and communication
- Video
- Control and data
- RF
- Loudspeaker
- AC power

Above all, keep microphone cables separate from others. Microphone cables are the most susceptible to interference because microphone signal level is so low.

Labeling cables before installation is also important because it helps you track the cables and make sure each gets where it needs to go. A label, like the one shown in Figure 19-22, ensures each end of the cable is terminated with the correct connector and plugged into the correct input or output. Once the cables are pulled and terminated, you will apply permanent labels. Permanent labels make it easier to service the AV system.

Typically, the system engineer will preassign cable-identification nomenclature. This would be part of a cable run sheet. However, you may be called on to create the labeling scheme in the field.

With cable preparation complete, you can begin pulling cable.

Figure 19-21

Creating a
pull-through
cable bundle

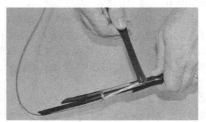

Figure 19-22
Cable label

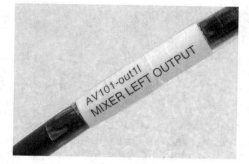

Pulling the Cables

Once you have identified the conduit route, you need to employ some type of tool, such as fish tape, to pull the cable(s) from one end of the conduit to the other. When multiple turns are involved, you will usually pull cables in steps, using the pull boxes along the conduit path.

Fish tape is long, flexible, metal or fiberglass tape, spooled like measuring tape, which AV technicians use to pull cable through conduit. It comes in a variety of lengths and has a hook on the end. First you feed, or "fish," the tape into the conduit from the far end until it reaches the end where the cable(s) will enter. Then you attach a pull string (or the cable itself, in simple situations) to the fish and pull the cable or cable bundle through the conduit. Roll up the fish tape immediately when you've finished. If possible, roll the tape back onto the spool as you're pulling it through the conduit because large amounts of loose fish tape can be hard to handle.

When necessary, you can use lubricants to reduce friction when pulling cable. If you are using cable lubricant, apply it constantly to the cable as you pull.

Be aware that it is possible to apply harmful stress to cables when pulling. Pulling tension is the force used when handling cable. Applying too much force can stretch a conductor, changing its electrical properties. Larger cables can typically handle more force.

While good cable specifications may indicate acceptable pulling tension, the number is not as important as experience. An experienced AV professional will learn, through practice, how to correctly estimate the pulling tension of a cable.

A cable's *bend radius* refers to the degree to which a cable can bend before its attributes change. Exceeding the cable's specifications can cause signal attenuation or permanent damage to the cable. Careful attention to bend radius is most important when using fiber-optic and coax cables. The tightest bend radius you can achieve without damaging the cable is usually about four times the cable's diameter.

NOTE When pulling cables, pay extra attention to bend radius. If a cable requires effort to bend, you're probably bending it too much. If you notice the cable changing shape, a wrinkly cable jacket, or bulging due to a distortion, the cable has definitely bent too much and should be discarded.

Best Practices for Pulling Cables

The following are some cable-pulling best practices recommended by AV installers:

- Observe safety precautions and safety code requirements.
- Coordinate with other contractors so that you can install plenum cable after electrical wiring, air ducts, and ceiling supports are in place, but before ceiling tiles. It can be hard to pull cable in a plenum environment with 2×2 ceiling panels while moving along access equipment.
- When you can, pull cables from their spools, rather than cutting them to length. This is the best way to ensure you have the necessary length of cable needed for terminating each end. It makes cable management easier and helps prevent tangles and cable damage. Take extra spools of cable with you to the job site.
- Use a cable spool/drum holder. This will keep your cables neat and orderly and allow for proper dispensing of the product. Do not lay the spool/drum on its side to remove the cable, as this will cause kinking, making it hard to pull, and may damage the cable.
- It's often easier to deliver the cable from the bottom of the cable spool.
- Consider color-coding cable to help maintain separation when you are bundling cables. Color-coding requires stocking more cable, but it can speed up installation of a big bundle. You should also always label cables.
- Calculate extra cable length. When you cut cable to length, verify how much cable is needed, and then add extra for proper termination and/or error.
- Maintain continuous communication between the technicians at both ends of a cable pull. Report progress and potential problems to each other via radios or telephones.
- Feed the cable into the conduit as straight as possible. Avoid twisting cable during installation.
- Avoid having the cables cross over each other when they enter the conduit.
- When cable gets jammed in a conduit or bunches up, pull it back toward the spool.
- If you're pulling the cable through a pull box, leave some slack. Be sure to identify the cable at the pull boxes to make sure it is your cable.
- Do not pull the cable too quickly. This can damage the jackets.

- When cables cross paths, they should do so at 90 degrees.

- Try to run low-level signals by the most direct route because they are easily susceptible to interference and loss.

- Maintain a clean work space. Make sure the conduit is free of debris prior to pulling and that all sharp edges on boxes and other cable contact points have been removed.

Mounting AV Components

Mounting AV components to the building's structure is a critical part of an installation technician's job. Installers are often required to permanently mount devices, such as flat-screen displays, projectors and screens, and loudspeakers to walls and overhead structures in a client space. Devices such as flat-screen displays can weigh more than 100 pounds (45 kilograms) and are often mounted above an audience, creating a potential safety issue. Installers must therefore understand how to safely and securely mount components to building structures.

The mounting location and approach are often defined in the AV design documents, based on a site survey completed earlier. The designer selects a mounting location based on the needs of the viewer or listener and the manufacturer's installation specifications.

The AV design documents may define a specific mounting approach, especially in locations that require special components to safely secure heavy equipment to the building structure. In cases where the AV installation technician is taking the role of designer, the technician will use typical mounting techniques and components to mount the equipment in the identified location. If there is a question about the safety of a mount, consult a professional engineer.

Hardware Load and Weight

The hardware used to mount components must meet all applicable building standards and local codes regulating composition, quality, and durability. Standards organizations such as the American Society for Testing and Materials (ASTM), International Organization of Standardization (ISO), and Society of Automotive Engineers (SAE) rate the strength and quality of mounting hardware such as bolts or fasteners. This rating helps installation technicians select appropriate mounting hardware.

Equipment and hardware intended for wall or overhead mounting have load limits. The load limit is the weight at which the mounting system will structurally fail. Part of the load limit is the safe working load (SWL), or a similar rating called the working load limit (WLL). This weight must not be exceeded. The load includes the weight of the equipment and all mounting hardware.

When assessing the weight of equipment that needs to be mounted, it is best practice to multiply the published design weight by five. If a loudspeaker and associated mounting hardware weigh 100 pounds (45 kilograms), you should use mounting criteria that will handle 500 pounds (225 kilograms). Five is considered the safety factor. (The *safety factor* is sometimes referred to as the *design factor*, *load factor*, or *safety ratio*.)

When two flat surfaces are fastened together, one takes on the integrity of the other. They must, therefore, be mated flat and tight. A CTS-certified professional should understand the mechanics and physics of stresses and loads.

Mounting Malfunctions

There are many ways a mount can malfunction. Here are some examples of what can go wrong:

- Improper calculation of loads
- Shear load, in which gravity pulls the mount down the wall, shearing off the bolts like a pair of scissors (see Figure 19-23)
- Tensile load, in which the load stretches the mounting fastener (see Figure 19-24), causing soft, inferior grade fasteners (like bolts) to stretch, distort, or break
- Pull-out, which refers to the fastener pulling out of the structure, wall, or ceiling from which it is mounted (see Figure 19-25)
- Shock loading, which refers to a sudden additional load, such as from an earthquake or a person hanging from a mount
- Bolts placed at the top of the mount carry more stress than those placed at the bottom (since not all fasteners carry the same the load in a mount), although following the manufacturer's instructions for placement should avoid this problem

Figure 19-23
Shear load
forces on a bolt

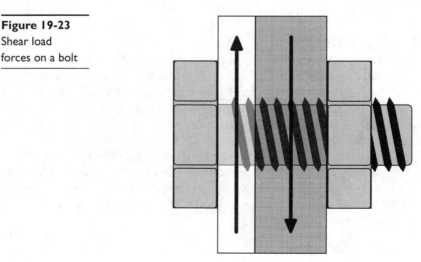

Figure 19-24
Tensile load
forces on a bolt

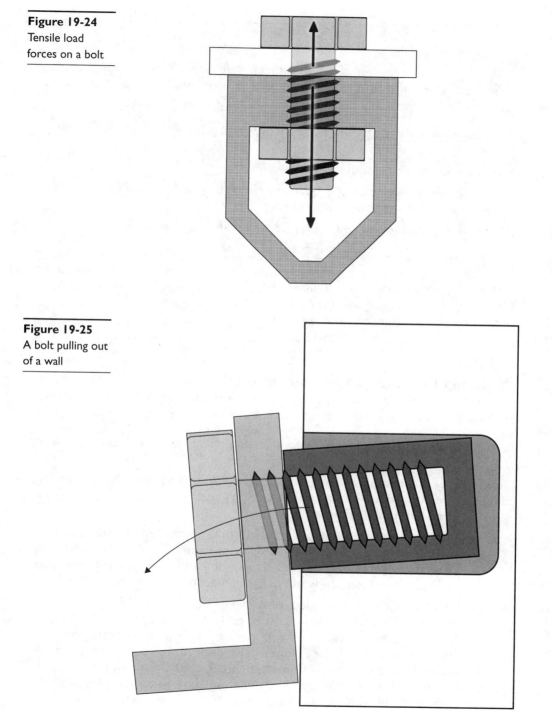

Figure 19-25
A bolt pulling out
of a wall

Mounting Safety Tips

Safety first! It is always important to define the elements of the mounting solution you plan to employ and to use rated hardware and components. Then do the following:

- Review the size and weight of what you will be mounting, including all brackets, mounting components, and hardware.
- Verify the exact mounting location.
- Inspect the structure to which you will be mounting the AV component and assess any potential problems.
- Review and follow all manufacturer guidelines and drawings.

A structural engineer should review all mounting plans and advise on any problem situations you may encounter. You should also have a licensed engineer sign off on any custom or modified mounting components you intend to use. When in doubt, always ask for help. Keep in mind that if you modify another contractor's work, you become responsible.

Preparing to Mount an AV Component

The general process for mounting AV equipment is similar, regardless of the equipment itself. To prepare for a standard installation, make sure to do the following:

- Determine the proper location for the component from the designer's drawings and specifications.
- Analyze the feasibility of mounting at the location, including locating any pre-installed structural blocking, cable pathways and terminations, and designated electrical and/or data outlets (if required).
  - If the AV design stage did not include a site survey to evaluate the structural integrity of the proposed mounting location, determine if the mounting location is suitable for the intended mounting solution. Consult a professional engineer if you have doubts about the strength of the structure.
  - If the installation is not based on a detailed site survey and system drawing, you, as an AV installer, may be called on to make decisions in the field. An installer must decide on the proper mounting method by inspecting the structure to which the mount will be fastened, and verify that the building structure will properly support the total weight of the equipment and its mount. The installer must also determine how to route cabling to the equipment, provide for any required electrical supply and/or data feeds, and evaluate issues such as how a service technician can gain access to the device for maintenance.

- Familiarize yourself with the mounting hardware and procedure. For example, a typical flat-panel display mount comes in two separate pieces: the wall bracket and the equipment bracket. These come together to hold the flat-panel display on the wall. Installers should use only premanufactured mount assemblies and should not attempt to create custom mounting devices in the field.

- Make a list of the tools and materials needed to mount the equipment at the determined location. Be sure you have everything before you start.

- Secure the area around the construction zone for safety. Clear ample space in which to perform the installation. Construct personnel barriers and remove furniture that is in the way or could be damaged. Cover other furniture or equipment to protect it from dirt, dropped material, and so on.

Each type of AV equipment has specific installation and mounting requirements. Here, we'll look at the installation of some common components.

Mounting Projectors

An AV technician must carefully position and align a projector relative to the projection screen, as shown in Figure 19-26. Use the information supplied by the projector manufacturer to determine if the desired mounting position is within range of where it should be located for correct display. The manufacturer's information will include measurement distance (or range of distances for a zoom lens) from the projection surface to a reference location on the projector. It will also provide information about the correct position for the projector relative to the top (or center or bottom) of the screen.

There are four possible orientations for a projector:

- A projector can sit upright on a table or cart in front of the projection screen.

- A projector can be inverted and mounted to the ceiling (on the same side of the screen as the audience). Figure 19-27 shows a technician installing a mounting bracket on a projector for mounting it to a ceiling.

- A projector can be located behind the screen (in a rear-projection configuration), in *either* a bottom (floor-mounted) or top (inverted) orientation.

Figure 19-26
Adjusting a projector

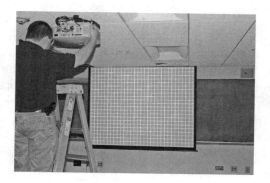

Figure 19-27
A projector with
a ceiling mount

Once a projector is physically positioned a certain way, it must be adjusted so it projects an image that appears in the correct orientation. You typically adjust projector orientation via a menu accessed through the projector's control panel.

Mounting Projection Screens

Projection screens should be located and mounted so that they will not cover any alarm indicators, thermostats, clocks, light switches, or power and data outlets. Additionally, a retractable screen should hang flat, clear of any wall-mounted furniture, marker trays, and other items in the area.

Electrically operated roller screens have limit switches, which are physical switches that turn off the motor. Make sure you follow the manufacturer's instructions for setting the up-limit switch, which prevents the screen from retracting too much, and the down-limit switch, which limits how far the screen extends.

Mounting Loudspeakers

Loudspeakers are available in models designed for mounting directly to a ceiling, or in enclosures with manufacturer-installed mounting points that enable them to be surface-mounted to an overhead structure or wall. When mounting loudspeakers overhead, use only loudspeakers with enclosures designed for that type of application.

Best Practice for Mounting Projectors

When mounting a projector, check the structure for vibrations by resting your hand on the surface to which the projector will be mounted. Sometimes, the HVAC unit is located on the roof above, creating vibrations that will transfer to the projector, causing images to appear unfocused. If you detect vibrations, you'll need to find a solution.

It's a good idea to check for structure-borne vibrations at the time of day when the AV system will typically be in use and with all normal building systems operating.

Mounting Flat-Panel Displays

Flat-panel displays are typically mounted to walls, as shown in Figure 19-28, or hung from an overhead structure using a mounting bracket. These displays can be heavy, so it is important to ensure that the surface on which they are mounted is strong enough to support the weight of the display and its mounting hardware.

Wall blocking by the builder is often required for mounting heavy displays. Blocking is the process of fixing heavy wood filling (or blocks) across two or more studs in the wall cavity prior to applying wall material and finishes. That way, the wall structure is strong enough to support the weight of the equipment. Blocking should be handled before the wall finish is applied, because doing so after may incur extra cost to take apart and repair the wall.

NOTE Blocking can be accomplished only through planning and coordination among the AV designer, installer, architect, and builder. This illustrates the importance of paying attention to the installation needs of an AV system *before* construction has begun. Ignoring such needs can be messy and costly.

Finalizing Mounting

Once you've mounted your client's AV equipment, finalize it as follows:

- Connect the appropriate signal, control cables, and electrical power (if required).
- Adjust the equipment to operate properly within the user's environment.
- Thoroughly clean up the site, including these tasks:
  - Remove all construction debris.
  - Remove all packing material from the client's premises.
  - Retain all installation and operational instructions for inclusion in the system documentation package.

Figure 19-28
Mounting a
flat-panel display

- Replace all equipment and furniture.
- Pack up and return all tools and leftover materials to where they belong.
- Remove personnel barriers.

Setting Up and Configuring AV Control Systems

Modern network-based control systems are powerful and complex. Control system installation follows preparation guidelines that are similar to those for other audio and video systems. However, the complexity of control systems requires special attention to system details, including functional design, signal-flow drawings, IP or other addressing information, and other technical documentation. In order to install the system correctly, the AV installer will often need to consult with the software programmer, system designer, and IT professionals involved in the project.

All AV control systems must be accompanied by a design that includes a button-by-button operational specification for all required touchpanel(s), as shown in Figure 19-29, and other user interfaces. As described in earlier chapters, the AV team should consider the users' needs and work with the programmer to identify how devices will be used and how they will connect to and interact with the control system.

AV professionals should do as much of the control system configuration and testing in the workshop as possible. Troubleshooting a control system on a job site can be very challenging. Verifying settings and the correct operation of the control system in the workshop allows you to interact with everyone involved in the project. You can verify touchpanel designs and troubleshoot issues in less time when configuring a control system in the workshop. Ultimately, there are fewer disruptions for programmers to work around, and clients get a functioning control system in a shorter period of time.

However, due to project deadlines and the need to test connections with the client's equipment and systems, working at the shop is not always possible. An AV professional will need to configure and test some items on the job site, although an efficient installation plan can keep this to a minimum. And when outside contractors are hired to program the system, close coordination is essential.

Figure 19-29
A control system touchpanel

Termination for Control Systems

To correctly terminate control system wiring, the installer needs technical documentation indicating which connector(s) to use, pinout and gender for the connector(s), and the type of signal the system will carry. Some protocols are associated with specific connectors, but that's not always the case. Improperly terminated pinout connectors can ruin an installation and compromise a design through loss of functionality.

Common signal types and related terminations for control systems include the following:

- Ethernet is a unique, flexible, bidirectional signal type. Wired Ethernet connections typically use either twisted-pair Category cable, terminated with an 8P8C or RJ-45 connector, or fiber-optic cable. Ethernet signals may also be transmitted wirelessly via RF. The base standard for wireless Ethernet connections, commonly known as Wi-Fi, is IEEE 802.11, with the specific version denoted by a suffix letter, which is usually b, g, or n.

- RS-232 is normally terminated with a DB-9 connector. It is often bidirectional and uses an unbalanced connection. You can use RS-232 to send commands from a control system or a computer to an AV device and receive feedback (responses) from the device. Bidirectional signal flow is preferred over unidirectional flow because it enables equipment to provide confirmation of its status and/or that an action has been performed. You need to configure RS-232 for proper speed and other connection parameters, which depend on the connected device.

- RS-422 is usually terminated with a DB-9 or DB-25 connector and uses a balanced connection. Operating over a balanced connection enables RS-422 to work over greater distances than RS-232 because balanced circuits benefit from common-mode rejection.

- RF is used mostly for user interfaces. It is susceptible to interference, so you should conduct on-site tests to confirm its reliability. This signal type can be unidirectional or bidirectional.

- Ramp voltage is an infinitely variable analog voltage for which there is no standardized connector type. It is mainly used for controlling things like projector lenses, pan/tilt/zoom on camera lenses, and volume controls.

- Relays and contact closures are binary controls, which are either on or off. A closed relay (with voltage) or contact closure (no voltage) passes a signal, and an open relay stops a signal. Relays are commonly used to execute simple on/off, open/close, or up/down commands, while contact closures usually monitor the status of something with only two states.

Assembling the Control System

Many devices in the AV industry, such as control systems and videoconferencing codecs, are also connected to corporate networks or other hybrid networks. Most of these devices communicate with each other using the Ethernet protocol and require IP addresses.

Setting up network-based control systems can be a challenge. For networked control signals to find their destinations, each device must have a unique identifier—either an IP address or a DNS hostname. You should work with the client's IT manager to determine which devices can be connected to the network. After that, you will continue to work with the IT staff to assign IP addresses or DNS hostnames to the devices.

Start by providing a list of network-connected AV and control system devices to the client's IT manager for review. The list should include the device, manufacturer, model number, MAC address, software and firmware versions, and intended location. It may also briefly explain why each device needs network connectivity. Leave space on this list for the IT manager to write in the IP address or DNS hostname of each device. Make sure the IT department doesn't have network security or bandwidth concerns with any of your gear. The IT manager will appreciate the advance notice, and you'll be able to work out any issues early.

Once the gear is approved, you can ask the IT manager to assign an address or DNS hostname to each device. If your devices require static IP addresses, ask the IT manager to reserve them and add the assigned addresses to the list. If your devices support DNS, you won't need permanent addresses. Instead, devices will find each other using their permanent DNS hostnames. They will be assigned an address dynamically whenever they need to connect. In this case, ask the IT manager to create a DNS record for each device, and add the domain name to the list of networked devices.

Once you've determined how devices will be identified on the network, you'll need to configure each device to respond to its new IP address or DNS hostname. There are several ways to assign addresses. DIP switches, jumpers, and software are three common methods. When in doubt, refer to the manual for each piece of equipment in the system.

After setting the address, mark the setting on the back of the device for future reference at the job site. Not only will this help if there are several pieces of the same model to install, but it also makes it easy to determine which settings have been established on each device without needing to search through menus.

After you have correctly configured all devices (including defining whether the IP address is static or dynamic), you should verify network connectivity. One way to do this is to open a DOS command prompt and use command-line text to ping the device. This will show if the target is online and accepting active connections. Other command-line prompts can help verify the network status and settings of a device.

To verify the IP address of a control system CPU, an installer can typically connect a computer to the CPU and perform checks per the manufacturer's instructions. Another method may be to select a menu option on the front of the CPU, which will display the IP address on the unit's screen.

Using Control System Software

A control system is similar to a computer: It must include software to operate. You can install hardware devices and wire the system, but until firmware and software are loaded, the system will not function.

Firmware is the CPU's operating system—a bit like DOS, Windows, Mac, or Linux. Most firmware is specific to the make of the control system, but a few use Linux or another multipurpose operating system.

The software is a unique application written for a specific AV installation. Some applications are preprogrammed, and others are programmed especially for the client. There are separate software programs for the controller and the touchpanels/button controllers (the interfaces). Each type of program must be installed on its respective device. The manufacturers of the devices may have proprietary computer programs for installing software on their devices. Refer to the control system's software manual for the installation process.

Don't forget to ask the software programmer if there are any "hidden pages" in a touchpanel GUI. Hidden pages are unseen hot spots on a touchpanel that allow the technician to configure passwords, autopowering, and other settings. Any hidden pages should be included in the documentation provided by the designer. Asking this question can save you time in configuring the system and training the client.

Testing and Installing the Control System

Once each device in the AV system has been inspected and configured, connect the control system and controlled equipment as completely as possible at the workshop. Test each button on every touchpanel for proper operation while referencing the button-by-button design documentation. Also, document any problems and report them to the software programmer or the designer. Any problems you can resolve in the workshop will make on-site installation easier.

It is imperative that the full functionality of the system be well known; otherwise, testing of the system may not be complete. It's possible that the software programmer has added a feature or has not included a feature.

After arriving on site, use the design documentation to confirm the locations of all the control devices in the room. Note any inconsistencies discovered during the installation process for future reference on drawings and/or for making software revisions.

Final Control System Testing

Once you've installed the control system at the job site, test it as follows to ensure proper AV system functionality:

- Verify that all cables are connected correctly and according to the drawings. Modify drawings or connections as needed.

- Power on the control system CPU, peripherals, and user interfaces, plus the attached AV equipment.

- Verify that all appropriate device-selection buttons are available on the user interface(s). For example, if you touch the button for a Blu-ray player, check that the switcher activates the correct input.

- Verify that the correct buttons appear for the operation of each device you select. For example, if you touch the button for the Blu-ray player, then the appropriate controls, such as play and pause, should appear on the touchpanel.

- Confirm that each device action button performs the specified action in accordance with the button-by-button specification.

- Test all wall buttons, control switches, motion and IR sensors, and all other devices that interact with the system. This process can take several hours. Remember that some control buttons may be programmed to perform more than one action.

After you've completed testing and resolved any issues, the client should have a reliable, fully functioning control system that performs as specified in the designer's documentation.

Configuring Networked AV Systems

We've touched on one AV application that commonly traverses an IT network: the control system. But these days, many more AV systems rely on IP and enterprise networks to function. Assuming you've coordinated the design of your AV system with the client's IT staff (see the section on integrating with IT networks in Chapter 16), you may need to take several steps to configure your AV system to operate on the client's network.

Keep in mind, what follows must also be addressed during the design phase, in coordination with the client's IT department. Issues regarding subnetting, IP addressing, VPN and VLAN requests, and LDAP integration will have been hashed out and settled before installation. But during installation, you will need to coordinate with the IT staff to ensure that networked AV devices are configured properly to operate on the network as they were designed to work.

When setting up an AV system to operate on a client's network, work with the IT department to configure subnetting, VLAN, VPN, and IP addressing, as necessary for the particular AV system design. Because networked AV systems are relatively new to AV professionals, it may help to review the network essentials covered in Chapter 6.

Subnetting

As networks get bigger, and the systems attached to them become more bandwidth-hungry, subdividing the network becomes important. Real-time media, particularly video, has an appetite for bandwidth. Fortunately, IP networks have a built-in subdivision mechanism: the subnet. IP networking uses two addresses to identify the system: the IP address and the subnet mask. By altering the subnet mask, a single physical network can be transformed into many logical ones.

Subnetting doesn't reduce bandwidth consumption, but it does help isolate it, limiting broadcast traffic to devices within the same subnet.

NOTE A *subnet mask* is a binary number whose bits correspond to IP addresses on a network. Bits equal to 1 in a subnet mask indicate that the corresponding bits in the IP address identify the network. Bits equal to 0 in a subnet mask indicate that the corresponding bits in the IP address identify the host. IP addresses with the same network identifier bits as identified by the subnet mask are on the same subnet.

VLANs

Sometimes clients want certain users to have access to their networked AV traffic, but not to their other resources. For example, a law firm may want to give its clients access to its videoconferencing system, but not its file servers. In such cases, you need to be able to subdivide your subnets. VLANs provide a means for this subdivision and allow the AV system to leverage an enterprise's existing network infrastructure, while still preserving its bandwidth and network security.

Because VLANs are not related to the network's physical topology, devices on the same VLAN can be located anywhere on the LAN. Also, while a device can belong to only one subnet, it may belong to multiple VLANs. On many devices, a VLAN can be configured with an access control list, which specifies the devices allowed to access the VLAN or the devices the VLAN is allowed to access. The former protects data on the VLAN; the latter keeps devices on the VLAN from being used as a launching point for attacks.

Implementing a VLAN can be labor-intensive because the VLAN broadcast domain must be segmented on every switch in the network. The network administrator must specify the MAC address, port, or IP address of every device on the VLAN, on every switch in the network.

A VLAN is a bit like a "friends and family" long-distance calling plan. Devices that need to "talk" frequently get efficient, unfettered access to one another. Their access to devices outside their VLANs, however, is slower and may be limited or even blocked. Because the communication among devices on a VLAN is typically switched rather than routed, it's efficient. Digital signage, for example, is well suited to a VLAN.

Requesting a VLAN

If you determine that an AV system should be segregated on a VLAN—for its own good or for the good of the network—you'll need to provide the network manager with the following information:

- What VLAN you want to create and why (for example, "I want to create an IPTV VLAN so that the network isn't flooded with streaming video traffic")
- Which devices should be included in the VLAN
- Whether any routing between the VLAN and other network locations should be required or permissible

NOTE A *broadcast domain* is a set of devices that can send Data Link layer frames to each other directly, without passing through a Network layer device. Broadcast traffic sent by one device in the broadcast domain is received by every other device in the domain.

VPNs

Almost any large or physically dispersed organization will require some way to hold managed communication between its LANs. This may also be true of networked AV systems that span multiple LANs, such as videoconferencing systems.

Services that are especially necessary for AV applications—such as QoS, low latency, managed routing, and multicast transmission—are impossible over the open Internet. Customers and service providers also commonly need a secure means of accessing systems remotely for monitoring, troubleshooting, and control purposes. VPNs provide a way.

A VPN may be able to secure a videoconferencing stream between offices, but be aware that using a VPN increases required bandwidth overhead. This is because an encryption and tunneling wrapper must be added to each network packet that travels across a VPN. This additional overhead may not be significant in terms of bandwidth requirements, but it can increase the Ethernet frame size to the point where packets must be fragmented before they can be sent across the network. Packet fragmentation can be disastrous for the quality of streamed video or conferences. Always be sure your frame size is set low enough to account for VPN overhead.

IP Addressing

When connecting an AV system to an IP-based network, you must carefully consider how networked devices will be assigned IP addresses. A control system, for instance, must be able to consistently locate each device it controls.

When installing your AV system, you may be limited to whatever addressing scheme the client already uses. The following are the main IP addressing schemes: DHCP, DNS, and static addressing, each of which has its own benefits and drawbacks:

- **DHCP** This approach allows network administrators to automate address assignment. When a device connects to the network, the DHCP service or server will assign an IP address to the device's MAC address. The pool of available IP addresses is based on the subnet size and the number of addresses that have been allocated. DHCP is easy to manage because no two devices get the same address, relieving potential conflicts. It also allows more devices to connect to the network. However, with DHCP, you never know what a device's IP address will be from connection to connection.

- **Static IP addressing** This approach assigns fixed IP addresses to networked devices. Many AV devices must always have the same address so control and monitoring systems can find them. In the absence of a DHCP server, you will need to manually set the IP address (which you must obtain from the IT manager), hard-coding it onto each device. To do this, you will need the IP address, the subnet mask, the device name, and the DNS server and gateway addresses.

- **Reserve DHCP** This is a hybrid approach to IP addressing. Using reserve DHCP, a block of statically configured addresses can be set aside for devices whose IP addresses must always remain the same. The remaining addresses in the subnet will be assigned dynamically. In order to reserve addresses for AV devices, you need to list the MAC address of each device that requires a static address.

 TIP If the client has a DHCP server, it's better to use a hybrid approach than manually assigning an IP address to every AV device.

- **DNS** This approach uses a hierarchical, distributed database that maps easily remembered names to data such as IP addresses. A DNS server keeps track of all the equipment on the network and matches the equipment names so they can be easily located on the network or integrated into control and monitoring systems. Many AV devices still don't natively or fully support DNS address assignment.

Producing AV Events

AV installers may also be required to set up and configure AV equipment in a temporary setting to support a presentation or performance. This task requires the installer to be able to do the following:

- Transport equipment to the site safely.
- Set up and operate the temporary AV system.
- Remove the system after the event.

Here, we will look at some of the main concerns when installing AV equipment to support a temporary event.

Preparing for the Event

Properly preparing for the event is critical. This includes the following tasks:

- Reference the system designs (sound, visual display, lighting, staging, and so on) and create a checklist of required equipment and cables.
- Allocate enough crew resources with the appropriate skills and credentials/ licensing for the project, as well as other resources, such as transport, additional equipment, materials to be hired, permits, and so on.
- Identify where you are going and where to park when you arrive. You may need to ensure access to a loading dock or service entrance.
- Create a realistic setup, testing, and rehearsal schedule so everything will be ready for the event start time.

Loading and Unloading a Truck

Working with equipment in and around trucks can be dangerous. Be sure to read your company's guidelines concerning safety and follow them.

Loading Equipment

The following are some of the typical safety guidelines when loading equipment on a truck:

- Do not move awkward, heavy items by yourself.
- Bend at the knees when lifting items, and follow proper lifting procedures.
- Follow all safety guidelines when using forklifts or other lifting equipment. Ensure that only appropriately licensed personnel operate these and other types of specialized equipment, as defined in applicable codes and regulations.
- Never leave an item in an unsecured position where it may slide forward and distract or injure the driver. Use load restraints, safe stacking practices, and proper truck weight distribution to prevent the load from shifting during transit.
- Do not leave road cases resting on their wheels. Turn the cases on their weight-bearing side to prevent the cases and the wheels from becoming damaged, and to make the load more stable. If it is not possible to leave a case on its side, use loading bars and straps to secure the case, and lock the wheels in place.
- The use of carts and dollies can make moving equipment safer and faster for everyone. Strap items to carts and dollies to prevent them from sliding off when being moved over bumps and along ramps. Do not overload the cart or dolly.
- Be familiar with the load limits of the truck or trailer you are loading and the total weight of the equipment you plan to load. There are laws regarding the safe operating weight of your vehicle. Know your company's rules to avoid liability issues.

Unloading Equipment

When you arrive at the venue for the event, consider these procedures when unloading a truck (see Figure 19-30):

- Identify the area where the truck will be unloaded and make sure you have approval to unload the truck in that area.
- If you are using a loading dock, use dock plates laid over the gap between the truck floor and the loading dock. You may also use dock lifts that raise or lower the dock to match the height of the truck, or lift-gate platforms installed on the back of a truck for lowering heavy items to the ground.
- If a loading dock is not available, use truck ramps to bridge the gap between the truck floor and the ground, the truck's lift-gate platform (if fitted), or forklifts to bring palletized loads from the truck to ground level. To prevent damage to equipment cases, the loads should be secured to pallets before being loaded into the truck.

Figure 19-30
Unloading a truck

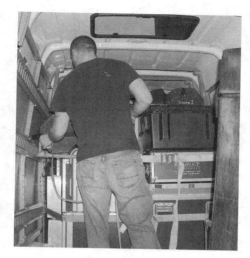

- Inspect the route you will use to reach the event space to make sure that large items will fit through doorways, hallways, and elevators.

- Determine if there are any special requirements to protect flooring, walls, and doors when moving equipment through the site.

Setting Up Equipment

An AV installer may be called on to set up equipment to support the event. This equipment may include pipe-and-drape backdrops, audio systems, regular projectors and screens, folding projection screens, and flat-panel displays.

Pipe and Drape

Pipe and drape is commonly used in an event space as a backdrop for a stage. This arrangement can hide unsightly areas where equipment cases and cabling are stored, add texture and interest to an event, or serve another purpose.

Setting up pipe and drape can be hazardous and requires at least two people (see Figure 19-31). The floor bases are very heavy and can injure you and damage floors. Poles can fall over and strike people and props. It is easy to trip and fall on the drape panels. The safety of all personnel involved should be a priority.

Sound-Reinforcement Systems

Sound-reinforcement systems include loudspeakers, amplifiers, and a mixer to blend and distribute various sources, including multiple microphones, DVD players, Blu-ray players, and computers.

Figure 19-31
Installing pipe and drape

Typical tasks for setting up sound-reinforcement systems include the following:

- Place or hang the loudspeakers on either side of the visual display area, taking care to put them in front of all microphones.

- Place each microphone directly in front of the positions from which persons will be speaking. To help minimize feedback, do not place microphones in the area in front of the loudspeakers. Microphones should be pointed toward the presenters' mouths, so they speak from above the microphone, not directly into it.

- Set up the mixer and outboard equipment on a table at the back of the room so that you can observe all the microphones and hear the mix in the same way as a member of the audience. Label inputs so that you know what each channel controls.

- Connect all cables and power on all equipment except the power amplifier. Set initial signal levels and test all inputs. Turn the power amplifier inputs to zero (fully attenuated) and power on the amplifier. Slowly bring the levels on the power amplifier up until the proper level is achieved.

- When testing and troubleshooting is complete, properly tape down all cables, following the best practices for laying tape over cables on the floor, as shown in Figure 19-32.

Projectors and Screens

You should set up projectors and screens to produce an optimal viewer experience. The size of the screen and the type of lens on the projector determine how far away the projector should be from the screen. This measurement is called the *throw distance*.

Figure 19-32
Taping cables to
the floor

Here are some guidelines for installing projectors and screens:

- Calculate the appropriate throw distance for your specific projector by multiply-ing the screen width by the lens zoom ratio. For example, if your screen is 10 feet (3 meters) wide and your lens ratio is 2.0, the distance from the screen to the projector is 20 feet (6 meters).

- If the projector will be floor-mounted (on a table or cart), check to see that no audience chairs are located immediately next to the projector's location. Projectors may generate large amounts of heat and fan noise, which is distracting for attendees.

- Verify that there is adequate space for the projector. Areas around the projector should have good ventilation and not trap heat. If a projector overheats, it will automatically shut down so it can cool.

- Place the projector table or cart at the desired location, and place the projector on the surface, securing it with a security cable to prevent theft, if necessary.

- Run all necessary cables to the projector's location and connect them. Note that some larger projectors require substantial power, so it may be necessary to use a separate power circuit, if possible.

- Send a test signal to the projector from a computer or video player.

- Turn on the projector. It may take several minutes before there is a usable image.

- Verify that you've selected the correct input on the projector. The projector may have an auto-detect or auto-image button, or you may need to toggle through several inputs.

- When the image is displayed, center and level the projected image on the screen by adjusting the projector's legs. Review the projector setup instructions for infor-mation about specific adjustments.

- Use the projector's zoom and focus features to fill the screen with the displayed image, at least from side to side. If you cannot place the projector in an optimal location, as a last resort, you may need to use the keystone features on the projector to make the image square on the screen.

- Walk into the seating area to see if the image appears to be focused and square from the audience's point of view.

Folding Projection Screens

Folding projection screens have a flexible fabric screen surface, which is snapped or laced to a rigid, four-sided metal frame. There are two types of screen fabrics. One type is used for front projection, when the projector is in front of the screen with the audience. The other type is used for rear projection, when the projector is behind the screen. Because the surface is a stretched fabric, folding screens are extremely flat, so creases should not be a problem.

The screen frame is a single, jointed piece that unfolds to form a rectangle of a fixed size. The frame is mounted on collapsible folding legs. These screens are commonly used with a black skirt and side curtains, giving the appearance of a portable theater environment. Folding screens are available in much larger image sizes than tripod screens.

Flat-Panel Displays

Flat-panel displays are typically set up on a stand or mounted to a vertical surface, such as scaffolding.

Here are some guidelines for installing flat-panel displays:

- Study the area where the flat-panel display will be located and verify that there is no restriction of airflow to the panel.

- Determine the safety of the audience and the equipment if it were placed in the proposed location.

- Ensure that the proposed location is flat and level.

- Verify that the audience can see the screen.

- Determine the possible pathway for running signal and power cabling and locate the designated electrical outlet.

- Assemble the stand or mount.

- Carefully lay the display face down on a flat, clean shipping blanket. Make sure there are no tools or other objects under the blanket to damage the panel.

- Connect the mounting bracket to the display. Use only the premeasured mounting bolts that are specified for attaching to that particular make and model of display. Using screws or bolts that are too long or have the wrong thread will cause damage, potentially destroying the flat-panel display.

- If you are using a base with a vertical post, set up the base following the manufacturer's instructions. If you are mounting the panel to scaffolding, make sure a certified rigger has constructed the scaffolding.

Figure 19-33
Mounting a
flat-panel
display on
a truss

- Mount the base bracket to the vertical post or scaffolding following the manufacturers' instructions (see Figure 19-33). Before tightening the hardware on the base bracket, ensure it is at the correct height and is level.

- With a partner, carefully raise the flat-panel display to the base bracket and connect the two brackets. During this process, avoid putting pressure on the delicate surface of the image area.

- Run all electrical and signal cables to the panel and connect them.

- Using a video test pattern, verify proper operation of the unit, making electronic adjustments as necessary.

- Clean up the area by picking up all tools, equipment, packing material, and storage cases.

Supporting AV Systems

In many cases, the level of expertise required to maintain a complex AV system is beyond the skills and knowledge of personnel in the client organization. In these cases, clients may be interested in obtaining ongoing support from the AV vendor.

AV vendors can typically provide several types of support services:

- **Maintenance and repair**   AV equipment or installation infrastructure may occasionally fail, requiring troubleshooting and repair services to return the system to full operation. The AV vendor that provided the installation services is often requested to supply ongoing maintenance and repair services, since this company has the best understanding of how the system was designed and installed.

- **System updates**   As the needs of the client expand or change, and the capabilities of AV technology increase (or reduce in cost), clients may want to update their AV systems. For some clients, updates can be on a regular, ongoing basis.

Keep in mind that offering support in each of these areas may be important to establishing a long-term working relationship with clients. They will come to view your company as a long-term partner in meeting their AV needs, rather than a one-time design or installation vendor. You will learn much more about supporting clients in Part V of this book.

Chapter Review

AV technical professionals must be able to install a wide range of AV equipment and systems at the client site. This task requires that the AV professional be able to read and understand AV system design documentation, know how to physically install AV equipment, and know how to configure the individual devices to operate as intended within the overall AV system.

In this chapter, you reviewed some of the basic tasks and procedures involved in installing an AV solution in a technically correct and professional manner, following design specifications created for the customer systems. You may need to interconnect devices and fabricate system elements, such as cabling, mounting components, and equipment racks. We discussed setting up and configuring individual AV devices, integrating them with a control system, and configuring them to operate on a network, where applicable. We also covered setting up AV equipment for temporary events and providing support for installed AV systems.

Review Questions

The following review questions are not CTS exam questions, nor are they CTS practice exam questions. These questions may resemble questions that could appear on the CTS exam, but may also cover material the exam does not. They are included here to help reinforce what you've learned in this chapter. For an official CTS practice exam, see the accompanying CD.

1. Which of the following is a sign of an improperly fabricated cable termination?

    A. Visible stray wires

    B. Visible solder

    C. Removed jacket or insulation

    D. Crimping

2. What type of termination does the following picture show?

    A. XLR

    B. 8P8C

    C. BNC

    D. RCA

3. Which of the following tests should be conducted to evaluate if a termination is properly transmitting a signal?

    A. Voltage test

    B. Continuity test

    C. Signal sweep test

    D. Isolation test

4. In order to communicate with devices installed onto a TCP/IP network, what must each device have?

    A. An IP address

    B. A wireless card

    C. A web address

    D. A TCP card

5. What is a drawback of using a rack-mounted RF receiver?

    A. The metal rack can shield the receiver from its intended signal.

    B. It relies on line of sight to the signal transmitter.

    C. It operates on different frequencies than its receivers.

    D. It is sensitive to fluorescent light.

6. Which of the following rack-ventilation methods uses a fan that draws air from the rack?

    A. Convection

    B. Pressurization

    C. Conditioning

    D. Evacuation

7. What does *signal separation* refer to within a rack layout?

    A. Separating cables according to signal strength

    B. Separating components according to signal type

    C. Running several cables carrying the same signal to separate components

    D. Splitting composite video signals into component RGB signals

8. When mounting heavy equipment, always mount to which of the following?

   A. The building's structural support or blocking

   B. Drywall

   C. Ceiling

   D. Any stud in the wall

9. Who should evaluate all mounting plans and advise the installation technician on difficult mounting situations?

   A. Structural engineer

   B. AV manager

   C. Mechanical contractor

   D. Client

10. What is the load limit?

    A. The maximum weight of an equipment rack

    B. The highest intensity a sound system can produce

    C. The weight at which the item will structurally fail

    D. The tendency for an equipment rack to tip over

11. Roof-mounted HVAC systems may produce vibrations that cause the projector to vibrate and cause an image to appear unfocused.

    A. True

    B. False

12. To what does the *permissible area* of a conduit refer?

    A. The locations within the ceiling where the conduit may be run

    B. The outer diameter of the conduit used for specific applications

    C. The amount of the inner diameter of a conduit that may be filled with cable

    D. Conduits must be rated as fireproof when used within ceiling (plenum) spaces

13. What is the main purpose for reviewing the system design when preparing for providing AV support for a live event?

    A. Allows the AV team to brief the presenters or performers on the capabilities of the AV system

    B. Ensures that the system meets building code or regulation requirements

    C. Helps the AV team integrate the live AV support with any broadcasting requirements

    D. Helps the AV installer determine the required equipment and crew resources

14. Why can implementing a VLAN be labor-intensive?

    A. Its broadcast domain must be segmented on every switch in the network.

    B. You may be limited to whatever addressing scheme the client already uses.

    C. It increases bandwidth overhead by adding an encryption and tunneling wrapper.

    D. You will need to manually set IP addresses for each device.

15. Which of the following objectives should the AV team target when locating microphones for a live event?

    A. Ensure that all microphones are behind loudspeakers

    B. Ensure that all microphone cables are no longer than 15 feet (4.5 meters) in length

    C. Ensure that all microphone cables are taped to the floor

    D. Mount all microphones on nonconductive stands

16. How far away from the screen should you place a video projector with a lens ratio of 2.0 to create a 10-foot (3-meter) wide image?

    A. 10 feet (3 meters)

    B. 15 feet (4.5 meters)

    C. 20 feet (6 meters)

    D. 30 feet (9 meters)

Answers

1. **A.** Stray wires that extend outside the termination are a sign of an improper termination. A stray wire can cause a short that will interfere with system operation.

2. **C.** The termination depicted in the picture is an example of a BNC connection.

3. **B.** The AV technician should test continuity of the cable to determine if it is properly transmitting signals or if it is shorted in some manner.

4. **A.** An IP address is used to give each component on a TCP/IP network a unique address that allows components to communicate.

5. **A.** The metal rack casing and components can interfere with an RF signal transmitted from within a rack.

6. **D.** A ventilation approach that uses a fan to draw hot air from inside the rack is called an evacuation method.

7. **A.** Signal separation within a rack refers to grouping cables according to signal strength, so that stronger signals will not create interference on cables carrying lower-level signals, such as power cables interfering with microphone cable signals.

8. **A.** Heavy equipment should be securely mounted, using either the building structural support elements or blocking installed behind the drywall specifically to be used as a mounting support.

9. **A.** Mounting plans for difficult mounting situations should be evaluated by a structural engineer familiar with the building's structural elements.

10. **C.** The load limit rating refers to the weight at which the item will structurally fail. For example, the load limit on a flat-screen mount is provided to indicate the maximum weight that can be supported using the mount.

11. **A.** It's true that HVAC systems can transmit vibrations that can adversely impact the quality of a projected image.

12. **C.** The permissible area refers to the amount of cable that can be run through a conduit, based on the outside cable diameter.

13. **D.** The installation team should review the system design to determine what equipment and materials are needed, along with the staff required to actually set up the system. It will also help the team determine how much time is required for system setup, which will aid in scheduling the installation tasks in a manner that meets the client deadlines.

14. **A.** Implementing a VLAN can be labor-intensive because its broadcast domain must be segmented on every switch in the network.

15. **A.** The AV team should strive to ensure that all microphones are mounted behind the loudspeakers to minimize any feedback.

16. **C.** You would place a projector with a 2.0 ratio lens 20 feet (6 meters) from the screen to create a 10-foot (3-meter) wide image.

CHAPTER 20

Managing an AV Project

In this chapter, you will learn about
- Managing project activities and resources
- Coordinating project activities among various stakeholders
- Providing project documentation

In today's AV industry, project management is a critical skill. In fact, many AV professionals, in addition to pursuing their CTS certification, also choose to study for their Project Management Professional (PMP) credential from the Project Management Institute (PMI). As AV systems grow more complex, more mission-critical to clients, and more integrated with IT systems, CTS professionals need to understand the basics of shepherding a project from start to finish, on budget and on schedule.

A project has five key elements, all of which are driven by the client's purpose:

- Scope, which covers deliverables and activities that must be accomplished
- Time, including its estimation and duration
- Cost, for human and material resources
- Quality, to make sure performance meets expectations
- Risk, including threats, opportunities, and response strategies

The person responsible for managing the AV project must understand how to coordinate the activities of company staff with stakeholders from the client side of the project, as well as with other contractors. The project manager must be able to direct project activities, effectively manage resources, and produce the documentation required for the project. But even if you're not going to be a regular project manager, management principles extend to all AV team members. It is important for you to understand relevant project issues and how to address them.

Domain Check

Questions addressing the knowledge and skills related to managing an AV project account for about 5% of your final score on the CTS exam (about five questions).

Managing Project Activities and Resources

Planning and management are important to the success of any project or business endeavor. You may be able to manage a simple project using an informal list of tasks, deadlines, designated staff, and equipment and supplies needed to accomplish the work. However, complex AV system installation projects or live events require a more detailed project breakdown and schedule for coordinating with other vendors performing work at the site. In these cases, the AV team often will use project management software to plan the work program and ensure task coordination.

In the AV industry, those responsible for managing AV installation projects or live events must be able to properly address a wide range of typical project management issues, including the following:

- Create a work breakdown structure
- Create a logic network for coordination of resources
- Manage the project budget
- Monitor work at the client site

Let's look at some of the issues involved with each of these areas.

Creating a Work Breakdown Structure

As mentioned in Chapter 18, the first step in scoping the labor required for a project is creating a work breakdown structure, or WBS, as shown in Figure 20-1. The objective of a WBS is to ensure that all staff members working on a project understand clearly what the client expects to receive. The functions of the WBS define a product or service, and are always expressed as nouns. A WBS does not address who, when, what, or how much.

The WBS breaks down the work into manageable pieces. It relates the elements of work to each other and to the total service or process. Each descending level represents an increasingly detailed definition of a project component.

The WBS facilitates project assignment and provides a basis for estimates. It also defines a baseline for performance measurement and change management.

Figure 20-1
A general work breakdown structure

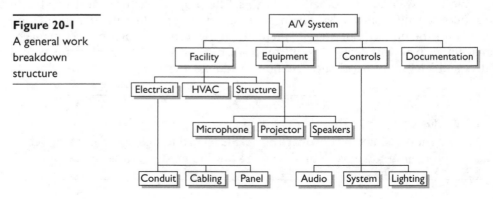

Creating a Logic Network for Coordinating Resources

Since your company will likely be involved with several projects at once, you will need to coordinate your project's requirements for staff, tools, equipment, vehicles, and other resources with other projects that have similar needs. In these cases, it is important to create a schedule that allows you to plan how these resources will be used and to identify potential shortages of equipment or staff that could impact the ability of your company to perform all agreed-upon work. You may need to obtain additional workers or equipment to address any resource shortages.

Once you're created a WBS, you can assign activities and milestones to the project's deliverables. A *deliverable* is an oriented grouping of project elements that will ultimately organize and define the total scope of the project. These elements form the basis for a *logic network diagram*, as shown in Figure 20-2.

The logic network diagram is a tool that helps sequence and ultimately schedule a project's activities and milestones. It performs the following functions:

- Lists activities and their sequential relationships
- Identifies the dependencies and their impacts
- Introduces activities into the network according to their dependencies, not on the basis of time constraints
- Identifies successors
- Identifies predecessors

Identifying Activities and Milestones

The logic network diagram includes the activities and tasks that are required to create the deliverable (see Figure 20-2), as they consume time and resources. The functions of a logic network are expressed as verbs, such as *Build, Test, Fabricate,* and *Develop.* Each activity should have an associated deliverable.

Milestones are key events in the project, such as the completion of a major deliverable or the occurrence of an important event. They can often be associated with scheduled

Figure 20-2

A logic network diagram

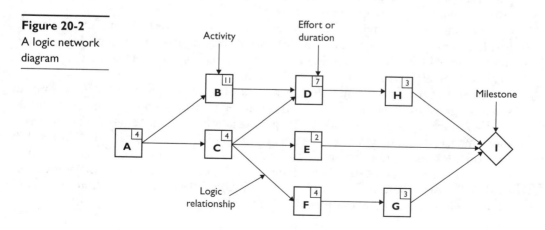

payments, client approvals, and similar events. Examples include *Materials Delivered*, *Racks Programmed*, and *Projectors Calibrated*. Milestones are typically shown as diamonds (as in Figure 20-2) or triangles in the logic network diagram.

Estimating Effort and Duration

Effort is how much time it will take to complete a task in its entirety. Many firms come up with an effort value through perfect-world estimating, assuming a best-case scenario where all needed resources are in place. This is an optimistic practice that usually increases risk and can lead to underbidding the labor component of an installation.

Mature organizations look at their resource requirements across all of their current projects before establishing timelines. Immature organizations make promises to clients before they have confirmed the availability of their resources. It may take only a few minutes to replace a faulty component, but if a technician must drive across town to get the component, those few minutes turn into hours.

Duration is how long it will actually take to complete a task (not just the time devoted to the task). It reflects a combination of the following:

- Effort (number of work periods)
- Resource availability (other projects, operations, and so on)
- Number of available resources (more is not always better)
- Personal/organizational calendars (weekends, holidays, and so on)

Managing the Project Budget

Each project should be based on an agreed-upon contract with the client for work and materials. Price should never equal cost. The fee that the client pays for a project is based on the cost to the company to provide the needed labor, supplies, and equipment, plus a markup that is intended to cover overhead costs (such as taxes, office rental, insurance, and staff benefits) and profit.

The project budget is composed of the labor, supplies, and equipment that the company estimates are necessary to complete the agreed-upon work. Figure 20-3 shows an example of an AV project budget estimate.

Assuming the AV team has accurately estimated the need for labor and materials, the role of the project manager is to manage how those resources are expended to ensure that the actual costs for performing the agreed-upon work do not exceed the amount that has been allocated to complete the project.

The project manager should also keep track of the activities of the staff and make sure the agreed-upon work is performed according to the standards defined within the work contract. If the project work is taking more labor and/or materials than estimated, the manager should try to determine the cause and make adjustments to get the project back on track. If any modifications or additions are required due to changes to site conditions or client requests, and these changes impact the project budget, the additional work and costs should be explicitly agreed to by both the AV company and the client. Chapter 18 covers creating estimates in more detail.

Estimated Budget

Item	Estimated Unit Cost	Estimated Quantity	Estimated Total Cost
Videographics projectors	$12,000	2	$24,000
Projector mirror assemblies	$3,500	2	$7,000
Audio/video switching systems	$3,500	1	$3,500
Surround-sound audio system	$3,500	1	$3,500
RGBS/audio switching system	$3,500	1	$3,500
Ceiling loudspeakers	$150	8	$1,200
Videoconference codec (384 Kbps, upgradeable) and integral I-MUX, with IP connectivity	$8,000	1	$8,000
Videoconference cameras	$4,000	2	$8,000
Cable television receiver	$500	1	$500
Document camera	$3,500	1	$3,500
Wireless mouse	$350	1	$350
Presentation monitors	$3500	2	$7,000
Computer video interfaces	$500	2	$1,000
Half-size equipment racks	$500	8	$4000
UHF wireless microphone	$700	1	$700
Audio reinforcement system	$11,000	1	$11,000
Boundary microphones	$500	2	$1,000
Audioconference interface	$2,000	1	$2,000
Audiovisual control system	$16,000	1	$16,000
Auxiliary audiovisual connections	$250	5	$1,250
Instructor workstation	$3,000	1	$3,000
Wall boxes	$500	4	$2,000
Miscellaneous hardware, cable, and connectors	$3,000	1	$3,000
Audiovisual contractor labor	$12,000	1	$12,000
Total Estimated AV Cost			$127,000

Figure 20-3 A project budget estimate

Monitoring Work at the Client Site

The amount of on-site management required for a project varies depending on the scope and complexity of the project. However, an AV project manager should have some common objectives for all projects, including the following:

- Ensure that the work is performed safely and conforms with relevant regulations and codes.

- Ensure that the work meets the company standards for quality.

- Ensure that the work performed conforms with what was agreed on in the work contract.
- Ensure that the work is completed according to the agreed-upon schedule.
- Address any problems or concerns that affect the project schedule or quality.
- Resolve conflicts that may arise as a result of the work activities.

Coordinating Project Activities

The manager of an AV installation project at a client site may be part of a larger construction effort that involves a number of other vendors and contractors. The following are some examples of other project stakeholders the project manager may need to coordinate with (other trade-specific contractors may be involved as well):

- General contractor
- Lighting contractor
- AV integrator
- Data/telecommunications contractor
- IT integrator/manager
- Mechanical contractor
- Security contractor
- Electrical contractor
- Life-safety contractor
- Plumbing contractor
- Structural contractor

Coordinating the work of the AV installation team with such a potentially large number of other stakeholders working at the project site can be a complex task. The manager of the AV portion of a project must understand the concerns of each of these stakeholders in order to effectively communicate and coordinate the project activities. A considerable amount of coordination is often needed during construction to avoid or resolve conflicts in the design documents or issues that may arise due to conditions in the field.

Depending on how the project is contracted, the general contractor and/or the construction manager are responsible for coordinating the trade contractors during their work on-site. The architect, owner, and various trade consultants are also involved to help resolve issues as they come up. However, conflicts that are caught and resolved prior to starting work are certainly the easiest to resolve, so close coordination and communication are critical.

Coordinating tasks with other stakeholders is also important for live-event AV projects. These stakeholders may include the client or promoter, the building management staff, building services staff, staging vendors, and on-stage presenters or talent. Each of these stakeholders may have specific issues that can impact how AV systems are installed or configured to support the event, and the manager must be aware of their tasks and concerns.

Let's focus on how to coordinate with other stakeholders on an AV installation project through meetings and scheduling.

Holding a Project Kickoff Meeting

A project kickoff meeting is typically held prior to beginning work. The objective of the kickoff meeting is to allow all relevant team members to get acquainted, discuss the project, and establish how project communications should occur.

Depending on the scope and type of project, and how the contracts are structured, the owner's project manager, the construction manager, or the general contractor may preside over this meeting. Stakeholders from all contributing teams that affect the outcome of the project should be in attendance. At a minimum, the architect, general contractor, electrical contractor, owner representative, IT integrator/manager, and AV providers should be present.

During the kickoff meeting, logistical protocols and communication channels should be established (if they are not already defined in the project specifications). These include proper forms, authorizations (financial and design), process for payment, process for handling requests for information, and proper lines of communication. Unless clarified or revised along the way, these protocols must stay consistent throughout the life of the project.

The following are other issues that should be addressed during the kickoff meeting:

- Coordination and sequencing issues specific to the AV integrator's work

- Any owner concerns or project politics that may need to be considered during the construction phase

- Contract issues needing review, such as contractor billing procedures, warehousing and bonding requirements, and project schedules (particularly as they relate to any penalty clauses)

- AV equipment delivery and installation logistics, including delivery of owner-furnished equipment, site cleanliness requirements, and equipment security concerns

- Scheduling of the "clean and clear dates" when AV equipment can be installed in finished spaces

Attending Project Status Meetings

Coordination with other trades is essential to the success of a project. This coordination is typically accomplished through weekly construction/coordination meetings.

During weekly project meetings, each team's project manager provides a status report of that team's progress and any issues that may impact other teams. Information is exchanged, schedules updated, and coordination issues resolved. The AV professionals may be asked to review trade-shop drawing submittals for coordination purposes.

AV providers may not need to be at every weekly meeting, particularly at the beginning of construction, when they have little or no involvement on-site. They should start to visit regularly as the construction of their spaces and rooms are built to check for field changes or irregularities that might affect their work. The AV system work begins in earnest with the cable installation, at which point, the AV integrator brings equipment and personnel on-site for system installation.

The weekly meeting organizer should keep minutes of each project meeting, recording all issues discussed, as well as any conflict resolutions and action items that may be required by members of the project team. This record should be distributed to all participants as soon as possible after the meeting. Each trade team leader may also maintain a record of meeting proceedings as they pertain to the leader's own team for its internal use.

Creating Project Schedules

To facilitate coordination, a general project schedule is typically established during the design and bid phases. It's finalized by the owner's project manager when work begins. Each trade or discipline must have a corresponding schedule of work (developed by each team's project manager), which is incorporated into the main project schedule.

The most common way to present a project schedule is a Gantt chart, which depicts the timeline for tasks and subtasks as horizontal bars. It shows the sequence in which tasks should be performed and any project milestones, such as the completion of major tasks. Most managers of large projects use project management software to create these Gantt charts, as shown in the example in Figure 20-4.

Because some tasks must be completed before others may start, the Gantt chart identifies these types of dependencies in a manner that clearly shows their sequence. For example, walls within a room must be finished and painted prior to mounting sensitive and fragile AV components, so the start dates of these AV tasks will be identified as dependent on the end dates of the room-preparation tasks.

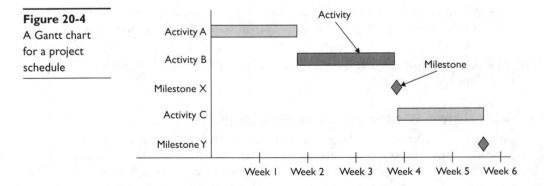

Figure 20-4
A Gantt chart for a project schedule

Coordinating AV Work with Other Trades

Like the work of other trades, the installation of AV systems must be integrated into an overall schedule. A successful AV project depends on the ability of the AV professional to dovetail the AV work with the other building trades.

The sequencing and duration of tasks are important. For instance, conduit and back-boxes for AV cabling must be in place before cabling can be installed, but the cable should be pulled before installing the ceiling grid where cabling runs overhead in rooms and corridors.

The AV team is most involved during the middle and final portions of the construction schedule, after the AV cable pulls, structural rough-in, and other AV-related infrastructure items are installed. What makes AV installation different from most other trades is that the bulk of the AV system (the electronics) must be installed after all other trades have completed their work and the site is clean, because the equipment is so sensitive to dust, moisture, and temperature.

Many different aspects of a construction project can impact AV installation, such as the following:

- Millwork that integrates with AV systems must be built to specifications and tolerance.

- Conduit, cable trays, cable paths, and electrical power must be located and sized correctly. The schedule of this work is also critical to the AV integrator.

- Lighting zones, fixture placement, and illumination properties should be checked for impact on displayed or projected images and cameras.

- Lighting dimming systems and any interfaces must be confirmed.

- Installation of HVAC ductwork and terminal devices must be monitored for conflicts with the AV installation.

- Dimensions of loading docks, elevators, doorways, corridors, and passageways must be large enough to allow delivery of equipment, particularly big projection screens and millwork that is fabricated off-site.

- Existing facility use must be considered, including any spaces that might be affected by work on the project. For instance, drilling or cutting may create noise on other floors, or equipment deliveries may be allowed only during off-hours.

- Ceiling layout and material, and the location of lights and sprinkler heads, must be coordinated with the installation of ceiling-mounted AV systems, such as loudspeakers, projectors, cameras, and projection screens.

- The scheduling of wall and ceiling installations must be monitored for sequencing of the installation of wiring, structural mounting requirements, and AV devices.

- Union and trade organization agreements must be honored.

- Acoustic treatments, including where and how they are mounted, can significantly affect the performance of audio systems, and they must be reviewed. Wall, floor, ceiling, window, furniture/upholstery, and other treatments affect the acoustic behavior of a space. (See Chapters 4 and 16 for information about acoustics.)

- Other work and treatments that may create a hazardous environment for AV equipment must be complete before equipment is delivered to the site. This work includes carpentry, sanding, painting, concrete work, and ceiling installation in areas where AV equipment is to be stored or installed.

- Data/telecommunication services must be installed and activated before functional testing of AV systems can occur.

This critical coordination of the AV installation with other elements of the project may be the responsibility of the AV company/consultant, the AV integrator, or both, depending on how the AV system will be delivered.

Providing Project Documentation

Consistent and accurate communication among all project stakeholders is crucial to a successful project. In many cases, it is also important to document any guidance received from a client or decisions related to the project. For example, decisions about changes to the project scope or specifications should be clearly stated, and the documents that identify these changes should be signed by all relevant parties to show that they understand and agree with the issues or terms described. In many cases, these documents become a formal part of the contracts or agreements among contractors, consultants, and owners.

Here, we will discuss some typical forms and agreements used by AV contractors and consultants to document decisions and agreements related to a construction project. These include letters of transmittal, requests for information, progress reports, requests for change, change orders, and construction change directives.

Regardless of the type of document, they all should include project-specific references, including the owner's name/address, specific installation location(s), contractor name(s), project numbers, and other appropriate contact or contract information.

 TIP Creating a set of standard approval forms that are project-specific and preaddressed at the beginning of the project can help save time and speed up the process.

Information-Related Documents

Examples of project documents associated with information include letters of transmittal, requests for information, and progress reports.

A *letter of transmittal* is used whenever documents, drawings, samples, or submittals are sent. It clearly indicates the addressee, sender, and contact information, and includes a list of what is being sent (including date or revision number) and any expected action to be taken by the receiving party. This form is used whether the items are sent by mail, courier, overnight carrier, or fax.

A *request for information* (RFI) is used to formally ask questions about a characteristic of a project. RFIs generally address three basic types of issues: a design issue, a site issue, or an owner change or request. This process is usually based on a paper or electronic form established for the project that indicates who the RFI is from and to whom it is directed, with spaces to enter the question and the response.

Some RFIs are resolved simply by a clarification from the RFI's recipient, without a change in anyone's contract. Others may need resolution through a change in the construction contract. In the latter case, other structured communications, such as change orders, may be generated, as discussed in the next section.

A *progress report* can be a formal document or an informal report that informs the client of the state of completion of the project as of a specific date. The progress report should include any issues or concerns that will impact the ability of the AV vendor to complete the project within the agreed-upon schedule and budget.

Change Documentation

Project documents related to changes include requests for change, change orders, and construction change directives. Figure 20-5 shows an example of the change process.

A *request for change* (RFC) is a common document used to change contractual obligations, equipment models, or specifications, or to modify a system design. When an RFC is generated (or answered) by the vendor, the pricing for the changes and their impact throughout the project must be included. The following are some issues that can prompt an RFC:

- Change in intended use of the system
- Discontinued product
- Architectural, mechanical, or millwork changes
- Discovery of system or product incompatibilities or functions

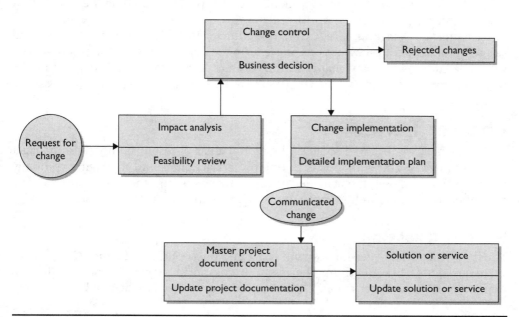

Figure 20-5 A change process

Any member of the project team can submit an RFC, although in an AV project, the integrator or consultant commonly creates it. An approved RFC then becomes a change order.

A *change order* (CO) is used to document agreed-upon changes to the project specifications or contract as the project progresses. COs are arguably the most important form used during the construction phase, because they can change the contract scope and pricing. Reasons for an AV system CO might include the following:

- Changes or clarifications in anticipated use by end users
- Architectural, millwork, finish, or other physical changes to the installation site
- Design conflicts, omissions, or errors
- Changes in product availability or specifications
- Availability of new products or technologies
- Discovery of hidden site conditions
- Budget adjustments
- Schedule changes and delays by others

The CO procedure and the people authorized to approve COs are established by contract at the beginning of the project. Since changes can dramatically alter the project schedule or budget, any necessary COs should be processed in a timely fashion.

COs almost always affect the overall value of the project. They can have an impact on financing, leasing, insurance, and bonding costs, so guarantors should be notified if a CO causes the project value to exceed the original budget. CO pricing can be included in the RFC or handled on a time-and-materials basis (pricing based on the actual labor and materials required for the work).

A *construction change directive* (CCD), sometimes called a *field order*, is usually issued as a result of time-critical conditions or events in the field, in order to trigger immediate work related to a necessary change in the contract. This form is typically used by the owner or the general contractor to bypass the more formal RFI–RFC–CO process.

Chapter Review

In this chapter, we reviewed some of the typical areas of responsibility for a manager of an AV project. These include directing project activities and managing company resources, coordinating the activities of company staff with other client stakeholders and other contractors, and producing the documentation required for a project.

Review Questions

The following review questions are not CTS exam questions, nor are they CTS practice exam questions. These questions may resemble questions that could appear on the CTS exam, but may also cover material the exam does not. They are included here to help

reinforce what you've learned in this chapter. For an official CTS practice exam, see the accompanying CD.

1. *Project budget* refers to which of the following?

    A. The total amount that the client has paid the AV company for the project

    B. The total cost of purchasing the needed materials and supplies

    C. The amount of labor and materials the project manager has allocated to complete the project

    D. The amount of labor and materials the project manager used to successfully complete the project

2. How should the AV company respond if a client requests changes in the project that end up increasing the cost of completing the project?

    A. Ask the client to agree to a CO that describes the changes to the project agreement and the additional costs.

    B. Perform the requested additional work in order to ensure that the client is satisfied.

    C. Never perform any work that was not agreed to within the original work contract.

    D. Remove another element of the project to keep the final cost the same.

3. What is a *dependency* within a project task schedule?

    A. A task that cannot begin before another task is complete

    B. A task that must be completed by a specific date

    C. A task that will be required if specific conditions occur at the project site

    D. An optional task that the client can determine is needed once the project is underway

4. Which statement best describes the relationship of the AV installation team with other vendors on the site?

    A. The AV team should focus on the AV installation; the project manager is responsible for coordinating work at the project.

    B. The AV team should review the project schedule; it defines all coordination necessary for the project.

    C. The AV team should actively communicate and coordinate with the project manager and other vendors to ensure that the AV installation is not impacted by other vendors working on the project.

    D. The AV team should negotiate work schedules directly with the other vendors.

5. Which of the following documents would an AV company submit in the event that a modification in the building design impacted the AV system installation requirements?

   A. Modification announcement (MA)

   B. Construction change directive (CCD)

   C. Request for change (RFC)

   D. Progress report (PR)

6. Which of the following defines project deliverables and relates the elements of work?

   A. Logic network

   B. Assumptions and risks

   C. WBS

   D. Gantt chart

Answers

1. **C.** The project budget is the amount of resources that has been allocated to perform the agreed-upon work.

2. **A.** The best approach is to document the changes and costs for the changes within a formal CO, and have both the client and company representative sign the CO to confirm that each side agrees to the changes to the original contract.

3. **A.** A dependency on a project task schedule refers to a task that cannot be started until another task has been completed. For example, the task of installing a wall-mounted flat-screen monitor cannot begin until the tasks of installing and painting walls have been completed.

4. **C.** The AV team should actively communicate and coordinate with both the project manager and other vendors to ensure that the AV installation is not impacted by other vendors working on the project.

5. **C.** An RFC is the appropriate document for the vendor to submit to the owner in the event that changes to the AV system installation contract are required.

6. **C.** The WBS (work breakdown structure) defines project deliverables and relates the elements of work.

PART V

After the AV Installation

435

Part V of the *CTS Exam Guide* picks up where an AV installation leaves off. Any AV professional who thinks the job is over once all the systems are plugged in is sorely mistaken. A CTS-certified professional knows that a client will need training and documentation on how to operate the installed system, assurances that the system will perform as designed, and help in troubleshooting and fixing the system should it malfunction. Moreover, AV companies are increasingly growing their businesses to offer ongoing support and maintenance services. In short, what you do after an AV system installation is often as important as what you do before.

Domain Check

Chapters 21 through 23 address tasks within Domain B (Operating AV Solutions) and Domain D (Servicing AV Solutions). They account for about 23% of the CTS exam.

CHAPTER 21

Operating an AV Solution

In this chapter, you will learn about

- Verifying the AV system is functioning properly
- Briefing the client on AV system operation
- Providing support, documentation, and training
- Documenting preventive maintenance on AV equipment

After installing an AV system, installers need to test it thoroughly to ensure it operates as intended and that all components are properly configured, calibrated, and functioning. An AV system set up at a temporary location to support an event or a performance must also be tested.

Once an AV system's proper operation is confirmed, the AV team must work with clients to ensure they can properly operate the relevant system components. Preventive maintenance must also be regularly performed to ensure the system continues to operate as intended.

The knowledge and skills required to accomplish these tasks include the following:

- Knowledge of AV system operation and maintenance
- Communication skills
- Skills in customer service

Domain Check

Questions addressing the knowledge and skills related to operating an AV system account for about 6% of your final score on the CTS exam (about six questions).

Verifying That AV Equipment Is Operating Properly

For many years, the term *commissioning* referred to a formal process for testing the elements of the AV system to ensure they operate as intended. Commissioning tasks could range from noting whether a component powers up and transmits a signal, to testing the signal output and calibrating individual components, to ensuring the output meets required performance standards. For example, video projectors have a number of functions that must be configured or adjusted once the projectors are installed at the site.

As AV has become more integral to the overall design and construction of buildings, it has become necessary to reconcile how the AV industry refers to certain processes with the way other trades refer to their procedures. In construction, commissioning refers to an ongoing process that lasts the lifetime of a built environment, from planning and delivery, to verification and operation. What AV professionals do when they commission an AV system is just one small step in what the construction industry considers commissioning. Therefore, InfoComm has come to refer to its role in the process as *systems performance verification*.

With the help of a volunteer task group, InfoComm has published the Audiovisual Systems Performance Verification Checklist. This is a detailed document of nearly 170 items that AV professionals can use to verify to clients that their installed AV systems work as they should. These include the items to check when performing systems performance verification so that you can document all AV systems tests, setup tasks, and other steps you've taken to demonstrate to clients that you've completed the job.

AV professionals use the checklist when "commissioning" AV systems to verify that they perform as intended. Along with the item identifiers and descriptions, a completed checklist also details who performed the test and its results. Where applicable, you can add notes about areas that require additional attention. The following is a small sample of the items on InfoComm's Audiovisual Systems Performance Verification Checklist.

Sample AV Systems Performance Verification Checklist Items

Item Number	Item Title	Description	Criteria	Responsible Party	Measurement/ Pass/Fail	Notes
VP-101	Projected Display Physical Alignment	Verify that the combined installation of projector and screen provides a displayed image that is correctly aligned to the active projection screen surface without misalignment unless an alternative condition is specified in the project documentation.				

Item Number	Item Title	Description	Criteria	Responsible Party	Measurement/ Pass/Fail	Notes
VP-103	Projected Display Brightness Uniformity	Verify that the combined installation of projector and screen provides a display to the viewer that meets the project requirements for uniformity of brightness across the area of the display.				
VP-110	HDCP Management Plan	Verify that the HDCP (High-bandwidth Digital Content Protection) management plan has been implemented per the project documentation.				
CABL-102	Patch Panel Labeling	Verify that all patch panels have been labeled as defined in the project documentation. Verify that all labeling is machine-printed, consistent, durable, accurate, and legible.				
CON-103	Control System Communications	Verify that all control communications are tested from endpoint to endpoint via the appropriate midpoint(s) for operation and functionality as defined in the project documentation.				
IT-104	AV IP Address Scheme	Verify and document that all network-connected equipment has the correct IP address, subnet mask, hostname, and gateway configuration as required for correct operation. This can be by static, DHCP, or reserved DHCP configuration. IP address scheme can be generated based on a list of equipment, location, VLAN requirement, and MAC address.				

The AV Systems Performance Verification Process

A formal systems performance verification (commissioning) process is a critical element of an AV system installation for several reasons:

- It provides a comprehensive step-by-step approach to testing all the elements of what may be a complex system.

- It provides a means for the AV company to objectively demonstrate that the installed system meets the system performance specifications that were defined in the system proposal and installation contract.

- It identifies any problems or issues that should be addressed and corrected prior to turning over the system to the client.

Systems performance verification confirms that the installation tasks are complete and the system is ready for operation.

For example, part of the testing procedure for a video display system is to connect the display devices to a test-signal generator at the beginning of the signal path and calibrate at various points along the path using a test pattern signal. The technician can evaluate the video signal components at the input of each piece of equipment in the signal path using a test instrument, such as an oscilloscope, and make adjustments as necessary to ensure the signal matches the reference values provided by the test pattern.

For video displays such as monitors and projectors, the technician views the output of the image and adjusts the monitor or projector controls until the images match a test-pattern reference. Technicians can use a variety of test patterns to calibrate various aspects of the image as it is displayed by the monitor or projector.

When checking networked AV systems, technicians must verify that there is enough bandwidth for control, audio, video, and data as part of either a shared or a dedicated AV network. They will do so by determining the network's peak load prior to bandwidth testing.

AV systems set up at temporary locations, for example, in support of live events, also require testing to ensure they are functioning properly. Technicians can assess the operation of temporary AV systems by confirming the proper signal flow through the system. The AV technician should determine that the source equipment is outputting the proper signals, that the signals are reaching the desired devices (including destinations other than displays), and that the destination devices are properly reproducing the content.

Checking Signal Flow

As you know, a signal originates from a source. It's then processed and sent to a destination device for output. For example, video and audio signals may originate from a media player. Video signals may be routed to a display or projector via a switcher. Audio signals may be routed to a mixer input, which controls the signals that are sent to loudspeakers.

To ensure and verify proper system operation, AV technicians must understand the following:

- How each device functions within the system. For example, is the device a source, a processor, or a destination?

- How a signal flows through the system. Understanding the path of the signal through the system allows the technician to properly connect devices and trouble-shoot problems.

- How to properly interconnect the devices. Selection of cable, correct termination, and proper interconnection of equipment are critical to ensuring system operation.

- How to properly configure, adjust, and calibrate the individual components. Many AV components require calibration on-site when they are installed as part of the system. You need to ensure these components are properly generating, processing, and receiving signals, as well as correctly reproducing images or sound.

In short, an AV technician needs to confirm that source components are generating the proper signals, the signals are passing through properly functioning processors, and they are reaching outputs that are properly calibrated. If you can confirm all of these functions, the AV system should be operating as intended.

Providing Support, Training, and Documentation

The success of a completed AV system lies not only in the design and installation, but also in the ability and willingness of the end users to use the system to its full potential. Once an AV system has been installed, clients often require operational support. In some cases, the client may want your AV experts to provide training to those responsible for maintaining the system, as well as to end users. In addition, you will be required to turn over documentation that describes how the system is configured and how to operate it.

Briefing the Client on AV System Operation

In most cases, the AV team will be required to brief the client and/or the end users on how to properly operate the system. The objective of the briefing is to ensure the client understands procedures for basic system operation, such as the following:

- How to start up and shut down the system
- How to display images from various sources
- How to select microphones or program sources
- How to change the volume of the sound system

This client briefing may be informal, such as giving presenters a demonstration of how to control the images and sound for their individual presentation. In other cases, the client briefing may be part of a formal training program that includes users and staff who will have responsibility for the ongoing operation and maintenance of the system.

In most cases, it is preferable to demonstrate operations by working with the installed AV systems. You want to give the client an opportunity to become familiar with the system by using the actual equipment.

Training Users and Support Staff

Beyond a simple briefing, you may be required to conduct more formal, in-depth sessions. Depending on the technical staff, you may provide only basic training on how the system is configured and operated, or you might conduct in-depth training that includes how to troubleshoot problems with the system. Comprehensive training serves many purposes:

- Teaches the end users how to use the system
- Improves the confidence of end users
- Reduces service calls resulting from user errors
- Reduces the chance of improper use and consequent damage of systems
- Enables in-house personnel to perform basic troubleshooting, possibly preventing system downtime, and prepares them for receiving assistance by telephone

You should identify specific training needs during the design phase and include training requirements in the contract. These may include duration, number of sessions, and anticipated number of trainees.

You should typically schedule training to occur after the system is complete and fully operational. It is not practical to offer training on an incomplete system. However, on a large project with multiple systems in various rooms, you may conduct training on smaller, stand-alone systems for users who do not need training on the overall project. It is usually most effective to train end users in small groups (up to six people). This allows all participants some hands-on time with the system.

Prior to training, the trainers should learn the backgrounds and abilities of the trainees in order to prepare suitable material. For example, someone who gives presentations in a media-rich space may need basic training on operating only limited functions of the system. Another user may routinely host videoconference calls, using various media sources and cameras, which requires more advanced training. There may also be a media department or technical staff that requires detailed training that addresses maintaining and troubleshooting the system.

Depending on the trainees' knowledge, it might be prudent to include education on AV basics. Reviewing nuances such as best seating positions, the unique features of various devices, and recognizing when a component isn't functioning properly may be appropriate for some support staff and/or users. During all training, the trainer should keep in mind the knowledge, background, and interest of the audience.

In preparation for end-user and technical training, it may be appropriate to "train the trainer," or teach select individuals how to train others so that training can continue within the organization. This person should be familiar with the design of the system, trouble-shooting techniques, and all possible scenarios in which the system might be used.

Documenting How to Operate the AV System

It is necessary to provide some level of documentation describing how to operate the system. Detailed operating procedures for every aspect of the AV system form one part of a total documentation package provided to the client at the end of the project (see the Final AV System Documentation Package Checklist shown in the next section).

Not all users require complete and detailed operating procedures. Usually, for typical system users, the level of detail can be reduced to a quick-reference guide. This document should focus on how to perform the most common system operations, such as the following:

- How to turn the system on and off
- How to switch between the various program sources
- How to turn on and mute microphones
- How to adjust volume levels of audio sources
- How to use the control system interface elements

When creating this type of document, keep in mind that many presenters will have only a moment to glance at the instructions—usually just prior to actually using the system—and they will want to see only the essentials. A best practice is to create a document that fits on a single laminated A4 letter-size sheet (double-sided if needed). Use diagrams, screenshots, and large text with simple instructions, preferably in bulleted point form.

For example, in AV systems where the control system does not control the shutdown sequence, the sequence in which the user turns off devices should be specified to ensure equipment is not damaged in the process. In the case of an audio system, the proper sequence is to shut down the power amplifier(s) prior to turning off devices such as mixers, to prevent transient noise peaks from damaging the loudspeakers. You can easily address this type of operation guidance via a numbered list that identifies the proper sequence in which to shut off the system.

Documenting the AV Systems Themselves

Providing operational documentation should also be part of the system specification and/or contract. This documentation, which you will submit to the client upon project completion, may include system-specific custom user guides, as-built drawings and schedules, and manufacturers' user and service manuals. Even if user documentation is not required by contract, you should still prepare training or job reference materials. Simple and practical explanations of basic functions are best, often in the form of small, laminated guides located with the equipment.

The system documentation may cover the following:

- Manufacturer manuals for all equipment, including instructions for operation
- System design and configuration documentation, including signal paths, to enable a technician to troubleshoot and correct problems with the system
- Cable and other schedules
- DSP and matrix switcher software configuration files
- Configuration of the control system, including DIP switch settings and IP addresses of individual components
- Operating instructions written for the AV knowledge level of the end user, using minimal technical language

At the end of the project, the items in the following checklist are typically provided to the client. The format(s), number of copies, and distribution should be determined in the contract.

Final AV System Documentation Package Checklist

The final project documentation package should include the items on this checklist.

As-built drawings, with final corrections, additions, and field-verified information in the following formats:

- ☐ Full-size paper
- ☐ Half-size paper
- ☐ Electronic DWG file
- ☐ Electronic PDF file
- ☐ Online system navigation drawing for larger or more complex projects
- ☐ Reduced as-built drawing sheets for use at equipment racks

Schedule and documentation of all physical system settings and adjustments, including the following:

- ☐ Signal gain settings
- ☐ DSP and other software settings
- ☐ Codec settings
- ☐ Projector settings
- ☐ All IP addresses
- ☐ Telephone and ISDN numbers

- ☐ All programming and equipment configuration files, including current versions of manufacturer's editing and loading software, on a CD-ROM

- ☐ Test reports in paper and electronic formats

- ☐ Manufacturers' users guides and manuals, alphabetized, bound in three-ring binders with an index, as well as a CD-ROM of any electronically available manuals

- ☐ Excel spreadsheet of all equipment provided, with all options and serial numbers noted

- ☐ System-specific custom operation guides, including basic setup and operational procedures

- ☐ Laminated flash card–style instructions for all basic system operations (e.g., playing a DVD, connecting a computer for display, volume and other audio settings, etc.)

- ☐ Basic troubleshooting guide in case of system malfunction, including common user errors and equipment failures

- ☐ List of consumable spare parts (lamps, filters, etc.)

- ☐ Key schedule with three duplicates of each key required for operation of the systems

- ☐ Description of recommended service needs and intervals

Warranty statement, including the following:

- ☐ System warranty start date, conditions, and term

- ☐ Summary of manufacturers' warranty coverage

- ☐ Description of extended warranties and service plans as purchased with the system

Documenting Preventive Maintenance

Some clients may enter into a contract with your AV company to provide long-term maintenance for their AV equipment and systems. It is very important to document all service and maintenance activities as they are being performed so that you have a detailed record of the work performed and issues identified. You can use your maintenance log to manage the long-term serviceability of the system.

In a system maintenance log, you should record the date of any maintenance activity, who performed that maintenance, and any problems identified. You should also note any corrective measures your technicians took to restore system performance, as well as any recommended subsequent steps (including whether or not those steps were taken). Table 21-1 is an example of a maintenance log with several detailed entries.

Date	Time	Initials	Work Performed
020313	10AM	AJB	Checked in with the on-site client before testing 1-year maintenance on Ballroom A. Tested for proper levels throughout the audio system using a meter and signal generator. All systems checked out OK. Noticed that fan in audio system rack was not operating. Recommended to client to call repair (800.123.4567) and speak with Tom for replacement. Before leaving at 3PM, secured all rooms and checked out with client.
060313	8PM	AJB	EMERGENCY CALL TODAY. DSP failed. Had to reference user's manual to get part number to replace the failed unit. Happen to have a spare device in the truck so replaced it. Reported to client status of problem. Will return to replace when part comes in. Left at 2AM.
060513	8AM	JAR	Normal tech (AJB) not available today. Replaced DSP after notifying client. Had to get key from security to get access. Lucky break—DSP data configuration file was on USB backup drive! AJB left directions on how to upload crosspoints on DSP. Verified that DSP was operating within normal parameters. No problems. Checked out with client at noon. Client said he was impressed with our work and has a friend who wants our company to do a retrofit install. Told him I'd have sales department contact him.

Table 21-1   Example of an AV System Maintenance Log

A high level of clarity and detail in the maintenance log is important, because it clearly communicates the status of the AV system in a manner that allows any technician to quickly understand previous system issues and maintenance activities. You will learn more about AV system maintenance in Chapter 22.

When you perform maintenance, log the details as follows:

- If you replace any equipment, such as upgrading a boardroom DVD player to a DVR, document the date of the upgrade, any changes to the signal path or types of signal, changes to control system connectivity and/or programming, and any other information that would be useful for ongoing maintenance and operation.

- When a device requires repair, document exactly what needs to be repaired, any options (such as replacing the device with an updated equivalent), estimated costs, and associated issues, such as system downtime. This provides your clients with the information they need to make decisions about the ongoing operation and maintenance of their AV systems.

Striking a Room

In cases when your company is providing AV for a live event, rather than installing a permanent system, there comes a time when the client has finished operating the AV equipment and you need to remove it. This is sometimes referred to as "striking the event" or "bumping out" (and setting up equipment is often called "bumping in").

Removing and storing equipment properly is just as important as setting up the equipment before the event.

Every company has its own method of striking an event. The following are some general guidelines to help you quickly and safely strike a room:

- Do not begin putting away the equipment until all the attendees have left.

- Create a plan of action in advance. This will save time and prevent the loss and/ or damage of equipment.

- Locate all storage cases and carts, and place them near the equipment to be packed. You will spend less time putting equipment back into storage cases if you don't need to walk all over the facility trying to find the cases that go with specific pieces of equipment.

- Strike equipment by category. For example, strike all of the video equipment first, then strike the audio equipment, and so on.

- Divide people into teams. For example, assign three people to the audio equipment, two people to video equipment, and four people to pipe and drape.

- Coordinate with in-house staff. This could be a brief phone call with the event or venue manager about space turnover or reserving the loading dock for a pickup. The facility may require that you strike the event within a certain time frame. Gaining a good understanding of the facility's situation can help you strike faster. Plus, the in-house staff members will appreciate your concern and respect for their space and time.

- Return any items you may have borrowed from the venue. Do not leave this equipment in unsecured areas. Find a person designated by the venue to take responsibility for the equipment.

- Make sure that if a particular piece of gear is associated with a particular road case (for example, by bar code or labeling), it is packed into the appropriate road case.

Clean up the space by removing tape or other materials that you brought in. Leave the space as you found it.

Chapter Review

In this chapter, you reviewed how the AV team evaluates the overall AV system to ensure it operates as intended and that all components are properly configured, calibrated, and functioning. You also examined how the AV team can work with clients to ensure that they can properly operate the system. Next, we discussed keeping a maintenance log as you conduct preventive maintenance to ensure that the system continues to operate as intended. Finally, we covered striking a room (removing AV equipment).

Review Questions

1. The maintenance technician for a system should document which of the following items?

    A. When preventive maintenance is provided

    B. When equipment is updated

    C. When another company's equipment fails

    D. All of the above

2. In the context of AV system installations, what is the systems performance verification, or commissioning, process?

    A. A process for registering the ownership of components

    B. A process for documenting that the AV system conforms with international standards

    C. A formal process for testing the elements of the AV system to ensure that they operate as intended

    D. Officially "launching" an AV system with users within the client organization

3. What is the objective of commissioning an AV system?

    A. Systematically test all components to demonstrate that the AV system operates properly

    B. Allow time to "burn-in" components to identify any potential failure points

    C. Document the delivery and installation of all system components to enable final billing for system installation

    D. Document how the user should operate the AV system

4. How does the AV technician use an understanding of system signal flow to ensure proper operation?

    A. To identify appropriate signal levels for each component

    B. To document the system during the commissioning process

    C. To calibrate AV system components

    D. To gain an understanding of overall system operation that will aid in identifying sources of problems

5. When briefing end users on how to operate the AV system, the AV team should do which of the following?

    A. Provide a detailed description of system components and operation

    B. Provide manuals for all system components

    C. Focus on how to perform basic presentation functions

    D. Use a schematic of the system to present how the signal flows through the system

6. What is a proper AV system shutdown procedure?

   A. Shut down system components in any order

   B. Shut down system components in a specifically defined order

   C. Turn off all system power at once

   D. Turn off sources, then processing, then display components

7. What is the main reason for carefully documenting AV system preventive maintenance tasks in a maintenance log?

   A. Provide detailed maintenance records that will aid in ongoing maintenance and repair

   B. Provide a record for client billing purposes

   C. Determine why a component failed

   D. Determine warranty coverage of individual components

8. Which of the following is *not* an objective of training end users on the operation of the AV system?

   A. Give users a higher level of confidence when operating the AV system

   B. Reduce service calls resulting from user errors

   C. Reduce the potential for damage due to improper use

   D. Eliminate the need for AV company maintenance and repair calls

9. AV companies are typically required to provide documentation of the AV system after a project is complete. What is that documentation typically composed of?

   A. Manufacturer manuals for all equipment that contain instructions for its operation

   B. System design and configuration, including signal paths, to enable a technician to troubleshoot and correct any problems with the system

   C. Configuration of the control system, including DIP switch settings and IP addresses of individual components

   D. Operating instructions written for the AV knowledge level of the end user

   E. All of the above

Answers

1. **D.** The maintenance log should document all relevant maintenance actions in a manner that aids in ensuring ongoing operation of the AV system.

2. **C.** Commissioning is a process that systematically evaluates the performance of the AV system and components, and documents that the performance meets the identified standards.

3. **A.** Commissioning is intended to demonstrate that the installed AV system operates as intended.

4. **D.** Understanding signal flow will help an AV technician to determine the source of any problems or performance issues.

5. **C.** Guidance for AV system users should focus on the basics, such as how to display an image, control sound, and turn components on and off.

6. **B.** The proper shutdown procedure for an AV system is typically based on turning off components in a specific order, to ensure that other components are not damaged by an improper shutdown sequence.

7. **A.** A detailed maintenance log provides valuable information that will aid in ongoing preventive maintenance.

8. **D.** End-user training will typically not address how to repair or maintain the AV system.

9. **E.** The AV system documentation is typically composed of manufacturer manuals for all equipment. These contain instructions for the system's operation and information about its design and configuration, including signal paths, to enable a technician to troubleshoot and correct any problems with the system. The documentation should also detail the configuration of the control system, including DIP switch settings and IP addresses of individual components. Additionally, the documentation should include basic operating instructions written for the AV knowledge level of the end user.

CHAPTER 22

Conducting Maintenance Activities

In this chapter, you will learn about
- Maintaining client AV systems
- Testing equipment required for maintaining AV systems
- Documenting maintenance

AV technicians must know how to perform preventive maintenance to help ensure that the client's AV systems continue to operate optimally. They must be able to identify faulty equipment and infrastructure, and to repair or replace components as necessary.

AV professionals also must document their maintenance activities carefully and clearly. Using this documentation, other technicians who visit the client site will be able to determine what maintenance and modifications have already taken place.

The knowledge and skills required to accomplish this task typically include the following:

- Skills in maintaining AV systems
- Knowledge of electronic testing procedures and equipment
- Knowledge of safety procedures

Domain Check

Questions addressing the knowledge and skills related to conducting maintenance on AV systems account for about 5% of your final score on the CTS exam (about five questions).

Maintaining AV Systems and Components

After the installation portion of an AV project is complete, systems verification finished, and users trained, your company's relationship with the client may shift to a service agreement. Such an agreement may include ongoing preventive maintenance and repair services, as well as addressing warranty-related issues.

Maintenance Warranties and Agreements

Three basic types of warranties or agreements are relevant to AV system maintenance and repair:

- Manufacturers' warranties for the equipment only, which may sometimes include a service plan option
- System warranties, covering integration and integrator workmanship during an initial period following installation of the AV system
- Service agreements, typically offered by the AV company to provide ongoing preventive maintenance, support, and repairs for the AV system installation, under agreed terms

The actual terms of warranties and agreements can vary, depending on the various manufacturers' terms, as well as those negotiated as part of a contract.

Manufacturer's Warranty

In most cases, manufacturers' warranties cover the repair of specific equipment by an authorized service center. A manufacturer's warranty may cover parts and labor over different periods of time. Removal and replacement of installed equipment, and transportation to and from a service center, are not typically covered by the manufacturer. A few manufacturers offer extended warranty plans, including optional multiyear service plans. Typically, these extensions are at additional cost to the owner.

Optionally, the AV integrator may, through a service agreement and at additional cost to the client, extend a manufacturer's warranty to cover the equipment at the installed location, or to include parts and/or labor and/or shipping.

At system sign-off, the AV company is responsible for delivery of documentation describing any manufacturers' warranties for equipment furnished by the integrator, as well as information about any options for extending the warranties.

System Warranty

As a best practice (and typically required by the client RFP and system installation contract), the AV company warrants all installation materials and workmanship against failure for a period of time after the installation is complete, usually for 6 to 12 months. This time frame should be defined in the original project scope, specification, or quotation.

A system warranty typically includes an extension of any manufacturers' warranties that are less than the duration of the system warranty.

Service Agreements

The owner may wish to contract for continued support of the AV system after the system warranty expires. AV companies may also enter into service agreements for existing systems they did not install, but this occurs less frequently.

A service contract will typically cover the following:

- **Extension of manufacturers' warranties**   This sometimes includes on-site service.

- **Emergency and nonemergency service visits**   Such agreements often specify a maximum response time. A contract might include provisions for a one-hour telephone response during business hours, same-day emergency response, and routine site visits by the end of the next business day. Some client sites are mission-critical and may require a 24-hour, 7-day-per-week (24/7) response.

- **Preplanned schedule of preventive maintenance visits**   Depending on usage and other site specifics, preventive maintenance might be performed monthly, quarterly, or every six months.

Contract Changes

The system may require modifications and adjustments after it has been installed and verified. Modifications may include changing settings and alignments based on unforeseen needs or preferences of the system's users. Some changes may be significant enough to require an addition to the contract. These changes are typically covered under the service agreement.

Adjustments and modifications made by the owner, an end user, or another party may void an existing warranty or service contract. To avoid breach of this contract, the owner should obtain written permission from the AV service provider before making such changes.

The warranty and/or service agreement may include software and firmware updates. In general, these updates should be performed only if the update improves the performance or reliability of the system or provides required additional functionality. Consider the situation carefully prior to installing updates, including those to owner-furnished equipment (for example, computers and network services), because an update to one system may trigger unexpected results from associated equipment or systems.

Performing AV System Maintenance

The maintenance you will typically perform on AV equipment includes repairs due to failure and scheduled replacement of components that have a limited life, such as lamps and filters.

In all cases, you should perform all maintenance and/or repair activities in a manner that complies with relevant safety guidelines for working with electronic equipment and electricity. You'll need to follow job-site safety procedures when working in locations that present physical hazards, such as climbing ladders to inspect installations in a ceiling space or lifting heavy components (for example, flat-panel monitors or loudspeakers).

Whenever you work at the client site, you should behave and dress in a professional manner. You will also need to minimize noise and disruptions to the client's business activities. Be sure to protect furniture and the work area from damage, and leave the client site the way you found it.

Repair and Replacement of Components

As an AV service technician, you may be called to a client site when a system exhibits obvious problems or some component has failed completely. Repair service visits are often considered emergencies, requiring a timely response to ensure that the system experiences minimal downtime and continues to meet client needs.

When you receive a client request for an emergency repair, get as much information as possible prior to traveling to the client site so that you have a better understanding of the problem and can prepare to address it effectively. Any information you obtain will allow you to conduct initial troubleshooting, which will help you figure out the best way to return the system to full operation quickly and efficiently.

For example, a problem with a monitor may be as simple as a failed input cable, or it might be impossible to fix, requiring complete replacement of the monitor. If the client states that the display monitor is not working properly, ask the client to describe the symptoms in as much detail as possible. Ask a series of questions such as the following:

- Is the monitor working at all?
- Does it appear to power on?
- Does it display any error messages?
- Is the problem occurring for all sources or just a specific source?

Answers to these types of questions help technicians narrow down the problem prior to leaving the shop and suggest solutions they can test at the site. We will review the process for troubleshooting in Chapter 23.

Once you arrive at the site and review the problem with the client, you will be in a better position to determine how to fix it. At the very least, you should be able to implement a work-around to provide as much functionality as possible until a permanent fix can be implemented.

In many cases, you will not be able to repair equipment at the client site; you will need to bring the component back to the workshop or send it to the manufacturer. In these cases, you will need to remove the device from the client system. Depending on the client's service agreement, you may need to provide a temporary replacement so that the system will continue operating while the original device is out of service.

In some cases, the problem is less complex and you can resolve it on-site. For example, you can easily replace a projector lamp, a ceiling loudspeaker, or cabling between devices, as long as you bring along the proper equipment and supplies. This often requires the same skills and tools you employed when installing the system, such as knowledge of how to properly terminate AV cables. You will learn more about repairing AV equipment in Chapter 23.

Preventive Maintenance Tasks

During a preventive maintenance site visit, an AV system technician should test each system and subsystem for proper operation, make adjustments to optimize performance, and review system performance with end users to identify any issues or problems. You should schedule preventive maintenance visits for times when you can fully assess the system without interrupting normal operations. The specific maintenance tasks to be performed during each visit will normally be determined prior to the visit, and should be defined in the service agreement with the client.

The following are examples of typical preventive maintenance tasks:

- **Replace system components that have limited, defined life cycles**   You may need to periodically replace certain components, such as air filters in rack-ventilation systems and projector lamps. You can usually predict the useful life of such components and develop a replacement schedule to ensure you replace them prior to failure. For example, the useful life of a projector lamp is typically rated for a specific number of hours. The maintenance schedule should track lamp usage and ensure the lamps are replaced prior to their rated limit so that they do not fail while in use. Through practical experience, you can fine-tune the time between replacements, which could be less (or more) than the manufacturer's specification based on conditions and usage patterns at the client's site.

- **Calibrate or adjust system devices**   Some equipment may require periodic adjustment or calibration to ensure it operates at optimal performance. In other cases, users may make adjustments to equipment that adversely impact the performance of individual devices and/or the overall system. For example, users misadjusting audio equalizer settings can result in feedback at lower audio levels than necessary.

- **Physically adjust equipment**   Over a period of time, users interacting with AV equipment may shift the position of devices. The maintenance technician should inspect all equipment to determine if its mounting or position has been affected and make appropriate adjustments.

AV System Testing Equipment

Whether conducting preventive maintenance or repairing equipment, an AV technician must understand the procedures for working with specific devices and how to use the appropriate tools and test equipment. Test equipment may include the following:

- Multimeters
- Oscilloscopes
- Sound-level meters
- Cable testers
- Impedance meters

- Audio signal and video-pattern generators
- Spectrum analyzers
- Light meters
- Computer network testing devices

AV system testing may also require other specialized testing tools for specific types of equipment, applications, and signal types. You will need to have all of the appropriate tools and use the appropriate tool to perform each job. For example, use proper cable-stripping tools, the correct size screwdrivers, and so on.

Documenting Maintenance Activities

You should maintain service records describing all maintenance activities. These records can be as simple as pages in a three-ring binder that you keep in the equipment rack, or as sophisticated as a database system maintained by the AV service provider or facility owner.

Each maintenance service record should contain the following:

- Identification of the service technician
- Date and time of the service
- A description of the specific service performed, including repairs, replacements, and modifications to devices and the system
- System downtime, if any (a service record that shows minimal downtime also shows that the service mission was properly planned and executed)
- End-user observations or reports of system issues or problems
- Any follow-up service or maintenance necessary to complete the maintenance or repair action

Whenever you make changes to a system, you should update any documentation associated with that change, including drawings of record (as-built drawings), equipment manuals, and operating procedures. This is crucial to the ability to manage, update, troubleshoot, and maintain the system over time. For many systems, maintaining documentation is not an insignificant task.

Following a preventive maintenance visit, you should send a report to the owner describing the findings and adjustments made during your visit. Any recommendations for changes to the system or operational suggestions should be included in this report.

Chapter Review

In this chapter, you reviewed the various types of preventive and corrective maintenance that AV professionals should perform to help ensure that a client's AV system continues to operate optimally. In addition, you reviewed how to document these maintenance activities in a manner that allows other technicians to determine the work that has been performed on the system.

Review Questions

The following review questions are not CTS exam questions, nor are they CTS practice exam questions. These questions may resemble questions that could appear on the CTS exam, but may also cover material the exam does not. They are included here to help reinforce what you've learned in this chapter. For an official CTS practice exam, see the accompanying CD.

1. Which of the following types of maintenance agreements commits an AV company to providing ongoing preventive maintenance for a client system?

    A. System warranty

    B. Service agreement

    C. Manufacturer's warranty

    D. Preventive warranty

2. When a new AV device fails within the first few weeks after installation, under which type of warranty is the repair typically addressed?

    A. System warranty

    B. Service agreement

    C. Manufacturer's warranty

    D. Preventive warranty

3. What is the first step that an AV technician should take when a client reports a problem with an AV system?

    A. Obtain more information about the problem, such as the symptoms of the system failure

    B. Locate the detailed system documentation

    C. Travel to the client site to repair the system

    D. Contact the manufacturer of the component to assist in troubleshooting

4. Which of the following is an example of preventive maintenance?

    A. Replacing system cabling that is causing static within the system

    B. Upgrading a display to a greater resolution to meet client needs

    C. Replacing a projector bulb that is nearing the end of its operational life

    D. Upgrading the programming on a control system to address newly installed components

Answers

1. **B.** A service agreement is a type of maintenance agreement that commits an AV company to providing ongoing preventive maintenance for a client system.

2. **C.** A new AV device that fails within the first few weeks after installation is typically repaired as part of the manufacturer's warranty.

3. **A.** The first step the AV technician should take when an AV system problem is reported is to obtain more information about the problem, such as the symptoms of the failure, so that the technician can begin to narrow down the cause of the problem.

4. **C.** The objective of preventive maintenance is to address system maintenance issues prior to a failure, to ensure that the system continues to operate. In this case, replacing a projector bulb according to a schedule and prior to the failure of the bulb is preventive maintenance.

CHAPTER 23

Troubleshooting and Repairing AV Systems

In this chapter, you will learn about

- The troubleshooting process
- Troubleshooting strategies
- Repairing common AV system problems and restoring systems to full functionality
- Tools for diagnosing network issues
- Recording maintenance and repair
- Billing for repair services

The process of troubleshooting and repairing an AV system requires a systematic, logical approach to identifying the problem source. Understanding and practicing this logical approach helps make you more confident when faced with actual system failure, especially in emergency situations.

When an AV system breaks down, an AV professional must be able to restore the system to full operation quickly and efficiently. AV professionals are expected to calmly and quickly diagnose the problem, and either fix it or apply a temporary work-around until permanent repairs can be made.

Inexperienced technicians (or more inexperienced end users) may randomly push buttons, make unrelated adjustments, or swap cables without any true insight into the actual cause of the problem. In most cases, this approach to troubleshooting wastes time and fails to achieve the desired result. As you will learn in this chapter, using a logical troubleshooting approach allows you to systematically identify and isolate the problem, even in a complex system.

Repairing AV systems requires an AV professional to understand how AV components operate and interconnect, how to address common AV system problems, and how to document repair activities.

Preparing to Troubleshoot

There's a problem with the AV system you installed, or perhaps it's a system that someone else installed but you're helping support. Now what?

Prior to attempting to troubleshoot a system problem, you need to do the following:

- Take steps to familiarize (or refamiliarize) yourself with the AV systems and equipment, either as part of a preventive maintenance schedule or prior to a live event. In many cases, the AV technician is under pressure to return a system to operation as quickly as possible, especially during a live event. The troubleshooting and repair process will be more successful if you have already reviewed the system design and operation, and know how individual devices should work, rather than learning about the system after you arrive on the scene.

- Have the project and/or system documentation available. System documentation will prove valuable if you need to examine drawings and trace component interconnections, or determine appropriate system and component settings. And you may need equipment manuals to look up specific troubleshooting and repair procedures.

- Bring the proper tools and supplies. In many cases, you will need to test the output of devices, open the cases of individual components, and so on. This will require tools, such as the following:
  - A multimeter
  - A signal generator and analyzer
  - An oscilloscope for video
  - Mechanical tools such as screwdrivers and spanners to open equipment cases or remove components from racks
  - Tools to install new terminations and connections
  - A supply of standard connectors

- Understand the warranty and maintenance agreements. The service and warranty agreements covering the AV system and its components may dictate how repairs or replacements should be addressed. In cases where warranties cover the components and/or the overall system—whether they are manufacturers' warranties or the AV company's warranty—you should repair or replace components according to the terms of those agreements. In cases where the client is paying the AV company directly for service and repairs, you may need to explain the issues to the client and obtain approval prior to addressing more costly system faults.

An Overview of the Troubleshooting Process

One way to figure out what's wrong with an AV system is to guess at the source of the problem, and then test each related component until you figure it out. Of course, this approach can take a long time and still not lead to a solution.

Proper troubleshooting follows a systematic, logical process to define the nature of the problem and narrow down the potential issues until you can identify the actual cause of the problem. The correct troubleshooting process offers the most efficient path to identifying and isolating failure points in a system while minimizing guessing.

The first step in the troubleshooting process is to clearly identify and understand the symptoms of the system's failure and determine how the system is malfunctioning. The next step is to select a troubleshooting strategy and use it to identify the cause of the problem.

 NOTE Remember that many of today's AV systems integrate with IT networks. When familiarizing yourself with a system that is having problems, make sure you understand which part of it—if any—traverses the client's network. That way, you can include possible network issues in your analysis and coordinate with IT staff as necessary. See the "Basic Network Troubleshooting Tools" section later in this chapter.

Recognizing Symptoms and Elaborating on Them

In order to clearly identify the problem, you must be able to differentiate "normal" system performance from "abnormal" performance. Understanding what the system *should* do when it's operating properly requires knowledge of the system's characteristics and design. To decide if the system is malfunctioning, you need to be able to recognize whether abnormal performance truly is a system problem or whether it is the result of user error.

You should also characterize the problem. For example, an intermittent problem—one that happens every now and then but is not persistent or permanent—may be more difficult to troubleshoot because you can't reproduce it. You can either observe the system, hoping it fails in your presence, or you can have the client contact you immediately when it happens. Either way, you will need to ask pertinent questions, especially in cases of intermittent system failures.

Asking questions helps define points of demarcation—before and after the abnormal performance began—to better develop a picture of what the system did or didn't do prior to the problem or failure. Here are some common questions for helping define a problem:

- When did it last work?
- When did it fail?
- Did any other equipment or anything related to this system change between when it last worked and when it failed?
- Who was using it when it failed?
- What were they doing when it failed?
- What happened when it failed?
- Was it a sudden or "catastrophic" failure, or did it gradually fail over time?
- What does it look or sound like when everything is "normal"?
- Can you reproduce the problem?
- Is there system documentation?

The next step is to elaborate on the symptoms by asking the following questions:

- Is it a system failure or a device failure?
- Is it turned on?
- Is it plugged it?
- Does it have a fuse or reset button?
- Does the electrical outlet have power?
- Can you reset the system or cycle the power?

Selecting a Strategy

The next step in troubleshooting the problem is to apply the appropriate troubleshooting strategy. The following strategies are used to identify faulty functions:

- **Swapping** While this may not seem like a formal, step-by-step process, swapping the suspected faulty component with a known good component can help identify problems. Prior experience with similar systems and their problems may quickly lead the experienced technician to a faulty component. For example, swapping out a small loudspeaker that isn't working with another small loudspeaker that you know is working divides the problem into smaller parts and indicates whether the original loudspeaker is the cause.

- **Divide and conquer** This strategy involves dividing a problem or system in half to define the half that is not working. Then divide the faulty half again and test for the failure point, repeating the process until you identify the source of the failure. In just a few steps, you can quickly identify a faulty component.

- **Divide the problem into subsystems**   Audio, video, or control? Divide the subsystem in half to identify the portion that's not functioning. Continue to divide the subsystems in half until you locate the fault.

- **Signal flow**   Start at one end of the signal path and continue to follow the signal path until you find the fault. With audio, you could begin at the source (such as the microphone) and continue all the way to the destination (such as the loudspeaker) until you find the faulty component (such as a cable, connection, or termination). Troubleshooting a video problem may begin at the output (the projector or display) and follow the signal path all the way back to the source. You could also do it the other way around.

A Troubleshooting Scenario

To demonstrate the troubleshooting process, let's walk through an example. Suppose that your client tells you that a microphone is not working. Assuming you know that the audio mixer, as well as the rest of the signal path, is connected properly and powered on (and you already verified this), begin by asking the following questions:

- Is the channel muted?
- Is the channel fader up?
- Is the microphone a dynamic or condenser model? If it's a condenser, is the phantom power switched on?
- Is it plugged into the correct input on the wall panel or snake/multicore?
- Is it plugged into the correct channel on the mixer?
- Is there a Mic/Line switch on the channel? If so, it is set for Mic or Line?

You can also swap the microphone for a microphone you've verified is working. If there is still no audio, you can assume it's probably not the microphone. However, while you have divided the system to try to identify the problem, only the microphone has been ruled out. Most of the rest of the system is still unknown.

Next, divide the system in half by determining whether or not you have sound out of the loudspeaker. You check this by connecting a known good input to the audio mixer, using something like an MP3 player. If you have sound from the loudspeakers, then you know the problem is somewhere between the microphone and the output of the audio mixer.

The next division is to take a known working microphone channel and plug the non-functioning microphone's line into it. If you have sound, then the problem is neither the microphone nor the cabling leading to the mixer.

Then take a known working microphone line and plug it into the nonfunctioning microphone channel's input. If you don't have sound, you know the problem is somewhere between the input and output of the mixer. You can then quickly follow the signal path through the mixer, checking adjustments and routings until you find the problem.

Best Practices for Troubleshooting AV Systems

The following are some basic troubleshooting best practices:

- Change only one thing at a time.
- Test after each change.
- Use signal generators in order to provide a known reference to measure against.
- Use signal analyzers to obtain quantifiable information.
- Document your procedures and findings.

An Overview of the Repair and Restoration Process

As described earlier, a successful troubleshooting approach should identify the cause of a system failure. The next step is to use that information to correct the problem. How you repair and restore an AV system depends on the nature of the problem.

In many cases, troubleshooting and repair are part of the same restoration process. For example, if, through troubleshooting, you've determined that a failed cable caused the problem, the repair process is simply to replace the cable. However, repairing the individual AV components of a larger system can be a complex task, depending on the nature of the component and the nature of the failure. Some companies have the capability to make corrective adjustments to a system and/or repair equipment in-house; others may contract with specialists to handle repairs. And in many cases, such as equipment under warranty, sending the device back to the manufacturer for repair is the best option.

You can remedy some system failures at the client site, such as problems related to faulty cable terminations or system configurations. Problems related to mounting or other physical issues are also best addressed on-site.

The key objective of any repair is to get the system working at full capability as soon as possible, with minimal disruption to the client's business. To minimize system downtime, it is often necessary to replace a failed piece of equipment with a working device of similar functionality. This ensures that the overall system continues to operate while the failed equipment is out for repair.

After the faulty equipment has been repaired, system repair and restoration are similar to the original installation process. The AV technician replaces or installs the new component, properly terminates connections, tests to ensure proper operation, adjusts the configuration settings, and verifies that the repair has corrected the identified symptoms and the underlying problem. The technician must therefore understand procedures associated with installation and testing to ensure the repair addresses the fault or problem.

Addressing Common AV System Problems

To illustrate issues in repairing AV systems, we will review how to troubleshoot and repair some common audio, video, and control system problems. What follows is not a comprehensive list of AV problems. It is provided to help demonstrate a general approach to isolating and addressing typical problems with AV components.

TIP When working with specific components, it's always a good idea to review the equipment manual or the manufacturer's website for suggested repair procedures.

AV Problems and Solutions

According to seasoned technology managers and live-event professionals, the following are examples of ten problems users describe having with AV equipment:

- I can't see an image.
- There is no sound.
- I hear feedback.
- My control panel is locked up.
- My video disc won't play.
- My CD/DVD won't run on my computer.
- I'm not getting an Internet connection.
- The projector keeps turning off.
- My microphone won't work.
- The audio sound is distorted.

Here, we will briefly review how to address each of these issues.

 NOTE The solutions described in this chapter work in many common situations. If they don't work in your specific situation, seek the guidance of an experienced technician to help you fix the problem. In general, you may need to engage the services of specialized technicians to address certain problems, such as a damaged cooling fan in a projector.

I Can't See an Image

If the user reports not being able to see an image, follow a troubleshooting process and pay particular attention to the following:

- Check the projector lens for a lens cap.
- Check the power connections, switches, and buttons, as follows:
  - Check the status lights on the source and the projector—do they indicate power?
  - Are any power strips plugged in?

- If the power strips have a switch, is it on?
- Do all the wall outlets you're using work? When in doubt, try different outlets or turning on light switches.

- Check the video, DVI, HD15, or other connections, as follows:
  - Are they pushed in all the way?
  - Wiggle the video, DVI, HD15, or other cable—is the image affected?
  - Look at the pins in the connector—are they bent or broken?
  - If you suspect a cable is bad, replace it.
- Some laptops will not allow the projection device and the local monitor to be displayed when the computer is running on the battery. If a laptop computer is the image source, plug the computer into an outlet to check whether this is the cause of the problem.
- If a laptop computer is the image source, check the following:
  - Did the screensaver come on?
  - Are the display settings correct?
  - Did it time off and go into standby or power-save mode?
  - Did you try toggling the display output?
- Check the status lights on the projector:
  - Is the temperature light on?
  - Is the lamp light on?
- Do you see any light coming out of the projector? If so, is the projector in standby mode? If not, does the lamp in the projector need to be replaced? If it gets dim and then bright, it may indicate lamp or image problems.
- Turn off the source device and projector, and then turn the projector back on. Now turn the source device back on.
- Find the sync or auto-image button on the projector and select it.
- Replace the cable with a cable that works.
- Replace the projector with a projector that works.
- Replace the source device with one that works with your projector.

There Is No Sound

Microphones and loudspeakers are the most obvious suspects when audio fails. However, cables and connectors fail more often than microphones and loudspeakers do. When you are going to help someone who is having audio problems, it is always a good idea to bring a spare cable. Simply replacing a cable is a quick fix, but not if your client must wait for you to run back to the office for a spare.

Troubleshooting audio is significantly easier when you use test equipment. However, if you're put in a position to solve an audio problem without a multimeter and SPL meter, here are some strategies:

- Make sure everything in the system is plugged in completely and powered on. Bring up the audio level slowly.

- Speak into the microphone to produce sound. While you are speaking, move the XLR connector, which attaches to the microphone, keeping the microphone and cable still. If you don't hear a change, move the cable around and listen.

- Move the connector on the other end of the cable and listen. If you hear any snaps or pops, then the connector or cable may have a problem. If you still don't hear a change, follow the troubleshooting process and begin replacing one component at a time, starting with the microphone.

- A simple way to determine if a microphone is failing is to switch it out for a different microphone. If the new microphone works, send out the old microphone for repair.

- When people use 1/8-inch or sub-miniature connectors on a regular basis and don't handle them carefully, they produce a variety of symptoms, including audio-popping sounds, static, loudspeakers not producing audio, or no audio at all. Users often report these problems as intermittent. To check for a problem with this connector, try moving around the cord and the connector while sending an audio signal to the loudspeakers. Listen for a change in the audio, such as popping, cracks, and audio cutting out on all or two speakers. If the problem is with the connector, you should hear a change in the audio while you manipulate the connector. The same is true if you manipulate the cord and hear a change in the audio. If you hear a change when you move around the cable, replace the cable with a functioning cable.

- If there is still a problem, the jack inside the device may be broken and require service.

I Hear Feedback

Feedback is the squealing or howling noise that drowns out the audio signal and annoys the listener. Feedback is generated between a microphone and a loudspeaker when the same tone is caught in a loop and amplified again and again. Feedback may stem from the audio system being too loud or the source (the person talking) being too far away from the microphone.

The best strategy is to place microphones near the origin of the sound to be reinforced and physically behind and as far from the loudspeakers as possible. The presenter should stand at an appropriate distance from the microphone. If the presenter is too far away, you may be tempted to turn up the volume of the microphone's output, which can cause feedback.

Feedback is best reduced through a combination of good loudspeaker placement and processing equipment. If you can't correct equipment placement, then turn down

Best Practices for Controlling Feedback

Best practices for controlling feedback include the following:

- Keep the microphone as close to the sound source as possible.

- Keep the loudspeakers in front of the microphones, and as far from the microphones as possible.

- Position the volume (gain) appropriately. Turn down all unused microphones.

- If your mixer has a mute button, mute the unused or temporarily unused microphones. The mute function will allow you to stop the microphones from picking up sound without disturbing the gain settings on the mixer.

the microphone or loudspeaker. Once the problem is corrected, the result isn't instantaneous; it might take a second for the feedback to complete a cycle and die out.

Feedback also occurs during audioconferencing and videoconferencing. It manifests itself in the form of an echo or other audio problems. The solution is often simple: Move the microphone away from the loudspeaker or turn down the incoming or outgoing audio.

My Control Panel Is Locked Up

Control panels may be wireless or hard-wired. If your panel is hard-wired, make sure all connectors are secure and undamaged. If you discover that a connector is damaged, seek the assistance of an experienced professional.

Touchscreens can be damaged by misuse. Control screens are often mounted in tabletops, which exposes them to potential damage. Customers may set heavy objects on the screens, not realizing what they are. Also, food or beverages may spill onto the equipment.

If a touchscreen stops working, inspect it for damage. Cleaning a touchscreen should be done only according to the manufacturer's instructions.

 NOTE A cracked or damaged screen can pose a health risk; do not touch a broken screen with your bare hands. Call an experienced professional to repair the panel.

Some touchscreens have a screensaver function, whereby the screen goes into a power-saving mode and won't display the control interface until someone touches it again. Users may incorrectly perceive a blank screensaver as a problem. Simply touching the screen may bring it out of power-saving mode.

Button panels don't have a sleep function. They often have lights that indicate whether they are receiving power and which button was last pushed. Buttons on these panels can become damaged through misuse. Pushing buttons too hard or spilling food or liquids into the panels can damage them. Follow the manufacturer's instructions to clean the interface.

My Video Disc Won't Play

The following are some steps to take when a user says, "My DVD (or Blu-ray disc) won't play in my player." After completing each step, try to play the disc. If it still won't play, move on to the next step.

- Make sure the power is on and the player is plugged in.
- Visually inspect the DVD/Blu-ray disc for scratches. Avoid touching the surface.
- Clean the disc. Use a cleaning cloth to wipe the disc from the center straight out to the edge. Do not use abrasive cleaners or cleaners designed for other media.
- Check the DVD/Blu-ray player. The disc should be inserted in the player.
- Insert the disc with the label side up and the playback surface down.
- Seat the disc in the circular depression in the tray.
- Check the cables. The video cable should be plugged into the video-out connector of the player and then into the monitor or projector.
- If you're using a television, it may need to be turned to channel 3 or 4.
- Projectors, televisions, and many types of monitors have several video inputs. What input is the DVD/Blu-ray player connected to? Use the channel or input button to select this input.
- If using a noncommercial DVD/Blu-ray disc, check if the presenter is certain that the files were burned correctly.
- If the presenter is using a noncommercial DVD/Blu-ray disc, make sure it is not, in fact, a CD.
- Many discs made in one country are not compatible with players from other countries. Ask the presenter where the disc was obtained.
- Has the DVD/Blu-ray player been in the cold air? Moisture may condense on the lens of the laser, making it unable to read the disc. Let the player adapt to the humidity and temperature of your current environment.

My CD/DVD Won't Run on My Computer

If you are a technology manager or live-event professional, you may be called upon to troubleshoot problems with any video or audio source, including a presenter's laptop. When troubleshooting problems with a computer's optical drive, check the following:

- The disc should be seated in the depression in the tray with the data side down.
- Look in the tray and make sure that there is only one disc in the tray.

- The CD/DVD/Blu-ray disc should be clean and free of scratches.
- Check the format of the disc, as some will not play in certain optical drives.
- If using a noncommercial disc, check if the presenter is certain the files were burned correctly.
- Not all laptops create CDs/DVDs that are compatible with all types of computers and operating systems. Ask the customer where the disc came from, as it may not have been finalized properly. Try to use the disc on a computer that most closely matches the computer on which the disc was created. Match the manufacturer, software, and operating system.
- If the optical drive is removable, turn off the computer, eject the drive, and put it back in. Follow the procedures in the operator's manual for your specific computer.

Figure 23-1 shows a troubleshooting and repair flowchart that illustrates the process for troubleshooting an optical disc-related problem.

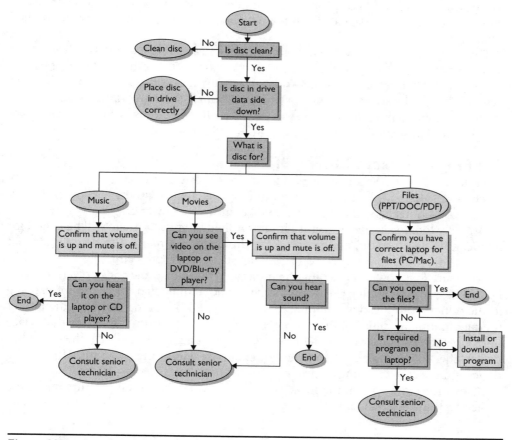

Figure 23-1 A repair process for an optical disc reader

I'm Not Getting an Internet Connection

If you are a technology manager or live-event professional, you may be called upon to troubleshoot network connections for presenters and facility guests. Start by checking the network to which you are connecting the computer. Making this connection isn't always as simple as connecting a computer cable. Networks are often secured and do not allow access to everyone. Do not attempt to connect to a network that you are not familiar with. Check with the person in charge of the network to learn how to set the computer to work with the network. You may need login information, an IP address, or some other configuration details.

Beyond that, the network cable should be plugged into a network wall jack, assuming you're using a wired network. Do not plug in the cable unless you are sure it is a network jack.

If the computer has a network card, it will often include small indicator lights near the network jack, on the network card RJ-45 (modular plug) adapter or on the network card. When these lights are illuminated, they indicate an active network connection.

There are two main types of wireless networks: one uses routers and access points to create a network, and the other uses computers to broadcast to one another. Access points send and receive signals from the network via wireless network cards installed in computers.

One common problem with wireless networks is losing a signal. This occurs when the user's computer is located too far away from an access point. Another obstacle for people wishing to connect to a wireless network is the network security system. Many wireless networks cannot be accessed without a password. Some users have wireless cards that are capable of using cellular technologies to access the Internet, thus avoiding the network security problems, but they still require a strong cell signal to obtain a connection.

The Projector Keeps Turning Off

Projectors will unexpectedly power off for many reasons. Digital projectors have safety mechanisms that may be triggered if conditions exist that could damage the projector. If these conditions persist, the projector will continue to power off.

Begin by looking at the projector's indicator lights. Typically, there will be a light for temperature, lamp status, and power. Read the operator's manual to determine what the lights on your projector tell you about the status of the projector.

To ensure that power is reaching the projector, check the projector's power indicator light. If the projector is plugged in and this light is not illuminated, check the power going to the outlet. Power-testing equipment is available, but if you do not have the proper testing equipment, plug a functioning electrical device into this outlet. If the device turns on, then you know the outlet is fine. Once you have confirmed that the outlet has power, check the power cord and the power supply within the projector. This can be done with the correct testing equipment.

Many projectors contain thermostats, which measure the air temperature inside the projector. If the internal temperature of the projector gets too hot, it will automatically go into a cooldown cycle. A red indicator light will illuminate, indicating that the projector has overheated. How do you fix this? Here are some suggestions:

- Clean the filters according to the operator's manual. This usually involves using a vacuum hose.

- Check the projector's air vents. Are they covered by something? Many people will place a projector on a padded chair, not realizing that the padding blocks the air, thus causing the projector to overheat.

- Listen for the cooling fan when you first turn on the projector. If you don't hear the fan, this might indicate that it has failed.

- Look for environmental factors that may contribute to the projector's overheating problems. If you suspect an environmental problem, adjust the environmental controls if it is in your power to do so.

- Check that all safety switches, covers, filters, and cables are properly connected and closed before you power on the projector. If the projector appears to turn on and then abruptly turns off, check the following:

  - Are the air filter and cover properly clipped into place?

  - Is the lamp properly installed in the projector?

  - Is the lamp cover properly installed?

  - Is the case of the projector cracked or damaged?

  - How many hours does the lamp have left? Some projectors will warn you when the lamp life is about to expire.

  - Has the lamp burned out? If so, have a technician help you replace the lamp or read the operator's manual and follow the instructions.

My Microphone Won't Work

Naturally, when a microphone isn't working, you need to check the audio system. Follow the troubleshooting steps outlined earlier to trace the signal and ensure that all the equipment is getting power. Systematically check the audio connectors and the cables and their connectors for damage.

As stated earlier, a simple way to determine if the microphone is failing is to replace it with a different microphone. If one microphone will not work with many systems, then there must be something wrong with the microphone. Many microphones require additional power to operate. Some have batteries; others get power from mixers. If you suspect a problem with a microphone, here are some possible solutions:

- Replace the batteries (if the microphone uses them). The battery polarity should reflect the directions on the microphone.

- Check for power switches on the microphone and turn them on.

- Look for mute switches on the microphone or transmitter and position them properly.

- Turn on phantom power switches on the mixer if the microphone requires phantom power.
- Many types of microphones have diaphragms. If they become torn or ripped, the microphone may buzz or not work at all. If you believe this has occurred, seek assistance.

Wireless microphones present additional issues, as follows:

- Wireless microphones transmit to wireless receivers. Transmitters must be set to the same frequency as the receivers or they will not work.
- If two microphones are running on the same frequency to the same receiver, they will interfere with each other. If your customer wants two wireless microphones in the same room, use two different sets of transmitters and receivers and two different frequency ranges.
- If two wireless microphone systems in proximity are running on similar frequencies, they will interfere with each other.
- You can use a transmitter only in proximity to a receiver. If the transmitter (wireless microphone) and receiver are too far apart, they will not work.
- Note that transmitters and receivers have volume levels that must be properly adjusted to allow the system to work and not interfere with other systems. Have an experienced technician set up these features.
- Transmitters have power and mute buttons that can cause the microphone to stop transmitting.
- Receivers should be set up so that their antennas are separated, pointing up, and not encased in metal. Try repositioning the antennas if you are not getting a signal. If antennas are located remotely, ensure that the coaxial cable is of the proper size used by in-line amplifiers.
- Some types of receivers are capable of both microphone-level and line-level audio output. This is typically controlled by a switch. Check this switch and make sure it has the proper output level selected.

The Audio Sound Is Distorted

Distortion is any change in the timbre (quality) of a sound resulting from internal interference. Distorted audio has several probable causes and therefore several possible solutions.

Overloading somewhere in the signal path causes distortion. One likely culprit is a mismatched signal level, like driving a −10 dBV input with a +4 dBu output without a pad. It may also be caused by an impedance mismatch. Frequency distortion happens when frequencies at the input are not present or equally produced at the output. Loudness distortion may result when the audio signal is prerecorded at higher than normal levels or reproduced at a level greater than the system can handle.

Check the output of each device in the system to see where the distortion becomes audible in the system. If the signal is distorted when coming out of the mixer, double-check the input gain of the distorted channel(s). Lower the input trim until the distortion goes away. If the problem is within prerecorded material, there is nothing that can be done to eliminate the distortion.

Poor gain structure is characterized by a constant high-frequency hiss in the system, or a waterfall sound. This is due to the inputs on the mixer being set too low and subsequent stages having their gain increased in an attempt to compensate. Double-check all microphone input gains and program-source input gains (DVDs, media players, and so on), and increase gain to the highest possible level without getting distortion or clipping.

Basic Network Troubleshooting Tools

Under most circumstances, if a networked AV system isn't working and the network is the cause, the client's IT staff or IT contractor will be responsible for identifying and rectifying the problem. Still, in an increasingly networked world, it will be important for AV professionals to acquire basic IT troubleshooting skills and/or tools.

For example, you should consider adding a network cable tester to your toolbox. Network cable testers are electronic tools used to verify Ethernet cable installation. Depending on the quality of the tester, it can test for simple connectivity, broken wires, crossed wires, and mismatched pairs. The tester can also identify attenuation—cable that is too long.

Here are some other networking tools:

- Network interface card (NIC) loopback test
- Network protocol analyzer
- Wireless network tester/analyzer
- Tone generator and tracker

Also, modern network devices themselves can hint at problems, as most have indicator lights that can provide immediate feedback. Light colors can vary by manufacturer, so read the technical manuals to translate light colors and patterns into usable diagnostic information.

In addition to hardware tools, you'll want to familiarize yourself with software utilities for troubleshooting IP networks, such as the following:

- Ping (packet Internet groper) can test the connectivity of networked AV devices by sending packets to their IP addresses.
- Traceroute and Windows tracert are diagnostic tools for checking the network path between two devices.
- Ipconfig and ifconfig are tools for diagnosing IP configuration issues.

As you learn more about networked AV systems, you will encounter many other tools for diagnosing and repairing network problems. But keep in mind that sometimes the best way to fix an issue with a network device is to turn it off, wait a minute, and then turn it back on.

Maintenance and Repair Records

In all cases, AV technicians should maintain detailed records of their actions to repair a system and the impact those actions have on the system. The repair records should be recorded in client system logs and sent back to the company for billing purposes, if appropriate.

The maintenance records should typically include the following:

- The nature of the problem and repair actions
- Parts and materials necessary to complete the repair
- Labor required to complete the repair
- Any follow-up actions required
- Warranty-related information (for example, if replacing a failed component, the warranty period for the new component should be recorded)

You should also identify repair-related issues that might impact long-term system maintenance and operation. For example, identify any reductions in system capabilities while components are out for repair or replacement.

Billing for Repair Activities

Depending on the nature of your contract with the client, you may need to invoice the client for repair activities. The information required for the accounting staff to bill a client for repair work typically includes the following:

- Labor time
- Travel time
- Component replacement
- Materials and supplies
- Costs for in-house repairs
- Cost for repairs completed by another vendor
- Other expenses

Chapter Review

In this chapter, you examined the troubleshooting approach typically used to identify the source of problems in an AV system. By following this systematic, logical trouble-shooting approach, you can narrow down potential causes until you identify the actual source of the problem. This step-by-step approach helps you find solutions more quickly, efficiently, and professionally.

We also reviewed examples of general repair steps AV professionals can take when a client identifies a problem with an AV system. The technician must understand how to restore the AV system to full operation in a rapid and efficient manner. This may require repairing equipment on-site, removing components for warranty repair, or replacing failed components with new components. We also introduced some basic tools for diagnosing network issues.

As you help customers with AV system problems, you will become more confident, and you will be able to resolve problems more efficiently. Just don't get overconfident. The AV industry is constantly changing, and technicians must be prepared to learn how to use new equipment and train others to do the same.

Review Questions

The following review questions are not CTS exam questions, nor are they CTS practice exam questions. These questions may resemble questions that could appear on the CTS exam, but may also cover material the exam does not. They are included here to help

reinforce what you've learned in this chapter. For an official CTS practice exam, see the accompanying CD.

1. What should be the first step in troubleshooting a failure in an AV system?

    A. Determine the possible failure points

    B. Determine which components are fully operational

    C. Review a system diagram depicting interconnections and signal flow

    D. Clearly identify the failure symptoms

2. What is an intermittent problem?

    A. A system failure that affects multiple components

    B. A system failure that is difficult to reproduce

    C. A disruption in signal flow affecting the system

    D. A failure of an interface component

3. Once the characteristics of the system failure are clearly identified, what is the recommended next step in order to return the system to normal operation?

    A. Examine the system to determine if the problem is simple to address and gather more information

    B. Logically divide the system in half, and determine which half has the failure

    C. Replace the component that has appeared to fail

    D. Ask the user to further describe the nature of the problem

4. What is typically the most efficient process for localizing the faulty function within a malfunctioning AV system?

    A. Test each of the components in the order of signal flow

    B. Test each of the components beginning at the final output, working backward

    C. Begin with testing the major functions, and then move to the minor functions

    D. Logically divide the system in half, and determine which portion has the failure, and then repeat for the failed half

5. When troubleshooting an AV system, should the AV technician consider user error as a source of the problem?

    A. No, because the users usually understand system operation

    B. No, because the users are not considered part of the system that has failed

    C. Yes, because users often do not understand how to properly operate the system, and may have changed settings or performed other actions that caused the failure

    D. Yes, because user error is typically considered the main cause of AV system failure

6. What is the best method to determine if a microphone is the source of a problem?

   A. Change the cord to the microphone

   B. Test the microphone with a multimeter

   C. Plug the microphone into a preamp and test

   D. Swap out the suspect microphone with a new microphone that you know is working and see if the system works properly

7. Which of the following is *not* an example of an issue that can cause a video projector to overheat?

   A. Brightness set too high

   B. Fan has failed

   C. Projector not properly positioned

   D. Air filters clogged with dust

8. Which of the following is the most likely source of an AV system failure or problem?

   A. Amplifier failure

   B. Projector failure

   C. Computer source failure

   D. Cable/connector failure

9. Which of the following is *not* a method of addressing feedback within an audio system?

   A. Adding compression to the microphone signal chain

   B. Keeping the microphone as close to the sound source as possible

   C. Keeping the loudspeakers in front of and as far from the microphones as possible

   D. Turning down or muting all unused microphones

10. An excessive amount of high-frequency hiss within an audio system is likely due to which of the following situations?

    A. Poor gain structure

    B. Insufficient compression

    C. Noise gate thresholds set too low

    D. Source output gain set too high

Answers

1. **D.** The first step in the troubleshooting process is to clearly identify the failure symptoms.

2. **B.** An intermittent failure is difficult to reproduce because it is not present all the time. This makes it hard for the technician to identify and isolate the problem.

3. **A.** Once the characteristics of the system failure are clearly identified, the next step is to examine the system to determine if the problem is simple to address, and if not, gather more information that will aid in identifying possible faulty functions.

4. **D.** Often, the most efficient process for localizing the faulty function within a malfunctioning AV system is to logically divide the system in half, and determine which portion has the failure, and then repeat for the failed half. Continue the process until the failure source is narrowed down to the faulty component.

5. **C.** The technician should consider user error when troubleshooting an AV system failure because some users do not understand how to properly operate the system. It's possible that a user has changed a setting or performed another action that caused the failure.

6. **D.** The easiest method to test whether a failed microphone is the source of a system problem is to swap out the suspect microphone with a new microphone that you know is working and see if the system works properly.

7. **A.** The brightness setting is not a typical reason for a video projector to overheat.

8. **D.** Cables and connectors fail more frequently than components. When cables and connectors are used on a regular basis and not handled carefully, they can be damaged and may present a variety of symptoms.

9. **A.** Adding compression to the microphone signal chain will not address feedback. The best practices for controlling feedback include keeping the microphone as close to the sound source as possible, keeping the loudspeakers in front of and as far from the microphones as possible, appropriately positioning the volume (gain), and turning down or muting all unused microphones.

10. **A.** Poor gain structure is a typical source of a constant high-frequency hiss in the system. This is due to the inputs on the mixer being set too low and subsequent stages having their gain increased in an attempt to compensate.

PART VI

The Business of Professional AV

A CTS-certified professional is by definition an expert in AV technology. But working in the professional AV industry is not only about designing great solutions, integrating cutting-edge components, and troubleshooting elaborate AV systems. As in any industry, a strong and growing AV company must also take care of business.

So-called "soft skills" describe various job functions a CTS-certified professional may need to perform that have little to do with technology itself. Granted, there are CTS-certified professionals who will enjoy successful careers without managing business operations, marketing and selling solutions, or managing personnel. However, InfoComm's volunteer advisory panel and certification staff have determined, through the regular JTA, that it is important for CTS-certified professionals to demonstrate a basic understanding of certain business skills.

Domain Check

Chapters 24 and 25 address tasks within Domain A (Creating AV Solutions) and Domain B (Operating AV Solutions). Although Domain A includes several other tasks, those included in this part of the *CTS Exam Guide* focus on marketing and selling AV solutions. The tasks in this part from Domain B involve managing an AV business, from tracking inventory to handling administrative work. All told, the tasks covered in Chapters 24 and 25 account for about 10% of the CTS exam. Chapter 26 is included in the *CTS Exam Guide* for your reference. Based on the 2012 JTA, this subject matter has been removed from the CTS exam.

CHAPTER 24

Managing AV Business Operations

In this chapter, you will learn about
- Working with vendors, suppliers, and customers
- Managing levels of stock and materials
- Maintaining appropriate security for company equipment
- Maintaining the professional skills and knowledge of your company staff
- Monitoring client sites from a remote location
- Performing administrative tasks

Managing the operations of an AV business is like managing any business. Management typically involves working with suppliers and vendors, maintaining relationships with clients, addressing issues such as security, and ensuring the AV staff is up-to-date on industry best practices and information.

This chapter reviews some of the typical tasks necessary for managing the operations of an AV company. You might not be responsible for performing some of these tasks within your organization, but it is important to have a general understanding of their scope and purpose. And it's possible you'll need to handle them at some point in your career.

The knowledge and skills required to perform these tasks include the following:

- Communication and customer service skills
- A general understanding of how to conduct typical business management
- A knowledge of administrative work, such as monitoring and maintaining inventories and training staff

Domain Check

Questions addressing the knowledge and skills related to managing AV business operations account for about 5% of your final score on the CTS exam (about five questions).

Working with Vendors, Suppliers, and Customers

When managing any type of business, you need well-developed communication skills. Managers spend much of their time communicating with staff, vendors, suppliers, subcontractors, clients, and users. Typically, such communication is conducted via telephone and e-mail, through documents such as purchase agreements and contracts, and in face-to-face discussions with staff and clients. Business managers must understand how to use all of these methods to ask the proper questions, make their needs and concerns understood, clearly and accurately communicate the desired message, and resolve problems and disputes.

A key concern of AV managers is maintaining a good working relationship with vendors and suppliers of AV systems and equipment. Because you need to work in close partnership with your vendors and suppliers to meet clients' needs, proper communication is important.

Here are some guidelines for working with vendors and suppliers:

- Try to meet face-to-face, at least initially. This will help establish a good relationship with their staffs by "putting a face" to the name of the person at the other end of a phone or e-mail message.

- Document transactions clearly and carefully in your communications to reduce the possibility of a misunderstanding about the nature of a transaction.

- Strive to resolve problems in a professional and courteous manner in order to sustain the working relationship.

- Make sure you communicate with the correct person at the vendor or supplier for the specific issue you need to address. In some cases, you may need to speak with a higher-level manager to prompt the company to address a specific need, such as speeding delivery of a special-order item.

Building a long-term partnership with your vendors and suppliers will enhance your ability to serve clients, as well as help build and maintain the ongoing success of your business.

Managing Levels of Stock and Materials

CTS-certified professionals must plan ahead to ensure they have the proper materials and supplies on hand to complete a given job task. For system integrators, this means the following tasks:

- Keep track of the inventory in a supply area or warehouse.

- Check in and out supplies and materials as they are used for client installations and maintenance activities.

- Order new supplies when stock levels reach a low point.
- Work with suppliers to obtain replacement stock.

Managing inventory in an AV rental and staging business can often be more complicated than dealing with a typical warehouse. In the rental and staging business, you need to track inventory as it is shipped and returned, rather than inventory that is simply shipped and never comes back. Maintaining an accurate inventory of equipment that comes and goes adds complexity to inventory tracking. For example, it's important to know the current location of equipment, but it's also important to track where equipment needs to go next.

Knowing what will be needed for future events helps planners determine if the supply in the warehouse is sufficient to meet the needs of those events. This knowledge allows you to procure equipment when necessary. There are various tools to help with this task. For example, you can use a software inventory database that manages inventory records and generates alerts when the supply of specific items is running low.

For technology managers supporting daily operation and maintenance of AV systems in a university, government office, corporation, or house of worship, inventory may include consumables such as extra lamps for projectors. Similar to rental companies, an end-user organization may need to inventory loaner or portable equipment, which should be checked in and out for resource tracking.

When it comes to checking equipment in and out of inventory, which is particularly useful for temporary events, you can use an inventory system that associates bar codes or radio frequency identification (RFID) tags with individual devices or crates. The bar code or RFID tag is associated with the item in a database. The unique identifier is physically attached to the equipment in the form of a tag or bar-code label, as depicted in Figure 24-1. The tag is read or logged in and out of the warehouse and venue, allowing the AV professional to know the equipment's current location. This system also serves as a scheduling tool. You can use it to create a schedule indicating where the equipment needs to be for upcoming jobs, and see whether there are overlaps where a particular item is required in two or more places at once.

Figure 24-1

A bar-code label used for inventory control of rental equipment

No. VIDEO RACK
Desc. Video Engineering Rack

Best Practices for Tagging Equipment

The following are some best practices for tagging AV equipment so you can track and monitor inventory:

- Tag equipment by serial number and use a call number to make the equipment easier to identify. For example, if you have ten LCD screens in your inventory, assign a call number to each screen from 1 to 10, and stencil or tag both the equipment and its case with the same number. This helps easily identify the item, and accurately ship and repack it.

- Standardize the placement of tags and bar codes on equipment to prevent people from needing to search for the tag. Use inspection ports on cases to speed up the process. This will allow technicians to scan equipment without removing it from the case.

- Use scan sheets or a list of the equipment and its bar code (its unique identifier) for items that are easily verified visually, but would be time-consuming to handle. For example, cables, accessories, and other kits containing miscellaneous supplies might be identified in this manner.

- Label equipment cases with the following general information prior to sending them to an event or venue:
  - All the company's contact information
  - The AV system with which the equipment belongs
  - A bar code or inventory-tracking number
  - The event's title, address, work order number, and specific room where the equipment needs to be placed
  - The weight of the case (if it always contains the same equipment)
  - Optionally (depending on your company's procedures), a contact name and number, room number, and date of use

Ensuring Security of Equipment and Supplies

A key aspect of managing your supplies and equipment is to take active steps to ensure they are safe from theft or damage. This is particularly important for AV equipment used at temporary installations or events. Cases containing valuable equipment can easily be removed from a site, either by accident (when another vendor is gathering its equipment) or on purpose (theft). Equipment and supplies intended for a permanent AV system installation can also disappear from a site in the same manner.

The AV manager should either personally monitor equipment and supplies or take specific measures to protect them, such as locking equipment cases in a secure room or vehicle.

Hiring Security Guards

Hiring on-site security guards can also reduce the possibility of equipment being stolen. Security guards can monitor entries, hallways, and loading docks. They are often instructed to inspect security badges, keep doors locked, and look out for suspicious activities.

To help security guards do their jobs, you should follow these practices:

- Create schedules of when areas should be secured, when areas are open to the public, when the crew can access areas, and when security guards should patrol.

- Make sure all employees wear appropriate ID badges to make them easily identifiable.

- Give security guards keys to lock and unlock doors. If the venue won't supply keys, give security guards the contact information for someone who has keys.

Deterring Theft

One step an AV company can take to reduce the potential for equipment theft is to mark equipment with company information. Stolen equipment is typically resold. Etching your company's contact information or branding on the equipment will make it difficult to resell. People interested in purchasing the equipment may notice the markings and notify your company and/or law-enforcement agencies.

 NOTE Etching or permanently tagging your AV equipment with your company's identification may be a good way to deter theft. But if your company sells its used equipment, be aware that such markings could decrease the resale value.

If your company rents equipment, it should take steps to properly identify the person taking charge of your equipment and determine if that person is a legitimate customer, instead of someone committing fraud to steal it. The steps to properly confirm that a rental customer is legitimate may include the following:

- Copy the driver's license of the person renting the equipment and the person picking up the equipment. This photographic identification needs to be corroborated by other pieces of identification, such as a passport or credit card.

- With a digital camera, take a picture of the person who picks up the equipment. The goal of this procedure is to verify that the person is picking up equipment for a legitimate rental. If you decide to go this route, you may need to advise your customers of this requirement and obtain their agreement in advance.

- Confirm the billing address with the credit card information the customer provides, and check to see if the credit card has been reported stolen.

- Record the locations where the equipment will be used, and then confirm the location information with the venue, if possible.

- Obtain a professional reference from a trusted source in the industry, including the reference's telephone number, address, and number of years the reference has known the person renting the equipment.

Keeping Up-to-Date Equipment Records

If equipment *is* stolen, it can be hard to determine exactly what is missing. Law-enforcement agencies need as much information as possible to identify stolen equipment. Your insurance company will also require this information if you make a theft claim. Keeping accurate records can help address both of these needs.

For example, the equipment schedule for an equipment rental should include the following:

- Where the equipment was last located
- The make, model, and serial number of each item
- Any inventory-tracking information associated with the equipment
- The value of the equipment
- The date the equipment was purchased and the purchase amount
- All accessories included with the equipment

This schedule should be kept up-to-date, including information about the equipment's current location at the site. You can create checklists to help the crew keep the records current. Paper checklists will allow the crew to take notes and access accurate information if something goes wrong. You may choose to scan the information on the paper, and enter it into an inventory management and billing system after the equipment is returned.

Maintaining the Professional Skills and Knowledge of Your Staff

Clients hire AV specialists because they expect the staff to be professional and up-to-date on the latest technology, standards, and industry best practices. In a fast-moving industry like AV communications, rapid innovation requires that staff members be proactive in keeping current on the latest information.

Resources for AV Staff

AV managers can help staff members maintain and upgrade their knowledge and skills by ensuring they are exposed to the latest technology and best practices. Here are some ways to support a staff's ongoing education:

- **Certification maintenance**   The CTS certification is an independent professional AV certification that requires holders to keep up-to-date by renewing their credentials on a regular basis.

- **Seminars and professional development courses**   Mastering new technologies and approaches may require formal training. Staff members should attend seminars and professional training courses to update their knowledge and skills in relevant areas.

- **Publications**   Subscribing to industry publications may help keep staff current with new technology and industry best practices.

- **Vendor presentations**   Vendors and suppliers are often good sources of information about the latest products and approaches. While you can expect vendors to focus on the specific benefits of their own products and services, they are also a good source of information about AV technology in general.

- **AV trade shows**   Trade shows are an excellent opportunity to experience the latest AV technology, equipment, design and installation approaches, and specialized training classes on AV technologies. These types of shows allow AV professionals to review a wide range of industry topics within a short period of time.

Training

Managers should constantly identify areas that require general ongoing staff training, as well as specialized training for certain professionals in the company. These training topics may be related to new technologies and techniques. They may also cover specific performance gaps in the company's collective skill set, such as ongoing problems with mounting techniques or AV networking.

In addition, some AV companies are called upon to provide training for client staff, to demonstrate how to operate or manage an installed AV system. In such cases, the company often develops its own standardized training materials, which can be adapted to meet the specific needs of individual clients. Client training should include guides that describe how to perform specific tasks, such as how to connect a PC to the system and switch between various inputs. These guides give users a reference that can supplement training.

Monitoring Client Sites

Some AV providers offer remote-monitoring services, in which they track real-time usage of the client's AV system via the Internet. Such monitoring systems allow staff to verify that AV systems are working properly without actually visiting the client's site. If the monitoring system senses something amiss, it will trigger an alarm that identifies the system or equipment involved.

Here are some features of remote-monitoring systems:

- They can track an item like a projector lamp's cumulative hours of operation and automatically generate a notification when the lamp is nearing the end of its useful life and requires replacement.

- They can trigger an alarm when something is removed from the system—as a result of vandalism or theft, or as an unauthorized modification to the system configuration.

- The system can generate alarms a variety of ways. Alarms may be audible and/or visual in the room itself, sent via Short Message Service (SMS) to an administrator, e-mailed to building security, and so on.

- Site-monitoring services may also include the ability to schedule the use of shared AV systems within the client organization. This can be accomplished via online scheduling tools or contacting a company representative by phone or e-mail.

Conducting Administrative Work

As you probably know, administrative work is critical to managing any type of business—AV or otherwise. Various personnel within an AV organization will be responsible for a wide range of administrative tasks, including the following:

- Invoicing clients for equipment and services
- Renting office and warehouse space and obtaining company vehicles, business insurance, office equipment, tools, and so on
- Purchasing AV equipment and supplies
- Handling staff needs, such as scheduling and tracking work hours, addressing insurance issues, issuing salaries or wages, and so on
- Hiring and paying subcontractors, vendors, and suppliers
- Providing accurate estimates or bids and quotes for projects
- Documenting and recording various business activities, such as work performed at client sites
- Handling the financial aspects of the business, such as managing cash flow and accounts, monitoring profit and loss, and managing the business's credit
- Conducting long- and short-term strategic business planning, including identifying potential product or specialization areas, identifying potential opportunities for pursuing new clients, planning for staff increases or reduction, and other strategies

Ensuring that these business administration tasks are performed properly is a key aspect of managing any type of business.

Chapter Review

In this chapter, you learned about typical AV business management and administrative tasks, such as working with suppliers and vendors, maintaining relationships with clients and vendors, managing inventories of equipment and supplies, monitoring security for company-owned equipment and materials, and ensuring staff are up-to-date on industry best practices and knowledge. Even if these tasks are performed by other people in your organization, it is important for everyone to understand their scope and purpose.

Review Questions

The following review questions are not CTS exam questions, nor are they CTS practice exam questions. These questions may resemble questions that could appear on the CTS exam, but may also cover material the exam does not. They are included here to help

reinforce what you've learned in this chapter. For an official CTS practice exam, see the accompanying CD.

1. Which of the following statements describes the typical relationship of an AV company with its suppliers and vendors?

   A. The AV company should identify the suppliers and vendors with the lowest current prices.

   B. The suppliers and vendors often become long-term partners with the AV company in meeting the client needs.

   C. The AV company should work with only one supplier or vendor to procure the needed goods and services.

   D. The AV company client will usually specify the vendors or suppliers to support the client's project.

2. Which of the following statements best describes how an AV manager should manage stock and inventory?

   A. Check in orders as they arrive.

   B. Ensure the proper amount of warehouse space is available for inventory.

   C. Ensure that the total value of stock and inventory does not exceed a specific amount.

   D. Keep track of the equipment and supplies the company has on hand, what is on order, and what is at the client site.

3. Why does an AV company use a unique bar code or other identifier on an equipment case for rental AV equipment?

   A. To identify a specific component and allow an inventory system to check the component in and out of the warehouse and venue site

   B. To allow the scanner to identify the price of the equipment for rental customers

   C. To identify the brand of the equipment or components

   D. To identify in which room the equipment should be placed when it arrives at the venue

4. Once equipment leaves the warehouse, who is responsible for ensuring security and preventing theft?

   A. Client

   B. AV company

   C. End user

   D. Whoever has taken control of the equipment

5. What is a key security concern when renting AV equipment?

    A. Properly identifying the person taking possession of the equipment from the AV company

    B. Ensuring that the renter knows how to properly operate the equipment

    C. Ensuring that the renter knows how to properly transport the equipment

    D. Ensuring that the renter has hired a security guard to protect the equipment once it has arrived at the site

6. Which of the following best describes the recommended approach to maintaining professional skills and knowledge?

    A. Take courses on state-of-the-art AV techniques at AV industry trade shows.

    B. Obtain and renew a professional certification.

    C. Obtain industry information from a range of sources, including courses, seminars, publications, vendor presentations, and professional certification courses.

    D. Take college-level AV technology courses.

7. Remote-monitoring services for client sites are intended to perform which of the following?

    A. Maintain security via the use of CCTV systems.

    B. Track system usage, identify system or component failures, and identify tampering.

    C. Track AV system operation costs.

    D. Track the type of program materials viewed via the AV system to ensure that inappropriate program materials are blocked.

Answers

1. **B.** The most desirable approach is for the vendors and suppliers to become long-term partners with the AV company, with all parties striving to meet the client needs.

2. **D.** The overall objective of managing stock and inventory is putting in place systems that will enable the manager to know the following:

    - The equipment and supplies the company has on hand at any time

    - The items that are on order and when they are expected to arrive

    - The specific equipment and supplies that have been sent to client sites, either for temporary use to support an event or for installation at the site

3. **A.** A unique bar code placed on an equipment case can be used to check that individual item of equipment in and out of a warehouse or venue.

4. **D.** Ensuring security for the AV equipment is the responsibility of whoever has control of the equipment at the time. This can be the AV company staff, the client, the end user, or security personnel. It depends on the use of the equipment and at which point in the event or installation process security is required.

5. **A.** The key initial concern related to security and theft prevention for rental equipment is ensuring that you properly identify the person renting the equipment. This reduces the opportunity for theft resulting from fraud, such as someone providing a false identification in order to take possession of your equipment.

6. **C.** AV professionals should strive to keep up-to-date on the latest technology and practices by obtaining industry information from a wide range of sources, including courses, seminars, publications, vendor presentations, and professional certifications.

7. **B.** Remote-monitoring systems are typically used to track system usage, identify system or component failures, and identify any instances of tampering or removal of system components.

CHAPTER 25

Marketing and Selling AV Solutions

In this chapter, you will learn about

- Defining your target market, audience, and value proposition
- Developing your marketing message and establishing a marketing budget
- Creating marketing materials and communicating your marketing message
- Developing proposals to address the client's AV needs
- Presenting the proposal to the client
- Creating a formal contract based on the proposal
- Maintaining a focus on customer service

Marketing is a critical aspect of any business. The objective of marketing is to bring in clients, because without clients, there's no business. Marketing is not sales. Marketing brings potential clients to your door and gives you an opportunity to make a sale. Marketing builds interest in your business by presenting your *value proposition*—the products or services that you offer that make your company stand out in the marketplace.

Marketing your business can be a time-consuming and expensive effort. You will need to understand how to target your marketing efforts in a manner that optimizes the effectiveness of your marketing program and your return on investment (ROI) in marketing activities. In this chapter, you will review some of the basic issues you should address to market your AV business.

Of course, successful marketing is measured in sales. Keep in mind that sales may happen at many stages in the design and installation process. Therefore, sales skills are needed at every level: consulting, design and integration, system commissioning, maintenance contracts and repair, rental and staging, and even the sale of a single device or piece of equipment. An AV professional should be able to develop a formal proposal appropriate to each of these types of sales opportunities.

In this chapter, you will also review the process of developing a proposal, presenting it to the client, addressing client concerns, and reaching a contractual agreement that defines which tasks you will perform and the associated costs.

The knowledge and skills required to accomplish these tasks include the following:

- Knowledge of client AV needs
- Knowledge of the AV market
- Skills in developing and implementing marketing campaigns
- Skills in presentation and selling

We will start with marketing.

Domain Check

Questions addressing the knowledge and skills related to marketing and selling AV solutions account for about 5% of your final score on the CTS exam (about five questions).

Defining Your Market, Audience, and Value Proposition

What is the market your AV company should serve? Who are the clients you should focus on? And what types of products and services do you intend to sell? Answering these questions can form the basis of your company's marketing message.

Defining the Market

Some may say their market is anyone with money who wants to buy an AV system. But defining a market in such a broad manner is usually not effective. Every AV company should form a strategy for capturing a specific portion of a well-defined market (or markets) by adjusting its products and services, processes, and marketing to attract those customers. Defining a market or markets allows the AV company to focus its resources in the most efficient and effective manner.

For starters, an AV company should carefully identify and target the type of clients and AV work that it wants to handle. A market may be industry-specific or technology-specific. Some typical markets are schools, universities, hospitals, corporations, corporate videoconference, government, auditoriums, houses of worship, and digital signage. The AV company determines which markets offer the best potential opportunity for making sales, and are therefore best to address with a marketing program.

Market Considerations

When defining a market, consider questions like the following:

- **Which services and products can your company provide?** For example, what type of installation services do you offer? What are your technical integration capabilities? What product lines do you carry? This gives you an initial basis for identifying your market.

- **What is your company really good at?**   Most companies are able to perform some tasks better than others. Identifying any special capabilities that make you stand out from your competition can help you target your marketing efforts. For example, are you really good at creating and installing videoconferencing systems, or have you developed skills in building AV systems for large auditorium spaces? If so, you have the makings of a marketing campaign that highlights your strongest skills.

- **Who are your potential clients?**   If you are offering installation services, you will likely need to service clients in your immediate geographical area. If you decide to specialize in specific types of services and clients, such as installing AV systems in religious institutions or schools, you will need to identify potential clients in those areas. Identifying and understanding the needs of your potential clients are critical to creating an effective marketing message.

- **Who are your competitors and what are their capabilities?**   Understanding how potential clients in a target market view your competition will help you target your marketing efforts. For example, do you know how you compare to your competitors on pricing, capabilities, and experience? How have they positioned their company in the market?

- **What market research is available?**   Having qualitative and quantitative data about market definition, economic data, and growth projections will bolster your business plan.

Addressing these types of questions will help you determine which types of clients and organizations offer the best opportunity for becoming potential clients.

Horizontal, Vertical, and Niche Markets

Another issue that impacts how you define your market is whether you decide to concentrate on horizontal, vertical, and/or niche markets. These are defined as follows:

- **Horizontal markets**   These address the needs of a wide range of clients by offering general categories of products or services. For example, a company that concentrates on offering flat-screen video display systems to any type of interested buyer is addressing a horizontal market. The company may offer a variety of products and provide installation services for any type of client interested in a display system.

- **Vertical markets**   These focus on the needs of specific businesses or customers. For example, a company may focus on serving the government market. In that case, the staff will have skills in addressing specific government procurement requirements, special rules, and government payment schedules. The AV company will understand the specific requirements of its government clients, such as the standard AV needs of law-enforcement agencies. Providing AV systems for live events, as shown in Figure 25-1, is another example of a vertical market.

Figure 25-1   Live-event AV system reinforcement is an example of a vertical market.

- **Niche markets**   These are usually a subset of a horizontal or vertical market. Companies that address niche markets focus on serving clients that have specialized needs requiring specialized products or services. For example, a company that focuses on creating digital signage systems in airports can be said to be addressing a niche market. Niche markets are smaller than horizontal or vertical markets, but they may offer an excellent opportunity for a business.

The final objective of defining your market is to identify your potential customers and gain a better understanding of their needs. This type of information is critical to the success of any marketing initiative.

Identifying Your Target Audience

Once you identify your target markets, the next step is to consider the specific audience for your marketing message within the organizations in your target markets. Who will receive and respond to your message? What are their job functions? Your target audience should be the people within your markets who will drive the purchase of your services.

It's also important to understand the process your potential clients will use for selecting and purchasing AV products and services within each of your target markets. Here are some examples of how that process may take place:

- The target company's technical staff may contact an AV company directly for support.

- The target company has a purchasing department that releases a request for proposal (RFP) to potential vendors, either to meet a current purchasing need or to select a number of preapproved vendors to be used for future purchases.

- A building manager or architect works as an agent for the corporation, and contacts the vendor.

- The potential client attends technical trade shows that exhibit equipment and services relevant to that client's market segment, and makes selections from those offerings.

Determining how your target markets and potential clients make their purchasing decisions will help you identify the most effective approach to get your marketing message in front of the right audience. Many times, you will determine that your company should incorporate several approaches into a single marketing plan, such as placing advertisements in technical magazines and websites, exhibiting at trade shows, responding to RFP announcements, or cold-calling specific potential clients. In each of these cases, the key is to determine the target audience for your message and how to best reach this audience.

Establishing Your Value Proposition

The term *value proposition* refers to the service or product offering that makes your company better than or different from a competing company. Your value proposition should become a key element of your marketing message.

Think about what your company offers that sets it apart and makes it better than or different from the competition. For example, a company may focus on configuring and installing systems for government command-and-control centers, as depicted in Figure 25-2.

The following are some typical examples of an AV company's value proposition:

- **Range of products** Do you offer such a wide range of products and choice of manufacturers that you can meet the special needs of any client?

- **Price** Do you beat the competition on pricing?

- **Reliability and reputation** Do you have a long list of satisfied customers that can vouch for your service?

- **Areas of specialization** Do you have special skills in areas of the AV business that set you apart, such as specialized installation or technical skills or extensive experience in a niche market?

Figure 25-2   Example of AV systems used within a command-and-control center

- **Design skills**   Do you create award-winning room and system designs that will capture the imagination of the client?
- **Specialized market experience**   Do you have experience or a capability within a particular market that sets you apart, such as holding an ongoing government contract agreement (a schedule) that would allow agencies to immediately procure your products or services without seeking multiple bids?

Keep in mind that one company will not be able to offer everything to everyone. The key is to know what your company can do very well and to highlight those areas when communicating with potential clients.

Developing Your Marketing Message

You have identified your markets and target audiences, and you've defined your value proposition. The next step is to develop the marketing message that you will use to promote your business. Your goal is to make your company's value proposition known to prospective clients so they will consider your company when the next AV project arises.

Your company should create a message that underscores what your company is about and gives clients a reason to contact you. The basic message should be relatively short, easy to understand, and memorable. You want to capture the attention of your potential clients.

When creating your marketing message, it may be helpful to focus on several areas:

- Capture the attention of your target audience by using a headline and/or images of your products or services in action, meeting the needs of the audience.

- Connect your product to the target audience by identifying how you can solve their problems or meet their needs.

- State the action that you want your audience to take, such as contacting your company for a demonstration or more information.

- Continually refresh your message in the marketplace. Highlight your company's excellent work wherever you can. Take photographs of your projects at multiple stages and include them in a portfolio, press releases, or case studies.

The key is to think about what would gain the attention of your target audience in a manner that will prompt them to take action.

Because you are an expert in AV technology—not marketing—it may be a good investment to hire a marketing professional to help you craft an effective and professional marketing message and to create marketing materials. In addition to creating professional materials, a marketing expert may also be able to identify methods for reaching your target audience that you had not envisioned.

Establishing Your Marketing Budget

Once you're determined a marketing strategy, you will need to determine what it will cost for you to conduct your marketing efforts. One approach is to plan a marketing activity, determine what it will cost, and ask yourself if it's worth doing. For example, a direct-mail campaign includes the costs of developing and producing the mailings, assembling the list of recipients, and postage. An advertisement in a trade publication includes the costs to create the advertisement and place the ad. In both cases, you can estimate the overall costs of the marketing effort, and determine if it is affordable and appropriate.

Another method, especially for established companies, is to set a marketing and advertising budget as a percentage of projected sales—for example, 3 or 5 percent—and plan a marketing campaign based on that amount.

Once you have estimated the cost of the marketing effort, you need to determine if it is affordable and represents a good investment.

Creating Marketing Materials and Communicating Messages

The specific format, style, and content of your marketing materials depend on the communication medium best suited to your message. Advertisements in a publication must immediately capture the attention of the reader; you have a limited space to get your value proposition across. A trade show display may allow you to use multiple approaches to capture customers' attention and communicate your message.

Because your overall marketing campaign may include a number of approaches and materials, you might need to employ the services of several vendors to create the materials. For example, you may need to hire a graphic artist to create images and a layout for a brochure or advertisement, a direct-mail company to create and send mailers, a marketing company to create T-shirts imprinted with your marketing message for distribution at a trade show, or a web designer to create a compelling company website. The specific mix of marketing methods should be based on an objective of reaching your target audience in the most efficient and effective manner.

A key element for success in any marketing campaign is ensuring the target audience is exposed to your message numerous times over an extended period. Your marketing campaign is intended to build awareness of your company and its services, which is unlikely to happen if the target audience sees your message only once or twice.

Check what your competitors are doing to get their message across. For example, you may want to advertise in some of the same publications. You may also want to determine if there is anything in their approach that provides guidance for your own marketing campaign. For example, you may discover that they are ignoring publications that are popular with your target audience, giving you a potential head start to reach that audience.

Keep in mind that marketing is not a one-time effort. It should be an ongoing part of your business, and your marketing campaign should evolve and change as the needs of your clients change. It is important to reevaluate your marketing message, audience, and approach on a regular basis, and work to ensure that it is effective.

Developing a Sales Proposal

You have begun marketing your products and services to potential clients. Great news: A few have asked your help. In response, you've worked with potential clients to determine their needs, evaluated site environments, and created an AV system design based on that information. In many situations, the next step is to create a sales proposal that will help persuade potential clients to give you their business.

The objective of your proposal is to document what needs to be accomplished to install an AV system that meets the client's needs. The size and scope of a proposal will vary depending on the complexity and scope of the project. Projects can range from a small installation that doesn't require a very detailed description to a complicated system within a large client space, which may also need room modifications by other contractors.

Proposal Contents

A typical sales proposal should include the following:

- A cover page with information that identifies the specific project by title, date, company name, etc. This also helps create a professional appearance.

- Optionally, a table of contents (for larger proposals).

- An up-front description of the overall project, often called an *executive summary*. This provides a brief description of the overall project. It may include a summary of the client needs and brief descriptions of the capabilities of each room/system proposed for installation.

- General project assumptions. These identify information such as who will obtain building permits, address zoning issues, and handle related issues, such as setting up the required electric utilities within the space.

- Room and equipment descriptions. For each room/system, the proposal should provide the following:

  - Architectural and technical AV drawings that depict the layout of the room and the AV components

  - A written description (narrative) of functionality, installation/configuration requirements, and the performance specification of the equipment and systems

  - A list of the specific AV equipment proposed for installation, including both major components and other required elements, such as cabling and mounting

  - Assumptions regarding the room, including use of existing equipment or facilities, work to be accomplished prior to installation by other contractors, elements of the system to be addressed by other contractors, acoustical and lighting criteria to be met, and so on

- Projected schedule and labor requirements.

- Information about warranties on your work and the equipment.

- Information about training of client staff to operate the AV system components, if that is defined as part of the project.

- The costs associated with the project elements. The client may ask for a breakdown of pricing by room or system. Major components may include individual price information.

- Information and cost for ongoing maintenance, if this task is included within your proposal. Note that an ongoing service agreement may also be provided as a separate proposal.

- An appendix that provides information about your company that the client can use to assess your ability to accomplish the job. This may include references or descriptions of recent similar projects, resumes of key staff (project manager, designer, etc.), company and/or employee certifications, legal statements required in the jurisdiction, and other company information.

Copying and Packaging the Proposal

Be sure to ask clients how many copies of the proposal they will require to complete their review. In many cases, the client will request several copies, such as when your bid must be evaluated by the general contractor or architect, the building manager, and the client.

It is also a good idea to spend some effort on packaging the proposal in an attractive and professional manner. For example, you might include color photos, an attractive cover page, plastic binding, and other enhancements.

Presenting the Proposal

Some clients will review the proposal and ask for written clarifications or changes. Others may ask that a company representative meet with the proposal reviewers to present the proposed plan and answer questions in person. That's when sales skills come in handy.

Salesmanship

Salesmanship is an important skill that helps move the client toward a commitment to the AV system purchase. Effective salespeople are able to establish credibility with the client. They are good listeners, responsive to client needs and concerns, and understand how to present the proposal to the client. Anyone in your AV organization who comes into contact with a client and other trade partner should be considered a salesperson. It is important to always act in a manner that will reflect positively on yourself and your company.

The following are key characteristics of good salesmanship:

- **Behaving in a professional manner at all times**   This includes proper dress and appearance, using professional business communication skills, and treating staff and clients with courtesy and respect.

- **Establishing credibility, both for your company and yourself**   Strive to present yourself and your company in a manner that reassures the client that you and your company are knowledgeable, competent, and honest, and that you will follow through on your commitments. You should be able to answer questions or commit to obtaining answers at a later date. And be sure to follow through on your commitments.

- **Understanding who makes the actual purchase decisions within the organization**   You want to focus on addressing the concerns of the decision maker(s). If you are asked to present your proposal, try to make sure you speak directly to the decision makers, rather than someone who will just relay information. This will allow you to address the decision makers' questions and concerns directly.

- **Focusing on addressing the client's needs, not the technology**   The client's main concern should be how well the AV system meets an identified need, not on the individual AV equipment components.

- **Having good listening skills**   Effective listening is a key skill for salespeople. Poor listening leads to incomplete or incorrect information, which can cause confusion and/or require you to do something twice. You should take notes, clarify anything that you may not understand, and summarize points at the end of the conversation to ensure that your understanding is accurate. Avoid arguing with, confronting, or interrupting the client. Avoid distractions and keep the discussion focused on the subject.

Remember that your role is to help clients make useful purchases; it's not simply to "sell" them. The sales discussion should focus on what the proposed AV system will do to meet the client's needs, rather than on technical specifications and operation. You should strive to link the system's performance to the client's environment. For example, instead of saying, "The control system will be custom programmed so that everything works from the keypad," a better approach would be to say, "Because some presenters will not have AV experience, all of the control functions will be programmed to display information in user-friendly language on the keypad menu. That way, you won't get dozens of calls from the presenters asking how to get the system to work."

As you interact with the client, ask questions to help foster a dialogue. Questions that serve to get the client involved in the process can be more persuasive than just statements. *Directive* questions can help you move the discussion in the proper direction. Continuing the preceding example, try asking, "Wouldn't it be great if we created a control system that would allow anyone in your staff to use the room with minimal training?" This kind of question directs the discussion away from technical issues and toward creating a system that will meet the client's needs, which is your overall objective.

Typical Sales Mistakes

There are a number of standard mistakes that you should learn to avoid during sales discussions, including the following:

- **Giving the client too many choices**   Most clients are best served by hearing your recommendation than by being informed of all the possible choices. You might present so many choices that the client becomes confused and decides not to act at all. Best practice is not to give more than two alternatives.

- **Not reaching the real decision maker**   People may lead you to believe that they have more authority than they actually do. You should gently attempt to find out who is the ultimate decision maker for the AV project and include that person in your discussions.

- **Presenting your proposal too late**   If the client has heard proposals and options from a number of other providers, it may be difficult for your proposal to stand out.

- **Not listening carefully enough**   Listening carefully will help you identify what is important to clients so you can be sure to target their actual needs. Don't assume you know their needs better than they do.

- **Failing to meet your commitments**   Missing a deadline or failing to deliver on something you promised as part of the sales negotiation can undermine the client's confidence in your ability to deliver on the actual project itself, costing you the sale.

- **Failing to identify an objection**   If your client is not responding to an offer or is not satisfied with your proposal, you need to strive to identify why. Ask questions and listen carefully to find out the actual objection, and then work to act on it.

- **Making a tactical, political, interpersonal, or technical mistake**   You will need to tread carefully when interacting with a client. You could say or do something that disturbs or offends a client and not know about it until long after the client is gone.

Creating a Formal Contract

Once you and your client have reached an agreement, and the proposal has been revised to reflect the final agreement, your company should create a contract that commits each side of the transaction to following through with the terms of the proposal.

Each organization handles contracts differently, but most contracts have similar terms and conditions, which define the nature of the transaction. These can include the following:

- The points of contact for the client and the AV company, such as the project manager and client representative, and the people authorized to accept or change the terms of the contract and approve the final work product

- The specific nature of the proposed work, which may be defined within the proposal, and/or attached to the contract, and may be a restatement of the project specifications that were in the proposal, adding detail to each element of the description

- The schedule for delivery/installation of each component of the project

- The cost for each component and associated payment schedule and terms

- Clauses that define what to do in the event of a dispute over the work or payment, such as an arbitration clause

- Other assumptions that affect the work, such as which elements of the project are to be addressed by other contractors, what to do in the event of a work delay due to weather, insurance requirements for workers at the site, and so on

The client's organization may also have some clauses that must be incorporated into the contract. These are usually defined in the initial RFP, and may address such issues as insurance requirements for the company and its personnel working at the client site.

Maintaining a Focus on Customer Service

Throughout the process of creating and presenting your sales proposal and contract, and during the installation of the AV systems, you must continue to build and maintain a positive relationship with the client—a relationship based on satisfying customer needs. This ongoing focus is critical to maintaining customer satisfaction with your company and your work.

While it is not always possible to meet every client demand or suggestion, you should strive to be responsive in some key areas that affect customer satisfaction. These can include the following:

- **Listening to the customer**   Again, listening skills are key to customer satisfaction. Proper listening helps you understand what the client wants from your company, whether the client is dissatisfied, and other issues.

- **Communicating with the customer**   Clients should expect that you will always return their calls or messages promptly, and that you will promptly address their issues and questions.

- **Dealing with problems or complaints immediately**   It is easier and more effective to address client complaints sooner rather than later. If not, you risk allowing the client to become angry while waiting for your response.

- **Properly addressing client suggestions or demands that you are unable to meet**   Suppose a client demands a change that is outside the scope of the agreed-upon contract, or wants to configure a system in a manner that is not technically feasible. You need to carefully explain why you are unable to meet the request and provide an alternative approach that addresses the client's concerns, if possible.

- **Following through on your commitments**   Clients need to trust that if you tell them you will do something, you will actually do it, within a reasonable time period.

- **Being competent in your job**   AV specialists should know the technology and be able to answer questions and propose solutions. If you don't know the answer to a specific question or issue, it is fine to tell clients that you will obtain the information and get back to them at a later date—just be sure that you *do* get back to the client with the information.

The bottom line is that you should treat the client in a manner that you would expect to be treated yourself if you were on the client end of the transaction.

Chapter Review

As you have learned, marketing consists of identifying potential customers and determining their needs, creating a message that communicates how you can address these needs, and getting the message to the right people. The objective of marketing is to build interest in your business by presenting your value proposition—the products or services that you offer that make your company stand out in the marketplace—and bringing potential customers to your door, giving you an opportunity to make a sale.

The process of selling a client your services requires you to address a wide range of issues, such as presenting the proposed AV system in a technically detailed and accurate manner, as well as salesmanship skills that help establish and maintain a positive client relationship. These elements form a foundation for moving forward with the project, and they are critical to ensuring that the final project is deemed a success—both by your company and the client.

Review Questions

The following review questions are not CTS exam questions, nor are they CTS practice exam questions. These questions may resemble questions that could appear on the CTS exam, but may also cover material the exam does not. They are included here to help reinforce what you've learned in this chapter. For an official CTS practice exam, see the accompanying CD.

1. Which of the following best describes the market that your AV business should strive to reach with its marketing campaign?

   A. Any persons or companies interested in purchasing AV equipment or services

   B. Persons or companies interested in purchasing AV equipment or services that are within your area of service

    C. Persons or companies interested in purchasing the types of AV equipment or services in which your company has special skills or capabilities

    D. Business executives who are the decision makers responsible for purchasing AV systems and equipment

2. An AV company that concentrates on serving the needs of retailers is focusing on which type of market?

    A. Integrated market

    B. Horizontal market

    C. Vertical market

    D. Niche market

3. What does the term *value proposition* refer to?

    A. A comparison of the price of your products and services with the competition

    B. The overall quality of the AV equipment carried by your company

    C. The overall client rating of your response to the company's RFP

    D. The elements of your company that give you an advantage over other companies

4. A proposal for an AV system installation should provide all *except* which one of the following?

    A. General project assumptions

    B. A detailed description of all of the components and accessories

    C. A detailed breakdown of the costs for components, materials, and labor

    D. A detailed description of the process used for identifying user needs

    E. Warranty information

5. What is the recommended role of the AV salesperson when meeting with the client to present a project proposal?

    A. To "sell" the client on purchasing the system

    B. To explain how the costs for the project were estimated

    C. To identify the needs that the system is intended to address

    D. To assist the client in making a purchase decision

6. Who is the person in the client organization that you should focus on meeting to present your proposal?

    A. The AV manager

    B. The IT manager

    C. The CEO

    D. The decision maker identified as ultimately responsible for the AV systems

7. What should the sales discussion focus on?

    A. How the proposed system will meet the client needs

    B. The technical specifications of the AV system

    C. The pricing of the AV system

    D. How the AV system was designed

Answers

1. **C.** Your marketing campaign should be based on the specific strengths of your company and its products—the products and services that your company excels at providing.

2. **C.** Companies that address the needs of specific types of businesses or customers are considered to focus on a vertical market. An advantage of specializing in vertical markets is that a company can develop special expertise in the specific needs of these types of customers, so it provides a better value than a company without this experience.

3. **D.** The value proposition is what your company offers that sets you apart from the competition, and makes you a better choice than the competition. The value proposition that your company offers should be a main element of your marketing message.

4. **D.** At the proposal stage, the client should already have an understanding of how the needs were identified. The proposal should focus on the specific components and labor required to install the system.

5. **D.** The recommended approach for the role of an AV salesperson is to assist the client in making a purchase decision by answering questions, identifying client concerns, and so on.

6. **D.** You should strive to identify which person is ultimately responsible for making the purchase decision, and try to meet with that person. In many cases, other company representatives, such as the AV manager or IT manager, may also participate in the meetings.

7. **A.** The salesperson will discuss all of the issues identified, but the focus should be on how each element of the system will meet the needs of the client and the client's organization.

CHAPTER 26

Managing AV Personnel

In this chapter, you will learn about
- Hiring AV staff
- Maintaining and improving staff skills
- Assessing staff performance

A key element of the success of any company is its ability to attract and retain high-quality staff. This chapter focuses on the steps for hiring the most qualified staff, including identifying the required skills, writing employment ads, and interviewing applicants. It also addresses how companies can help ensure that staff members keep up-to-date on skills and maintain their job performance.

Domain Check

Questions addressing the knowledge and skills related to managing AV personnel were removed from the CTS exam following the 2012 JTA. That said, InfoComm educators have deemed these skills very beneficial to CTS-certified AV professionals, especially as they grow and manage their own businesses.

Hiring AV Staff

Companies are only as good as the employees who perform the work. A systematic approach to identifying and hiring qualified and competent staff is critical to the company's success.

Hiring the appropriate staff is the result of a careful process consisting of identifying the job requirements, writing a compelling job advertisement, and carefully evaluating the job candidates.

Determining Job Requirements

The first step in the hiring process is to look at the needs of your company and its clients and identify the specific employee characteristics and requirements to address those needs. This typically includes the following:

- Determine the knowledge and skills required for specific tasks and technical areas. This will vary according to the type of job and level of employee. For example, entry-level assistants require less knowledge and skills and can be trained to take on higher-level work, while higher-level jobs require candidates with a background that includes specific education and work experience.

- Determine the specific work experience necessary to demonstrate that a candidate will likely be able to perform the job.

- Determine any education or certification requirements that may demonstrate knowledge of the job tasks.

- Determine the specific physical attributes that may be appropriate for staff in physically demanding roles, such as working from ladders or lifting heavy components.

- Assess any government requirements for the job role, such as authorization to work within a jurisdiction, obtaining clearances to work in particular secure or governmental facilities, meeting minimum age requirements, or holding special licenses for specific work tasks (such as electrician certifications).

Taking the time to assess the knowledge, skills, experience, and other characteristics the job requires will help you focus your search on the most appropriate candidates.

Writing Job Advertisements

A well-written job advertisement (a "help-wanted" ad) will lure qualified candidates and discourage ill-equipped candidates from applying for the job. The ad must present both the opportunities offered by the position and the specific requirements that applicants must meet to be qualified for the position.

Using some simple marketing techniques when composing job advertisements can make the difference between an ad that attracts highly qualified candidates and one that attracts few or no candidates. The key is to view your company's job opening as a product you are trying to sell, and to view potential candidates as the clientele you are trying to reach. To attract top people, your ad should sell candidates on the fact that working for your company will improve their quality of life and help their careers.

The first step in creating a job advertisement is to consider which elements of the job would interest the candidates you want to target. What kind of job do *they* want? What do *they* value? Where are *they* positioned along their career path? For example, consider the following workplace criteria and how your company might be attractive to the ideal candidate:

- Location
- Salary

- Training
- Room to grow
- Prestige of company brand
- Work environment
- Working independently
- Working in a team

If you can look at the job opportunity from the perspective of the candidate—the person you are trying to "sell"—then you are better positioned to write an ad that will attract the attention of the ideal candidate.

In writing your ad, the job title, which functions as a headline, is important. The job title is the first part of the ad to capture someone's attention. Titles that are exciting and a call to action stand out and make people want to read your ad.

The ad itself should begin by selling your company to the job seeker. Ask yourself "Why would talented people want to work for our company rather than for the competition?" List three to five selling points about the position and your company.

The body of your ad should be easy to read. Using a bulleted outline format, rather than one big block of text, makes it easier for candidates to quickly read the ad content. The following are some typical areas to address within the ad:

- **General scope of the work the applicant will perform**   Walk the applicant through a typical workday, describing the job duties.

- **Team management**   If the employee will be managing a team, how large is the team?

- **Your work environment**   Why would an employee want to work in that environment?

- **Career-advancement opportunities**   Candidates want to know how they can move ahead in your company.

- **Company benefits and perks**   These may include tuition reimbursement, free training, an outstanding benefits package, on-site daycare, an on-site gym, nearby walking trails, etc. Benefits and perks separate your company from the competition and help you attract the best candidates.

- **Job requirements**   Include only the "must-have" skills—usually education and experience. The more skills you list, the fewer candidates will apply for your job. If you can teach an employee some unlisted skills, you have just created a job benefit (free training) and increased the number of good candidates who will respond to your ad.

NOTE   Be careful not to scare away great candidates by listing dozens of skills they may never use. Don't request specific personality traits (such as "outgoing" or "detail-oriented"), because people are likely to come in for an interview and imitate those characteristics, whether they actually possess them or not.

In your ad, be specific. According to certain job surveys, job seekers list location (as specific as possible—suburb, neighborhood, subway stop, etc.) and salary (when possible, include it) as top criteria when scanning job ads.

When writing a job advertisement, follow these simple copywriting rules:

- Engage in a conversational tone (as opposed to being pretentious and wordy) and use "you."
- Keep it simple and concise, with useful information. Don't add any fluff.
- Use bullets, and space out paragraphs to make the ad easy to read.
- Use boldfacing to highlight important points, but don't overdo it.
- Be action-oriented by using action verbs. Instead of saying, "This is a great opportunity," use "Seize this great opportunity."
- Try to keep it short. List only the essential points.
- Always double-check for editing mistakes. Nothing looks more unprofessional than sloppy grammar and spelling mistakes.

Finally, specify how employees should contact you. Depending on the type of job (professional or nonskilled), you may want to have the person mail, fax, or e-mail a cover letter and resume, or simply call to set up an appointment to fill out an application.

The amount of detail you provide in your ad will depend on how the ad will be distributed. Online ads can typically be as lengthy as necessary. On the other hand, most newspapers and other forms of print advertising charge by the amount of space—or even the number of characters—that you use for your ad. Print-based ads are usually cut to fit a budget and include only the necessary information. You can keep down the length of an ad using the following techniques:

- Select one or two job responsibilities to explain what the person does on a daily basis.
- Use abbreviations where you can.
- List your top requirements, such as education and experience, noting whether they are required (*reqd*) or preferred (*prfd*).
- Include your contact information.

Assessing Job Applicants

Your job advertisements should yield a number of resumes that describe candidates' qualifications, experience, education, and other characteristics. The process of reviewing resumes and selecting candidates for interviews is called *short listing*.

The purpose of short listing is to determine if the applicant's knowledge and experience match the job competencies identified in your ad. In some cases, the candidates may specifically describe how they fulfill each requirement. In other cases, you will need to review general descriptions of their education and job experience to determine if they are a potential match.

Example of an AV Job Advertisement

AV Technician for High-Growth Company

Interested in joining a challenging, growth-oriented organization? ABC Integrated Solutions is an AV systems integration and service company located in Anytown, USA. Join a firm that provides tremendous training and career growth opportunities in an expanding industry.

As an AV technician, you will:

- Fabricate racks in-house

- Set up, test, operate, tear down, and troubleshoot AV equipment

- Work on a team with other skilled technicians

- Use the latest AV equipment in the industry

- Learn from seasoned AV professionals

Qualified candidates should know how to solder and read wiring drawings. Knowledge of DEF, GHI, JKL, MNO, and PQRS equipment a plus. 1–2 years AV experience and flexibility working with client scheduling requirements are required attributes.

AV technicians can expect:

- Compensation commensurate with experience

- Medical, dental, life insurance

- Short- and long-term disability

- Tuition reimbursement

- 401(k) plan

- Generous paid time off

- On-site training

Send your resume and cover letter to HR@ABCIntegrated.abc or fax to 555-555-5555.

Issues and questions that the short-listing process should address include the following:

- Do the knowledge and experience described in the cover letter and resume align with the identified job duties and requirements?

- Does the applicant appear to have the required level of job experience and/or education?

- Do the applicant's cover letter and resume look sufficiently complete, competent, and professional?

- Does anything in the response stand out in a highly positive manner? For example, does the candidate possess specific work experience, education, or certifications that would be highly desirable for the position?

- Does anything in the cover letter or resume stand out in a negative manner? Here are some examples of responses that provide hints of potential problem areas:

  - A job history with many unexplained gaps in employment
  - Several job terminations for cause
  - Indication of unwillingness to do elements of the job role
  - Elements of the applicant's qualifications that appear to be fabricated or suspicious in some manner

The objective of the short-listing process is to rate the applicants as either qualified or not qualified. Rank each of the qualified candidates according to how well that applicant appears to meet the requirements in each major area of the identified job responsibilities. Next, select the top candidates that you want to invite to interview.

Interviewing Applicants

The job interview is your opportunity to get to know more about the job applicant, clarify your understanding of the applicant's background and qualifications, and learn about the candidate's interests and objectives. It gives you an opportunity to ask questions to gain a clearer understanding of what each candidate has to offer.

The interview is typically a straightforward process. Having selected qualified applicants during the short-listing process, you or someone in your company should contact them and invite them for an interview. Face-to-face interviews are preferable to telephone interviews because seeing candidates during discussions offers a better chance to assess their truthfulness.

The interview should include several people from the AV company, such as employees who would work with the candidate and the potential supervisor. You might also include other staff members with specialized knowledge and skills to help assess the technical capabilities of the candidate.

It's important that your company actually prepare for the interview. The selected interviewers should review the candidate's resume prior to the interview to identify any specific issues to clarify or discuss. It is also a good idea to prepare some standard interview questions that will be asked of all candidates.

It is important to remember that laws and regulations that prohibit discrimination can be a serious issue when it comes to hiring. There are numerous subjects and topics that employers may be prohibited from discussing during job interviews under various antidiscrimination laws. When asking interview questions, keep in mind that something that would be considered improper on a written application would also be improper in a verbal interview.

Table 26-1 provides some examples of questions that may be allowable or prohibited, based on the jurisdiction in which the interview takes place.

Sample Initial Interview Questions
Name of Applicant _____
Position _____
Department _____

- What prompted you to apply to our company?

- What are some of the things that you value in an employer?

- How would you describe your work style and your work ethic?

- How do you stay current with trends in your field?

- What have you done in the last year to continue your learning/education?

- If you had only one word to describe yourself, what would it be? Why?

- What challenges do you foresee in this type of job, and how would you overcome them?

- What are the clues you have come to recognize to indicate that you are under too much stress?

- Would you feel comfortable with us contacting your bosses, peers, subordinates, or customers?

- What do you think your bosses, peers, subordinates, or customers would say about you?

- Describe an instance when you had to overcome a difficult situation. How did you do it?

- What do you like most about your current position? Why? What do you like least? Why?

- When did you last receive feedback at work that made you feel proud? When did you receive criticism that upset you?

- Describe a situation where multitasking was necessary to achieve a specific goal. How did you feel about that situation?

- Do you have any specific salary requirements?

- What resources do you use to manage your time? May we see them in a follow-up interview?

- What is your motivation to succeed?

- What do you need your next employer to provide for you to succeed?

(Continued)

- Rate yourself in the following areas (1 being weak, 10 being strong):
  Organization skills: _____
  Creative thinking: _____
  Analytical thinking: _____
  Interpersonal skills: _____
  Technical skills: _____
  Time-management skills: _____
  What have you done in the past year to improve the weakest of these skills?

Age	
Do	**Don't**
• Ask if the person is of legal age to work	• Ask how old the person is
	• Ask what the person's birthdate is

Sex, Family, and Marital Status	
Do	**Don't**
• State your company policy regarding work assignment of employees who are related	• Ask a female applicant's maiden name
	• Ask what kind of child-care arrangements he/she will have
• Ask if the person knows of any reason why he/she would not be able to get to work on time and on a regular basis	• Ask if his/her spouse expects him/her home at a certain time
• Ask (if pertinent) if he/she is available to work overtime and if there are any limitations on his/her ability to work overtime	• Ask what he/she will do if his/her children get sick
	• Ask how he/she will get to work
	• Ask how many children he/she has
• Ask if there any reasons why he/she might not stay if hired	• Ask if his/her spouse lives with him/her or contributes to his/her support
• Ask what his/her career objectives are	• Ask if he/she is married
• Ask if he/she intends to stay in the area	• Ask if he/she plans to get married
• Ask if he/she is willing to relocate to this or any other area	• Ask if he/she plans to have children
	• Ask if he/she is likely to quit if he/she gets married or has children
• Ask if there are any areas to which he/she would not be able to travel for business reasons	• Ask if his/her spouse is likely to be transferred
	• Ask if his/her spouse is from this area
• Ask if he/she is willing to entertain clients on weekends and evenings	• Ask if he/she gets along well with other men/women
	• Ask what the person's spouse's name is
	• Ask if he/she owns a home
	• Ask if he/she owns a car
	• Ask if he/she has any debts
	• Ask if he/she has any loans

Table 26-1   Interview Dos and Don'ts

Race and Color

Do

- Avoid changing the structure of the interview based on a person's race or color

Don't

- Ask whether it will create any difficulties if the person's supervisor is of a specific race
- Ask how he/she feels about having to work with members of a different race
- Ask if the person is a militant

Religion, National Origin, and Birthplace

Do

- Avoid changing the structure of the interview based on a person's origin or religion
- Be understanding if a person asks to reschedule an interview based on a religious observance
- Provide a statement of the employer's regular days, hours, or shifts to be worked
- Ask what other languages he/she reads, speaks, or writes, if this is applicable to the job
- Ask if, after employment, he/she can submit proof of legal authorization to work in the United States

Don't

- Ask what language his/her mother speaks
- Ask if the person was born in this country
- Ask any questions with regard to nationality, lineage, ancestry, national origin, descent, or parentage—about the applicant or any of the applicant's family members
- Ask if the person has people in the "old country"
- Say "That's an unusual name. What nationality are you?"
- Ask he/she to provide a photograph of himself/herself
- Ask what religious days he/she observes
- Ask if his/her religion prohibits him/her from working weekends or holidays
- Ask what language he/she uses most
- Ask how he/she learned another language other than English
- Ask if the person is a citizen of the country
- Ask the person to provide citizenship documents prior to the decision to hire the applicant
- Ask which church he/she attends
- Ask what his/her religion is

Physical and Mental Disability

Do

- Ask if the person can perform job-related tasks

Don't

- Ask if he/she is in good health
- Ask if the person has ever filed for workers' compensation
- Ask if he/she has any disabilities
- Ask if he/she has been treated by a psychiatrist or psychologist
- Ask if he/she has ever been treated for alcoholism or drug addiction
- Ask if he/she has ever been injured on the job
- Ask how many sick days he/she took last year

Table 26-1 Interview Dos and Don'ts

Example of a Candidate Evaluation Form

Name of Applicant _____

Position _____

Department _____

Rate the applicant in the following areas:

	Excellent	Good	Fair	Poor	N/A
Education					
Job Experience					
Supervisory Experience					
Technical Skills					
Interpersonal Skills					
Motivation					
Time Management Skills					
Attention to Detail					

Strengths

Comments _____

_____

Weaknesses

Comments _____

_____

Overall Ranking

[ ] Excellent

[ ] Meets job requirements

[ ] Does not meet job requirements

[ ] Not applicable for this position

Comments _____

_____

_____

Salary Expectations _____

Date Candidate Available to Begin Work _____

Interviewer _____

Date of Interview _____

Keep in mind that in some cases, a company may be required to collect certain types of personal information for government reports through the use of nonidentifying demographic questionnaires. However, companies cannot ask about this information in an interview.

Interviewers should take notes during the interview and discuss their impressions of a candidate once the interview is complete. You can use the evaluation form found on the previous page to record interviewers' impressions of individual candidates.

Maintaining and Improving Staff Skills

The AV industry is defined by rapid technical innovation. To keep up, AV professionals must take steps to upgrade their knowledge and skills throughout their careers. An AV company can assist in this process by arranging for staff to receive training classes and to obtain and maintain professional certifications (such as CTS). AV professionals can keep up-to-date by participating in technical seminars provided by trade organizations and vendors, reviewing trade publications, and attending trade shows.

Employers may also want to conduct regular in-house training sessions to address the following issues:

- Upgrade staff skills in specific areas in order to make the company more competitive.

- Provide cross-training for staff so that the company does not rely on one or two key staff members for performing critical tasks.

- Resolve identified work performance problems.

Employers may also consider developing evaluations or tests as part of a certification process for new staff or for employees working to upgrade their skills. In some cases, special certification may be required for people working in specific job roles or with specific types of technology. Such certifications may be dictated by legal requirements, such as needing to use a licensed electrician, or they may be required by the client to provide assurance that the workers are qualified to perform the work tasks. An AV company should identify these types of opportunities for staff to improve their professional skills and prompt staff to take advantage of them.

Assessing Staff Performance

All companies should monitor the performance of their staff and provide employees with feedback that will aid in aligning staff performance with the needs and expectations of the company. Feedback can come from the following sources:

- **Supervisors** Feedback should be given on a regular basis. It shouldn't come only as part of a once-a-year evaluation to justify wage increases or promotions. Feedback should give staff members targets or other performance suggestions they can use to guide their work. This also gives the staff member's supervisor something specific to evaluate at the next feedback session.

- **Peers**   Supervisors should seek performance feedback from the staff member's peers in the organization. Because they work closely with the staff member, such peers may be able to provide insight into the person's capabilities, strengths, and weaknesses. This information may help the supervisor identify areas that require additional effort or training. (Of course, the supervisor should keep in mind that there may be personal issues between the staff members, such as friendships or disputes, which may color peer assessments.)

- **Senior staff or experts**   Feedback should be obtained from other supervisors or experts in specific tasks or technologies related to the employee's evaluation. Senior staff can observe the performance of specific tasks to ensure that employees are able to meet the required standards.

Promoting Staff

Promoting staff to new positions may be necessary because your company is growing, or because an employee has matured in a position or gained new skills. Most staff members want to move up within an organization as they develop their skills. If your company does not provide promotion opportunities, employees may leave the company to advance in their careers.

If appropriate positions are not available to provide room for advancement, consider increasing pay. Also consider cross-training to provide people with opportunities to gain fresh knowledge or move into other positions (such as an installer moving to an administrative or sales position).

Disciplining Problem Staff

An AV manager may find it difficult to deal with staff performance or behavior problems. To address such issues, a supervisor should do the following:

- Clearly identify and document the nature of the problem.
- Document instances when the staff member has displayed the inappropriate performance or behavior.
- Clearly define what the staff member needs to do to correct performance problems.

Supervisors should strive to be professional in these interactions and remain objective. Once the staff member has been informed of the problem and advised how to correct it, the supervisor should monitor the staff member's subsequent behavior and follow up to provide feedback on whether the staff member has adequately addressed the problem.

Chapter Review

This chapter focused on how to hire the most qualified staff by identifying the required skills, writing employment ads, and interviewing applicants. And because staff members should strive to maintain and improve their skills, this chapter also reviewed how companies can prompt employees to obtain continuing education and professional certifications. Finally, we looked at how to assess staff performance and provide feedback.

Review Questions

The following review questions are not CTS exam questions, nor are they CTS practice exam questions. They are included here to help reinforce what you've learned in this chapter. For an official CTS practice exam, see the accompanying CD.

1. What should be the first step in the hiring process?
    A. Determining if the company already has someone in-house who can perform the work
    B. Identifying the specific job requirements
    C. Identifying the skills, knowledge, and experience needed by prospective employees
    D. Determining the required education and certifications for the prospective employee

2. Which of the following best describes how a job advertisement should function?
    A. Describe the job requirements in as much detail as possible to ensure that the applicant understands the requirements
    B. Provide only a few details, so that the applicant contacts the company to learn more
    C. Provide the main details about the position in a manner that "sells" the company and position to the prospective employee
    D. Promote the company in a manner that convinces the applicant that it would make an attractive career choice

3. When evaluating resumes of prospective job applicants, how should the reviewers decide who to invite for interviews?
    A. Only applicants who have previously worked with well-established companies should be selected for an interview.
    B. Reviewers should call current employer and references of applicants to determine if the resume is accurate prior to selecting applicants for interviews.
    C. The reviewers should evaluate candidates against the stated job requirements, and create a short list of qualified candidates to invite for interviews.
    D. Reviewers should select only candidates whose resumes exactly match the stated job requirements.

4. Which of the following best describes the interview process?

   A. The interviewers are legally allowed to ask only standard questions that focus on the person's work experience.

   B. The interview should be an informal process of getting to know the candidate in order to determine if the candidate's personality will be a fit within the organization.

   C. Interviewers should focus on technical questions intended to assess the candidate's knowledge of the industry.

   D. Interviewers should ask a range of prepared questions intended to assess the candidate's knowledge, skills, and work experience, as well as overall personality, behavior, and attitudes.

5. What is the main purpose of providing staff with feedback on job performance?

   A. Once-a-year feedback is used to determine salary increases and promotions.

   B. Regular specific feedback should be provided to give staff members guidelines for improving performance.

   C. Feedback should be provided when a complaint is received about an employee.

   D. Feedback is used to identify areas that require additional training.

Answers

1. **B.** The first step in the hiring process is to identify the specific job requirements that you want the employee to fulfill.

2. **C.** A job advertisement should be written in a manner that provides the main details and requirements of the position, presented in a way that helps sell the company and position to the prospective employee.

3. **C.** Reviewers should evaluate each of the candidate resumes against the stated job requirements and create a short list of the most qualified candidates who will be invited for interviews.

4. **D.** In general, interviewers should ask job candidates a range of prepared questions that are intended to assess the candidate's knowledge and skills; relevant work experience; and overall personality, behavior, and attitudes.

5. **B.** The main purpose of feedback should be to give staff members specific information that will provide targets or guidelines for improving their performance.

PART VII

Appendixes

Future Trends in Professional AV

Throughout the *CTS Exam Guide*, we emphasize that technology changes quickly, and the AV industry must constantly adapt. This is one reason that CTS certification must always be maintained and reinforced with new knowledge and training. This appendix introduces some of the new areas of expertise you can expect to learn about as you maintain your CTS credentials in the years to come: security for networked AV systems, sustainability of AV systems, and the use of smart building technology.

Secure AV

Networked AV systems and information technology (IT) issues are discussed in Chapters 6, 16, and 19 of this book. If you've spent time working in IT or collaborating with IT staff, you know that network security is a top consideration of all IT professionals. As you design and integrate AV systems to run over IT networks, their security concerns must also be yours.

Some markets for AV systems require a heightened level of security, most notably government, health-care, and financial services. In government installations, especially in the federal government, AV systems are often deployed in sensitive compartmented information facilities (SCIFs). And even when they are not installed in SCIFs, networked AV systems usually must adhere to a variety of government standards for protecting information, most notably the Federal Information Processing Standard (FIPS), overseen by the National Institute of Standards and Technology (NIST).

Other markets have unique security requirements, such as the need for financial services providers to meet standards to ensure the protection of consumers' financial information. The health-care industry operates under legislation such as the Health Insurance Portability and Accountability Act (HIPAA), which regulates, among other things, patients' privacy and health records. For example, there is a growing trend in acoustic solutions for health-care facilities to ensure patient conversations that are protected under HIPAA may not be overheard.

As AV professionals deal more with clients that require secure, networked AV systems, they will begin to think differently about security. They must think in terms of an organization's security posture: what it needs to protect, how important it is to protect those resources, and how its organizational culture affects its tolerance for risk and its response to risk. A customer's security postures may limit what equipment can be designed into an AV installation, and it will surely affect how the overall system is designed and configured.

When identifying the types of security measures required for AV systems, it is important to address confidentiality (who has access to information), integrity (who can do what with the information), and availability (how long the customer can be without the information, in the event of a technology failure or other disruption). From there, those who work in sensitive markets must address the different kinds of security, including the following:

- **Operational security** This includes everything from controlling physical access to secure facilities, to designing systems for continuity of operations (COOP) in the event of system failure or natural disaster. AV professionals must constantly ask questions about operational security, such as whether the architecture of a secure room could inadvertently allow the transmission of sound beyond the room.

- **Communications security** This involves protecting information exchange, such as during a videoconference, as data passes over a network. This type of security often involves some type of data encryption, which can vary based on the type of network that is handling the transmission.

- **Emissions security** This deals with the potential that information could be compromised through the electromagnetic emissions surrounding a cable or wire transmitting secure AV data. It also applies to wireless signals. Emissions security is often addressed through cable shielding or the use of fiber-optic cables, which don't create magnetic fields.

- **Application security** This refers to encryption or security measures implemented at the Application layer of the OSI model (see Chapter 6). Users needing to log in with a secure password to access a network AV system is an example of application security.

- **Network security** This comprises many of the best-known technologies used to control unauthorized access to a network, such as firewalls and intrusion detection devices. An AV professional designing or installing a system will not necessarily be responsible for network security. But in order to make a networked AV system functional, the AV professional may need to coordinate with the IT staff to ensure that appropriate AV system traffic can pass through the organization's network security.

As more AV systems integrate with clients' IT networks, and as AV companies perform more work with clients that have unique information security requirements, secure AV will be a growing requirement. AV professionals will need to know how to do the following:

- Assess and mitigate risk.
- Coordinate with IT and other security professionals to ensure AV systems adhere to all security requirements.
- Identify and document AV equipment, systems, and designs that comply with clients' security standards.

Sustainability

Whether you call something "green" or "sustainable," it means the same thing. It's a sign that the product, system, or building in question consumes less energy than others of its kind and has a minimal impact on the environment. For example, hybrid cars and Energy Star appliances are considered sustainable products.

In 1998, the United States Green Building Council (USGBC) began its Leadership in Energy and Environmental Design (LEED) program. LEED was created to give architects, builders, building owners, and building operators a framework for implementing sustainable design, construction, operation, and maintenance solutions.

By the following decade, LEED had grown popular with technology solution providers. Those who worked in LEED-eligible buildings felt their own commitment to sustainability and green systems, and saw an opportunity to contribute to the LEED movement. After all, a new office with a well-integrated videoconferencing system, for example, could reduce a company's carbon footprint by eliminating travel and generating other sustainable benefits. A control system that can turn down lights or monitor a building's projector systems could save energy.

In 2008, BICSI, an association that supports the information transport systems (ITS) industry, created the Green Building Technology Alliance (GBTA), which included InfoComm International, the Telecommunications Industry Association (TIA), and the Continental Automated Buildings Association (CABA). The goal was to write technology-related credits that could be incorporated into the LEED rating system and make technology systems, including AV, part of sustainable building. The U.S. Green Building Council (USGBC), which must consider a wide variety of criteria in sustainable buildings—from the building materials to its proximity to public transportation—chose not to adopt any of the GBTA's proposals.

But AV and other like-minded industries remained interested in making their systems energy efficient and sustainable, in part because clients were asking about it. A public that has been exposed to a steady stream of green messaging has come to wonder how the AV systems they use can be greener.

For AV professionals, a couple of new initiatives will help guide the industry to sustainable solutions. The first is a standard: ANSI/InfoComm 4: 2012, Audiovisual Systems Energy Management. Approved by the American National Standards Institute (ANSI) in 2012, the standard spells out processes and requirements for the ongoing power management of an AV system. It prescribes control and continuous monitoring of electrical power for AV systems, encouraging conservation whenever possible, and operating AV components at the lowest possible power-consuming state without compromising performance.

The second initiative for promoting sustainable AV systems is the Sustainable Technology Environments Program (STEP). STEP emerged from a pair of task forces established by InfoComm. It was written by a diverse group of industry professionals, including several LEED Accredited Professionals (LEED APs), to help promote sustainable technology. Today, the program is administered by the STEP Foundation, an independent nonprofit organization founded by InfoComm, BICSI, TIA, and the Computing Technology Industry Association (CompTIA).

The STEP Foundation (www.thestepfoundation.org) oversees the new STEP rating system, which is a voluntary system for rating the sustainability of information communications technology systems in the built environment. In many ways, it is like LEED for technology.

The STEP rating system is broken down into five phases: program, architectural and infrastructure design, system design, systems integration, and operation. Each phase has credits associated with it. The more credits a project earns, the higher its eventual STEP rating. The current, nearly 200-page STEP Reference Guide details every current STEP credit. But this number will grow as STEP Foundation members from outside the AV industry add their credits to the rating system.

Currently, several STEP pilot projects are underway by AV integrators and consultants. The Randal A. Lemke Center for Professional Development at InfoComm headquarters in Fairfax, Virginia, is the first completed STEP project.

Smart Buildings

The AV industry often sits at the crossroads of significant changes in building design and technology. Today, AV/IT network convergence, sustainable technology, and something called *integrated project delivery* (IPD) are changing the way people design, inhabit, and use buildings, from schools to offices. The result is smart building technology (SBT), a growing force within the design and construction industries that benefits from a strong AV industry contribution.

Building owners increasingly want their facilities to be "intelligent"—to respond to the ways people behave in the building and to changes in the microclimate. The goal of SBT is to integrate disparate building systems—automation systems, management systems, HVAC, life safety, security, AV, and lighting—in the same way that AV professionals integrate AV systems in a room or venue.

SBT integration comes down to the following:

- **Physical integration** This means putting many building systems on a common structured cabling system and consolidating cable pathways and equipment rooms throughout an entire building. Physical integration recognizes the physical convergence of systems and offers an opportunity to use one cable contractor, which should reduce the cost of installing the various systems.

- **Network integration** This reflects more than just AV and IT convergence. Modern building automation and management systems use protocols such as BACnet, Modbus, and LonWorks, which communicate via IP over an IT network. Increasingly, AV systems can speak those building system languages.

- **Application integration** This allows different building systems to work together and provide extra functionality. For example, an employee security card that is embedded with identifying information for access control could also trigger HVAC, AV, and lighting systems to adjust when that employee enters a room. Application integration also means facility managers can better monitor and manage a building's performance, from air quality to projector bulbs.

AV professionals are well suited to integrating SBT. They are widely recognized as early adopters of new technologies. When clients want the latest systems in their boardrooms, lobbies, hospitality suites, and other venues, it often entails an AV experience, and members of the AV industry are called upon to make it work. Also, in recent years, there has been considerable emphasis on AV system ease of use and ease of operation. In response, AV programmers, consultants, and integrators have developed unique skills for creating intuitive, user-friendly tools and control interfaces to mask the connections among complex systems that don't normally communicate with one another. That's the challenge of integrating disparate building systems.

InfoComm has spent considerable time and resources studying SBT and SBT opportunities for AV professionals. It has sponsored a pair of volunteer task forces dedicated to quantifying the requirements and capabilities of smart buildings and identifying business opportunities and solutions. Because SBT is bigger than one industry, InfoComm coordinates with other groups, such as BICSI, the International Society of Automation (ISA), and the American Society of Heating, Refrigerating and Air-Conditioning Engineers (ASHRAE).

STEP and the STEP rating system offer one path for helping building owners implement and measure the benefits of SBT. Ultimately, what building owners and developers say they want in smart buildings should be achievable through STEP, and confirmed via the STEP rating system.

InfoComm Standards

The American National Standards Institute (ANSI) is the official U.S. representative to the International Organization for Standardization (ISO). InfoComm International's Certified Technology Specialist (CTS) certification exam, for which you are studying, is ANSI-accredited under the ISO and the ISO/IEC 17024:2012 Conformity Assessment—General Requirements for Bodies Operating Certification Schemes of Persons standard.

In addition, InfoComm is an ANSI-accredited Standards Developer (ASD), developing voluntary standards for the commercial AV industry. Accreditation by ANSI signifies that the processes used by standards development organizations (SDOs) to develop ANSI standards meet the ANSI's requirements for openness, balance, consensus, right to appeal, and due process. Subject matter experts work cooperatively to develop voluntary ANSI/InfoComm standards. There are currently more than 450 volunteers involved in task groups working on industry standards for audio, video, control, documentation, automation, and sustainability.

ANSI/InfoComm standards are not product-specific standards. They are system performance standards, management standards, documentation standards, and verification standards. These standards provide guidance for AV system performance. They take into account technology, physiology, architecture, and other variables in determining the best way to design, implement, and manage the performance of all types of AV systems. You can keep up with news about current standards and those in development by visiting InfoComm's website (www.infocomm.org/standards).

Don't be surprised if you see references to some of these standards on the CTS exam. It is important for CTS professionals to be able to recognize and understand relevant standards. As a CTS-certified AV professional, you should consider standards when designing, installing, and managing AV systems (although implementing ANSI/InfoComm standards is not a requirement for being CTS-certified).

In this appendix, we offer a brief synopsis of the existing ANSI/InfoComm standards.

ANSI/INFOCOMM IM-2009, Audio Coverage Uniformity in Enclosed Listener Areas

One of the fundamental goals of sound system performance for both speech reinforcement and program audio is the delivery of consistent audio coverage in the listening area. A well-executed audio system design is one that allows all listeners to hear the system at approximately the same sound pressure level (discussed in Chapter 16 of this book) throughout the desired frequency spectrum range, no matter where they're positioned in the designated listening area. The ANSI/INFOCOMM 1M-2009, Audio Coverage Uniformity in Enclosed Listener Areas standard provides a procedure to measure this spatial coverage, and criteria for use in the design and evaluation of audio systems.

ANSI/INFOCOMM 2M-2010, Standard Guide for Audiovisual Systems Design and Coordination Processes

A successful professional AV system installation depends on the clear definition and coordination of processes, resources, and responsibilities among AV design and installation project teams. A properly documented AV system provides the information necessary to understand and implement the system goals and project requirements in a logical and efficient manner. The documentation should complement and coordinate related architectural, engineering, and construction documentation.

The ANSI/INFOCOMM 2M-2010, Standard Guide for Audiovisual Systems Design and Coordination Processes standard outlines a consistent set of the standard tasks, responsibilities, and deliverables required for professional AV systems design and construction. It provides the basis for several of the domains/tasks on the CTS exam, including much of the planning detailed in Part III of the *CTS Exam Guide*.

ANSI/INFOCOMM 3M-2011, Projected Image System Contrast Ratio

The ANSI/INFOCOMM 3M-2011, Projected Image System Contrast Ratio standard, which applies to both permanent and temporary installations, defines requirements for minimum contrast ratios for both front- and rear-projection AV systems. It also provides metrics for measuring and validating the contrast ratios as defined.

ANSI/INFOCOMM 3M-2011 puts the needs of the viewer ahead of all other considerations. The system contrast ratio refers to the image as it is presented to viewers in a space with ambient light. The standard's requirements go beyond the individual performance factors of a projector and a screen because the contrast ratio they ultimately deliver is affected by the light in the space.

The standard defines four contrast ratios, based on the type of content an audience needs to view: passive viewing, basic decision making, analytical decision making, and full-motion video. It provides metrics for evaluating, planning, and designing projected-image system installations.

ANSI/INFOCOMM 4:2012, Audiovisual Systems Energy Management

An AV system that is designed to minimize electrical power consumption uses power monitoring and automated component control to achieve its goal. And this applies not only when the AV system is in operation, but also when it's in standby mode or isn't used at all. Design of the technical architecture of the AV systems and components, implementation based on design documentation, and thorough testing procedures of installed systems are critical to the success of an energy management program.

The ANSI/INFOCOMM 4:2012, Audiovisual Systems Energy Management standard defines and prescribes processes and requirements for ongoing power-consumption management of an AV system. It identifies requirements for the control and continuous monitoring of electrical power for AV systems, whereby power is conserved whenever possible and components operate at the lowest possible power-consuming state without compromise to the system's performance.

AV systems in conformance with the standard meet the defined requirements for automation, measurement, analysis, and training.

AV Math

How long has it been since you solved a word problem or used a math formula? Many CTS exam candidates have not done math in a formal setting in years. You may be familiar with the skills and tools in this section, but then again, you may need a refresher.

Using the Proper Order of Operations

Most AV math formulas use only the four common operators: add, subtract, multiply, and divide. However, some formulas require a solid foundation in the order of operations. The order of operations helps you correctly solve formulas by prioritizing which part of the formula to solve first. It is a way to rank the order in which you work your way through a formula. This section will review how to apply the proper order of operations.

This is the order of operations:

1. Any numbers within parentheses or brackets
2. Any exponents, indices, or orders
3. Any multiplication or division
4. Any addition or subtraction

If there are multiple operations with the same priority, then proceed from left to right: parentheses, exponents, multiplication, division, addition, and then subtraction. Several acronyms can help you remember the order of operations: PEMDAS, BEMDAS, BIDMAS, and BODMAS.

Steps to Solving Word Problems

All math formulas summarize relationships between concepts. Word problems are designed to test how well an individual can apply that relationship to a new situation.

By following a few basic steps, you can turn a complicated word problem into a few straightforward steps. This section provides a structured approach to solving problems.

Within this structure, you will find many strategies for solving different types of problems. This strategy is based on *How to Solve It: A New Aspect of Mathematical Method*, by G. Polya (Princeton University Press, 1985).

Step One: Understand the Problem

As typical within the AV industry (and in general), the first step is to understand the problem you're trying to solve. Here are the tasks to complete for this step:

- Read the entire math problem.
- Identify your goal or unknown. What information are you trying to determine?
- Identify what you have been given. What data, numbers, or other information in the problem can help you determine the answer?
- Predict the answer if you can. What range of values would make sense as an answer?

Example: Calculate the current in a circuit where the voltage is 2V and the resistance is 8 ohms.

First, identify your goal. What are you trying to solve for? When you see the word *calculate*, generally the word that follows is your goal. Other words that identify the goal include *determine*, *find*, and *solve for*. Your goal in this problem is to calculate current.

An easy way to identify your given information is to find the numbers in the problem. In this example, the numbers are 2 and 8. Look for context clues or units to identify what those numbers represent. "The voltage is" identifies 2 as a voltage. The "ohms" after the 8 identifies 8 as the resistance.

Sometimes it's unclear what each number represents. In that case, drawing a diagram can help you make sense of what the problem is trying to say. You may want to make a chart of your given and unknown information for quick reference. For more complex problems, tables of given information can be extremely helpful.

Givens and Goal	Values
Current	?
Voltage	2V
Resistance	8 ohms

Step Two: Create a Plan

The second step in this process is to translate the words in the problem into numbers you can enter into a formula. Begin with the following:

- Assign appropriate values to the goal and given information.
- Determine a formula that describes the relationships between your variables.

If there is a single formula that has all your variables in it, move on to the next step.

In some cases, you may be unable to identify a formula that will determine your goal based on your given information. If you're stuck, consider the following strategies:

- Use an intermediate formula to solve for the information you are missing.
- Use an outside reference, such as a chart or graph, to find information not listed in the problem.
- Use a strategy that has worked to solve similar projects in the past.
- Diagram the scenario described in the word problem. Use the diagram to keep track of the relationships between values. For instance, as listeners move farther away from a sound source, you know to expect a loss of sound pressure.

Example: Calculate the current in a circuit where the voltage is 2V and the resistance is 8 ohms.

Again, start by assigning variables to your given and unknown information. Note that some items are represented by different variables in different contexts. If you are having trouble determining which variables to use, consider drawing a diagram and labeling it with your given information, like this:

Givens and Goal	Values	Variables
Current	?	I
Voltage	2V	V
Resistance	8 ohms	R

Once you have assigned variables, think about the relationship between the information. There should be a formula that describes the relationship. For complex problems, you may need to use several formulas.

You may know formulas from memory. If you don't, look them up (we've provided several useful formulas at the end of this appendix). It is helpful to think about a time when you solved for the value previously, or a similar problem you may have solved in the past. Try to think of a problem that used the same givens and unknowns.

For example, you might not remember how to solve for current using voltage and resistance. But if you remember how to solve for voltage using current and resistance, you can solve this problem. Voltage is equal to current multiplied by resistance.

Step Three: Execute Your Plan

The third step is to put your plan in action, as follows:

- Write the formula(s).
- Substitute the given information for the variables.
- Perform the calculation.
- Assign units to your final answer.

Example: Calculate the current in a circuit where the voltage is 2V and the resistance is 8 ohms.

Once you have determined the appropriate formula, write it down, and then replace the placeholders with the numbers for this problem. People who skip this step are prone to making mistakes.

Formula: $V = I * R$

Substitution: $2 = I * 8$

To solve this equation, you need to get the I by itself. The 8 is currently being multiplied. To move it to the other side of the equation, perform the opposite mathematical function. In this case, that function is division:

$2 / 8 = (I * 8) / 8$

$2 / 8 = I$

$0.25 = I$

You need to assign units to your answer before it is final. Because I represents current, you would assign 0.25 the unit for current, which is amps (A).

$I = 0.25$ A

Step Four: Check Your Answer

Your final step is to make sure the numbers you've calculated still make sense when translated back into words. Compare your answer to the scenario described in the problem. Is the result reasonable? Is it within the range you originally predicted?

For example, suppose that you are calculating the voltage present in a boardroom loudspeaker circuit. A result of 95V is probably a reasonable answer. A result of 50,000V indicates that you made a mistake in your calculations.

If you have an incorrect answer and use it to solve other parts of a process, it will result in cascading problems.

Example: Calculate the current in a circuit where the voltage is 2V and the resistance is 8 ohms.

In this example, is less than 1 amp a reasonable answer? Understanding the problem is essential here. A D battery has a voltage of 1.5V. When a D battery is attached to a circuit, there is not much current. So the small number of 0.25 amps is a reasonable answer.

Rounding

Many of the results listed in workbook answer keys have been rounded to the nearest tenth. When solving multistep problems, you may be tempted to round at each step. The earlier or more often you round in a multistep problem, the less accurate your result will be. Only round your final result.

AV Math Formulas

This section presents some common math formulas that may be useful for AV professionals.

Aspect Ratio Formula

The formula for finding aspect ratio is:

$$AR = W / H$$

where:

- *AR* is the aspect ratio.
- *W* is the width of the displayed image.
- *H* is the height of the displayed image.

Note that *W* and *H* are measured in inches or millimeters.

Screen Diagonal Formula (Pythagorean Theorem)

The diagonal of a screen can be calculated using the Pythagorean theorem:

$$A^2 + B^2 = C^2$$

where:

- *A* is the height of a screen.
- *B* is the width of a screen.
- *C* is the diagonal length of a screen.

Estimating Throw Distance Formula

The formula for estimating throw distance is:

$$Distance = Screen\ Width * Throw\ Ratio$$

where:

- *Distance* is the distance from the front of the lens to the closest point on the screen.
- *Screen Width* is the width of the projected image.
- *Throw Ratio* is the ratio of throw distance to image width.

Refer to the owner's manual of your projector and lens combination to find an accurate formula for your specific projector.

Determining Image Height Formula

The formula for determining image height based on a certain viewing task is:

$$S_H = D / S_L$$

where:

- S_H is the height of the image on the screen.
- D is the distance to the farthest viewer from the screen.
- S_L is the level of screen detail:
  - $S_L = 4$ for inspection detail
  - $S_L = 6$ for detailed use
  - $S_L = 8$ for general viewing

Note that the unit of measurement for S_H will be the same for D, whether inches, feet, or millimeters.

Determining Farthest Viewer by Task Formula

The formula for finding the farthest viewer by task is:

$$D = S_L * S_H$$

where:

- D is the distance to the farthest viewer from the screen.
- S_H is the height of the screen.
- S_L is the level of screen detail.
  - $S_L = 4$ for inspection detail
  - $S_L = 6$ for detailed use
  - $S_L = 8$ for general viewing

Note that the unit of measurement for S_H will be the same for D, whether inches, feet, or millimeters.

Determining Viewer Distance by Text Height Formula

The formula for calculating viewer distance by text height is:

$$D = T_H * 150$$

where:

- D is the distance to the farthest viewer from the screen.
- T_H is the height of the text on the projection screen.

Note that all units in this formula must be identical. Avoid mixing inches with feet or millimeters with meters.

Decibel Formula for Distance

The formula for decibel changes in sound pressure level over distance is:

$$dB = 20 * \log (D_1 / D_2)$$

where:

- dB is the change in decibels.
- D_1 is the original or reference distance.
- D_2 is the new or measured distance.

The result of this calculation will be either positive or negative. If it is positive, the result is an increase, or gain. If it is negative, the result is a decrease, or loss.

Decibel Formula for Voltage

The formula for determining decibel changes for voltage is:

$$dB = 20 * \log (V_1 / V_R)$$

where:

- dB is the change in decibels.
- V_1 is the new or measured voltage.
- V_R is the original or reference voltage.

The result of this calculation will be either positive or negative. If it is positive, the result is an increase, or gain. If it is negative, the result is a decrease, or loss.

Decibel Formula for Power

The formula for calculating decibel changes for power is:

$$dB = 10 * \log (P_1 / P_r)$$

where:

- dB is the change in decibels.
- P_1 is the new or measured power measurement.
- P_r is the original or reference power measurement.

The result of this calculation will be either positive or negative. If it is positive, the result is an increase, or gain. If it is negative, the result is a decrease, or loss.

Current Formula (Ohm's Law)

The formula for calculating current using Ohm's law (see Figure C-1) is:

$I = V / R$

where:

- I is the current.
- V is the voltage.
- R is the resistance.

Figure C-1
Simple Ohm's law
formula wheel

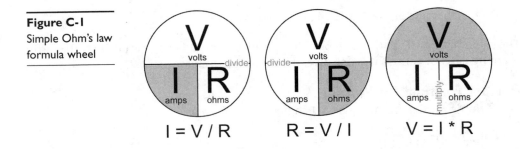

Power Formula

The formula to solve for power (see Figure C-2) is:

$P = I * V$

where:

- P is the power.
- I is the current.
- V is the voltage.

Figure C-2
Simple power
formula wheel

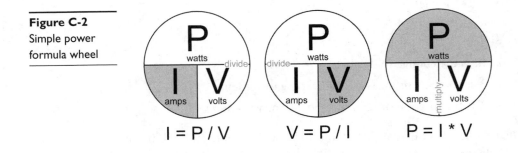

Series Circuit Impedance Formula

The formula for calculating the total impedance of a series loudspeaker circuit is:

$$Z_T = Z_1 + Z_2 + Z_3 \ldots Z_N$$

where:

- Z_T is the total impedance of the loudspeaker circuit.
- Z_x is the impedance of each loudspeaker.

Parallel Circuit Impedance Formula: Loudspeakers with the Same Impedance

The formula to find the circuit impedance for loudspeakers wired in parallel with the same impedance is:

$$(Z_T) = Z_1 / N$$

where:

- Z_T is the total impedance of the loudspeaker system.
- Z_1 is the impedance of each loudspeaker.
- N is the number of loudspeakers in the circuit.

Parallel Circuit Impedance Formula: Loudspeakers with Different Impedances

The formula to find the circuit impedance for loudspeakers wired in parallel with differing impedance is:

$$Z_T = \frac{1}{\dfrac{1}{Z_1} + \dfrac{1}{Z_2} + \dfrac{1}{Z_{3\ldots}} \dfrac{1}{Z_N}}$$

where:

- Z_x is the impedance of each individual loudspeaker.
- Z_T is the total impedance of the loudspeaker circuit.

Series/Parallel Circuit Impedance Formulas

Two formulas are used to calculate the expected total impedance of a series/parallel circuit.

First, the series circuit impedance formula is used to calculate the impedance of each branch:

$$Z_T = Z_1 + Z_2 + Z_3 \ldots + Z_N$$

where:

- Z_x is the impedance of each loudspeaker.
- Z_T is the total impedance of the branch.

Then the parallel circuit impedance formula is used to calculate the total impedance of the series/parallel circuit:

$$Z_T = \frac{1}{\dfrac{1}{Z_1} + \dfrac{1}{Z_2} + \dfrac{1}{Z_{3\ldots}} \dfrac{1}{Z_N}}$$

where:

- Z_x is the total impedance of each branch.
- Z_T is the total impedance of the loudspeaker circuit.

Heat Load Formula

The formula for calculating heat load is:

Total Btu = W_E * 3.4

where:

- W_E is the total watts of all equipment used in the room.
- 3.4 is the conversion factor, where 1 watt of power generates 3.4 Btu of heat per hour.

This formula does not account for the heat load generated by amplifiers.

About the CD-ROM

The CD-ROM included with this book comes complete with a MasterExam practice exam and the electronic book in PDF format. The software is easy to install on any Windows XP/Vista/7 computer and must be installed to access the MasterExam feature. For more practice exams and CTS exam preparation resources, visit www.infocomm .org/certification.

System Requirements

Software requires Windows XP or higher and Internet Explorer 8 or above and 200 MB of hard disk space for full installation. The electronic book requires Adobe Acrobat Reader.

Installing and Running MasterExam

If your computer CD-ROM drive is configured to auto run, the CD-ROM will automatically start up upon inserting the disk. From the opening screen, you may install MasterExam by clicking the MasterExam link. This will begin the installation process and create a program group named LearnKey.

To run MasterExam, select Start | All Programs | LearnKey | MasterExam. If the auto run feature does not launch your CD, browse to the CD and click the LaunchTraining. exe icon.

MasterExam

MasterExam provides a simulation of the actual exam. The number of questions, the types of questions, and the time allowed are intended to represent the exam environment. You have the option to take an open book exam (including answers), a closed book exam, or a timed MasterExam simulation. The practice exam questions included with this edition were subject to the same item development process as questions on the actual CTS exam, but they may or may not reflect the actual exam experience and neither McGraw-Hill nor InfoComm make any guarantees that using this practice exam will ensure passing the actual CTS exam.

When you launch MasterExam, a digital clock display will appear in the bottom-right corner of your screen. The clock will continue to count down to zero (unless you choose to end the exam before the time expires).

Help

You can access the help file by clicking the Help button on the main page (in the lower-left corner). An individual help feature is also available through MasterExam.

Removing Installation(s)

MasterExam is installed to your hard drive. For best results removing programs, use the Start | All Programs | LearnKey | Uninstall option to remove MasterExam.

Electronic Book

The entire contents of the book are provided in PDF format on the CD. This file is viewable on your computer and many portable devices. Adobe's Acrobat Reader is required to view the file on your PC and has been included on the CD. You may also use Adobe Digital Editions to access your electronic book.

For more information on Adobe Reader and to check for the most recent version of the software, visit Adobe's website at www.adobe.com and search for the free Adobe Reader or look for Adobe Reader on the product page. Adobe Digital Editions can also be downloaded from the Adobe Web site.

To view the electronic book on a portable device, copy the PDF file to your computer from the CD and then copy the file to your portable device using a USB or other connection. Adobe does offer a mobile version of Adobe Reader, the Adobe Reader mobile app, which currently supports iOS and Android. For customers using Adobe Digital Editions and the iPad, you may have to download and install a separate reader program on your device. The Adobe Web site has a list of recommended applications and McGraw-Hill Education recommends the Bluefire Reader.

Technical Support

For questions regarding the content of the electronic book or the MasterExam questions and answers, please visit http://www.mhprofessional.com/techsupport. For customers outside the 50 United States, e-mail international_cs@mcgraw-hill.com.

LearnKey Technical Support

For technical problems with the software (installation, operation, removing installations), please visit www.learnkey.com, email techsupport@learnkey.com, or call toll free at 1-800-482-8244.

3.5 mm mini   A connector that is similar in appearance to a 1/4-inch phone connector, but much smaller. It measures 3.5 mm in diameter.

8P8C (eight position, eight conductor)   A modular Ethernet connector. It is attached, or terminated, to the cabling. It is commonly referred to as an RJ-45 connector, which is technically the term for a different connector (8P2C).

AC   See *alternating current.*

acceptable viewing area   A viewing range for a screen suggested as a 45-degree line extending outward from the left edge and right edge of a displayed image.

acoustics   The science that explains and quantifies the interaction of cyclical mechanical compression and rarefaction of a medium, typically air, occurring within a commonly accepted frequency range of 20 Hz to 20 KHz (the audible spectrum) and the physical environment in which those waves occur.

AES   Audio Engineering Society.

AHJ   See *authority having jurisdiction.*

alternating current (AC)   An electric current that reverses its direction periodically.

ambient light   All light in a viewing room produced by sources other than the display.

ambient noise   Sound that is extraneous to the intended, desired, and intentional audio; background noise.

amplifier   An electronic device for increasing the strength of electrical signals.

amplitude   The strength of an electronic signal as measured by the height of its waveform.

analog   A method of transmitting information by a continuous but varying signal.

angularly reflective screen   A screen that reflects light back to the viewer at a complementary angle.

ANSI   American National Standards Institute.

aperture   An opening in a lens regulating the amount of light passing through the lens to the imager.

artifact   A small disturbance that affects the quality of a signal.

aspect ratio   The ratio of image width to image height.

attenuate   To reduce the amplitude (strength) of a signal or current.

audio processor   An electronic device used to manipulate audio signals in some manner.

audio signal   An electrical representation of sound.

audio transduction   Converting acoustical energy into electrical energy or electrical energy back into acoustical energy.

authority having jurisdiction (AHJ)   1. An organization, office, or individual responsible for enforcing the requirements of a code or standard, or for approving equipment, materials, an installation, or a procedure. 2. The entity responsible for interpretation and enforcement of local building and electrical codes (*BICSI Information Transport Systems Dictionary*).

balanced circuit   A two-conductor circuit in which both conductors and all circuits connected to them have the same impedance with respect to ground and all other conductors.

balun   1. Short for *bal*anced to *un*balanced. 2. A transformer used to connect a balanced circuit to an unbalanced circuit. For example, a transformer used to connect a 300-ohm antenna cable (balanced) to a 75-ohm antenna cable (unbalanced).

band   A grouping or range of frequencies.

bandwidth (BW)   1. A range of frequencies. 2. In terms of a circuit or equipment, the range of frequencies that the circuit or equipment can reliably pass. 3. In terms of a spectrum, defines the range of frequencies used or allowed. 4. In terms of networking, the available or consumed data communication resources of a communication path, expressed in terms of bits per second (bps). It is also called *throughput* or *bit rate*.

bandwidth limiting   The result of encoding a higher-quality signal into a lower-quality form, such as RGB converted into S-Video.

baseband   A video signal that has not been modulated.

bend radius   The radial measure of a curve in a cable, conductor, or interconnect that defines the physical limit beyond which further bending has a measurable and/or harmful effect on the signal being transported.

bidirectional polar pattern   The shape of the region where some microphones will be most sensitive to sound from the front and rear, while rejecting sound from the top, bottom, and sides.

bit Shortened form of binary digit, symbolized by ones and zeros. A bit is the smallest unit of digital information.

bit depth The number of bits used to describe data.

block diagram An illustration of the signal path through a given system from source(s) to destination(s).

blocking Pieces of wood that have been inserted between structural building elements to provide a secure mounting point for finish materials or products.

BNC connector A professional type of video connector featuring a two-pin lock. The BNC is the most common and most professional coaxial cable connector because of its reliability and ruggedness. BNC connectors are used to transport many different types of signals, such as radio, frequency, component video, time code, sync, and power.

boundary microphone A microphone design where the diaphragm is placed close to a sonic "boundary," such as a wall, ceiling, or other flat surface. This prevents the acoustic reflections from the surface from mixing with the direct feed and causing phase distortions. This microphone is commonly used in conference and telepresence systems.

branch circuit The circuit conductors between the final overcurrent device protecting the circuit and the outlet(s).

breaker box See *panelboard*.

broadcast domain A set of devices that can send Data Link layer frames to each other directly, without passing through a Network layer device. Broadcast traffic sent by one device in the broadcast domain is received by every other device in the domain.

buffer amplifier An electronic device that provides some isolation between other components.

bus Also *buss*, a wiring system that delivers power and data to various devices.

busbar An electrically conductive block or bar of metal, typically copper or aluminum, that serves as a common connection for two or more circuits.

buzz A mixture of higher-order harmonics of the 50 or 60 Hz noise (hum) originating from the AC power system and audible in the sound system.

BW See *bandwidth*.

byte (B) An 8-bit word. The abbreviation for byte is B.

cable An assembly of more than one conductor (wire).

capacitance The ability of a nonconductive material to develop an electrical charge that can distort an electrical signal.

capacitive reactance (XC) Opposition a capacitor offers to alternating current flow. Capacitive reactance decreases with increasing frequency or, for a given frequency, the capacitive reactance decreases with increasing capacitance. The symbol for capacitive reactance is XC.

captive screw connector Sometimes called a Phoenix connector, a molded plastic connector. Termination requires a wire to be stripped and slid directly into a slot on the connector. A set screw then pushes a gate down to hold the wire in place.

cardioid polar pattern A heart-shaped region where some microphones will be most sensitive to sound predominately from the front of the microphone diaphragm and reject sound coming from the sides and rear.

carrier Modulated frequency that carries video or audio signal.

Category 5 (Cat 5) The designation for 100-ohm unshielded twisted-pair (UTP) cables and associated connecting hardware whose characteristics are specified for data transmission up to 100 Mbps (part of the EIA/TIA 568A standard).

Category 5e (Cat 5e) An enhanced version of the Cat 5 cable standard that adds specifications for far-end crosstalk (part of the EIA/TIA 568A standard).

Category 6 (Cat 6) A cable standard for Gigabit Ethernet and other interconnections that is backward-compatible with Cat 5, Cat 5e, and Cat 3 cable (part of the EIA/TIA 568A standard). Cat 6 features more stringent specifications for crosstalk and system noise than the other types.

CATV Community antenna television system or cable television. Broadcast signals are received by a centrally located antenna and distributed by cable.

CCD See *charged-coupled device.*

CCTV See *closed-circuit television.*

center tap A connection point located halfway along the winding of a transformer or inductor.

charged-coupled device (CCD) A semiconductor light-sensitive device, commonly used in video and digital cameras, that converts optical images into electronic signals.

chassis Also called a *cabinet* or *frame,* an enclosure that houses electronic equipment and is frequently electrically conductive (metal). The metal enclosure acts as a shield and is connected to the equipment grounding conductor of the AC power cable, if so equipped, in order to provide protection against electric shock.

chassis ground A 0V (zero volt) connection point of any electrically conductive chassis or enclosure surrounding an electronic device. This connection point may or may not be extended to the earth ground.

chrominance   The color portion of a composite or S-Video signal.

clipping   The deformation of an audio signal when a device's peak amplitude level is exceeded.

clock adjustment   Also called *timing signals,* used to fine-tune the computer image. This function adjusts the clock frequencies that eliminate the vertical banding (lines) in the image.

closed circuit television (CCTV)   A system of transmitting video signals from the point of origin to single or multiple points equipped to receive signals.

CMRR   See *common-mode rejection ratio.*

coaxial (coax) cable   A cable consisting of a center conductor surrounded by insulating material, concentric outer conductor, and optional protective covering, all of circular cross-section.

codec   An acronym for *coder/decoder.* An electronic device that converts analog signals, such as video and audio signals, into digital form and compresses them to conserve bandwidth on a transmission path.

collision domain   A set of devices on a local-area network (LAN) whose packets may collide with one another if they send data at the same time.

color difference signal   A signal that conveys color information such as hue and saturation in a composite format. Two such signals are needed. These color difference signals are R-Y and B-Y, sometimes referred to as Pr and Pb or Cr and Cb.

combiner   In a process called *multiplexing,* puts signals together onto one cable, constituting a broadband signal.

common mode   1. Voltage fed in phase to both inputs of a differential amplifier. 2. The signal voltage that appears equally, and in phase, from each current carrying conductor to ground.

common-mode rejection ratio (CMRR)   The ratio of the differential voltage gain to the common-mode voltage gain; expressed in decibels.

compander   A device that combines compression and expansion.

component video   Color video in which the brightness (luminance) and color hue and saturation (chrominance) are handled independently. The red, green, and blue signals—or more commonly, the Y, R-Y, and B-Y signals—are encoded onto three wires. Because these signals are independent, processing such as chroma-keying is facilitated.

composite video signal   The electrical signal that represents complete color picture information and all synchronization signals, including blanking and the deflection synchronization signals to which the color synchronization signal is added in the appropriate time relationship.

compression   The action of the air molecules moving closer together.

compressor   A device that controls the overall amplitude of a signal by reducing the part of the signal that exceeds an adjustable level (threshold) set by the user. When the signal exceeds the threshold level, the overall amplitude is reduced by a ratio, also usually adjustable by the user.

condenser microphone   Also called a *capacitor microphone*, a microphone that transduces sound into electricity using electrostatic principles.

conductor   In electronics, a material that easily conducts an electric current because some electrons in the material are free to move.

cone   The most commonly used component in a loudspeaker system and found in all ranges of drivers.

conferencing systems   The technology by which people separated by distance come together to share information. Conferencing systems may include projection systems, monitor displays, computers, satellite connections, video, audio playback devices, and many more items.

continuity   The quality of being continuous (as in a continuous electrical circuit).

control track   The portion along a length of a recorded tape on which synchronization control information is placed. A control track is used to control the recording and playback of the signal.

CPU   Central processing unit. The portion of a computer system that reads and executes commands.

crossover   Used to separate the audio signal into different frequency groupings and route the appropriate material to the loudspeaker or amplifier. This ensures that the individual loudspeaker components receive program signals that are within their optimal frequency range.

crosstalk   Any phenomenon by which a signal transmitted on one circuit or channel of a transmission system creates an undesired effect in another circuit or channel.

current   The amount of electrical charge that is flowing in a circuit, measured in amperes.

curvature of field   A blurry appearance around the edge of an otherwise in-focus object (or the reverse) when the velocity of light going through the lens is different at the edges than at the center of the surface. Curvature of field is due to the lens design.

DA   See *distribution amplifier*.

Dante   A proprietary digital audio Network layer protocol designed by Audinate. Dante sends audio information as Internet Protocol (IP) packets. It is fully routable over IP networks using standard Ethernet switches, routers, and other components.

Dante traffic requires no separate infrastructure; it can coexist with other data traffic. Dante controller software manages data prioritization and audio routes.

dB   See *decibel*.

dB SPL   A measure of sound pressure level represented in dynes per centimeter squared. Its reference, 0 dB SPL, equals 0.0002 dynes/cm². dB SPL is used as a measure of acoustical sound pressure levels, and is a 20 log function.

DC   See *direct current*.

decibel (dB)   A base-ten logarithmic relationship of a power ratio between two numbers. Used for quantifying differences in voltage, distance, and sound pressure as they relate to power.

dedicated ground   An ambiguous term that refers to an insulated equipment grounding conductor.

dedicated power   An ambiguous term that refers to one of more individual branch circuits with a supplemental grounding conductor.

deflection coil   A uniform winding of wire used to electromagnetically direct an electron beam to draw an image on a CRT.

delay   An audio signal-processing device or circuit used to retard the speed of transmission on one or more audio signals or frequencies.

demodulator   An electronic device that removes information from a modulated signal.

depth of field   The area in front of a camera lens that is in focus from the closest item to the camera to the item farthest away.

differential mode   1. Voltage fed out of phase to both inputs of a differential amplifier. 2. Signals measurable between or among active circuit conductors feeding the load, but not between the equipment grounding conductor or associated signal reference structure and the active circuit conductors.

diffusion   The scattering or random redistribution of a sound wave from a surface. Diffusion occurs when surfaces are at least as long as the sound wavelengths, but not more than four times as long.

digital   A method of transmitting information by discrete, noncontinuous impulses.

digital-to-analog converter   An electronic device that converts digital signals into analog form.

D-ILA   JVC's Digital-direct Drive Image Light Amplifier projection system.

DIN connector   For *Deutsche Industrie-Norm*, a connector that follows the German standard for electronic connections.

direct current (DC)   Electricity that maintains a steady flow and does not reverse direction, unlike alternating current (AC). It is usually produced by batteries, AC-to-DC transformers, and power supplies.

direct sound   Also known as *near-field*, sound that is not colored by room reflections.

dispersion   An effect that can be seen when a white light beam passes through a triangular prism. The different wavelengths of light refract at different angles, dispersing the light into its individual components.

distributed sound   A sound system using multiple loudspeakers separated by distance. It typically operates at a lower sound pressure level than a high-pressure system. The loudspeakers are most often suspended over the heads of the listeners.

distribution amplifier (DA)   An active device used to split one input into multiple outputs, while keeping each output isolated and the signal level constant.

DLP   Digital Light Processing by Texas Instruments. A projection system that has technology based on the digital micromirror device (DMD). It uses thousands of microscopic mirrors on a chip focused through an optical system to display images on the screen.

document camera   An imaging device used to create a video image of printed documents or three-dimensional objects.

Domain Name System (DNS)   A hierarchical, distributed database that maps names to data such as IP addresses. A DNS server keeps track of all the equipment on the network and matches the equipment names so they can easily be located on the network or integrated into control and monitoring systems. (See also *reserve DHCP.*)

dome   A type of loudspeaker driver construction. Fabric or woven materials are used to create a dome-shaped diaphragm, and the coil is attached to the edge of the diaphragm.

driver   In audio, an individual loudspeaker unit.

D-subconnector   A generic name for a *D*-shaped serial connector used in data communications.

DTV   Digital television. A signal transmitted digitally.

DVD   Digital video disc or digital versatile disc. An optical storage medium for data or video.

DVI   Digital Visual Interface. A connection method from a source (typically a computer) and a display device that can allow for direct digital transfer of data. The digital signal is limited to 5 meters.

DVI-D   One of two common multipin connectors available for DVI signals. The DVI-D carries only digital information; no analog video information is sent. The digital signal is limited to 5 meters.

DVI-I   One of two common multipin connectors available for DVI signals. The DVI-I adds analog video to the connection, permitting greater distances than the digital signal limit of 5 meters.

Dynamic Host Configuration Protocol (DHCP)   An IP addressing scheme that allows network administrators to automate address assignment.

dynamic microphone   A pressure-sensitive microphone of moving-coil design that transduces sound into electricity using electromagnetic principles.

dynamic range   The difference between the loudest and quietest levels of an audio signal.

early reflected sound   Sound created by sound waves that are reflected (bounced) off surfaces between the source and the listener. The sound waves arrive at the listener's ear closely on the heels of the direct sound wave.

echo cancellation   A means of eliminating echo from an audio path.

EGC   See *equipment grounding conductor*.

electromagnetic interference (EMI)   Improper operation of a circuit (noise) due to the effects of interference from electric and/or magnetic fields.

emissive technology   Any display device that emits light to create an image.

encoded   A signal that has been compressed into another form to reduce size or complexity, as in a composite video signal.

equalizer   Electronic equipment that adjusts or corrects the frequency characteristics of a signal.

equipment grounding   The connection to ground (earth), or to a conductive body that extends that ground connection, of all normally noncurrent-carrying conductive materials enclosing electrical conductors or equipment, or forming part of such equipment. The purpose is to limit any voltage potential between the equipment and earth.

equipment grounding conductor (EGC)   The conductive path installed to connect normally noncurrent-carrying metal parts of equipment together and to the system's grounded conductor, to the grounding electrode conductor, or to both.

equipment rack   A centralized housing unit that protects and organizes electronic equipment.

Ethernet   A LAN system used to transmit data at 10 MBps, 100 MBps, or 1 GBps. Ethernet signals are transmitted serially, one bit at a time, over the shared signal channel to every attached station.

expander   An audio processor that comes in two types: a downward expander and as a part of a compander.

fc   See *footcandle*.

F connector A threaded connector that is used in transmission applications such as cable television. The cable's center conductor also serves as the connector's center pin.

feedback 1. In audio, unwanted noise caused by the loop of an audio system's output back to its input. 2. In a control system, data supplied to give an indication of status, such as on or off.

fiber-optic A technology that uses glass or plastic threads or wires to transmit information.

field In video, one half of a video frame containing every other line of information. Each standard video frame contains two interlaced fields.

filter Removes or passes certain frequencies from a signal.

firewall Any technology, hardware, or software that protects a network by preventing intrusion by unauthorized users and/or regulating traffic permitted to enter and/or exit the network. A firewall controls what traffic may pass through a router connecting one network to another. Firewalls control access across any network boundaries, including between an enterprise network and the Internet, and between local-area networks (LANs) within an enterprise.

fixed matrix A type of display that has a fixed grid on which it re-creates an image.

FL See *focal length*.

flex life The number of times a cable can be bent before it breaks. A wire with more strands or twists per inch will have a greater flex life than one with a lower number of strands or fewer twists per inch.

focal length (FL) The distance, in millimeters, between the center of a lens and the point where the image comes into focus. This is the value given to a lens, stated in inches or millimeters. The shorter the focal length, the wider the angle of the image will be.

focus The act of adjusting a lens to make the image appear clear, sharp, and well-defined.

footcandle (fc) An English unit of measure expressing the intensity of light illuminating an object. A footcandle equals the illumination from one candle falling on a surface of 1 square foot at a distance of 1 foot.

footprint 1. Indicates where possible mounting points are to join two pieces together, the total contact area, and how they may fit together. 2. Space required to house an equipment rack or device. 3. Coverage area of a communications satellite.

frame 1. An individual segment of film. 2. A complete video picture or image of odd and even fields. Two fields equal one frame.

frequency   The number of complete cycles in a specified period of time. Formerly expressed as cycles per second (cps), now specified as hertz (Hz).

frequency response   The range of frequencies within which a microphone is sensitive.

Fresnel lens   A flat glass or acrylic lens in which the curvature of a normal lens surface has been collapsed in such a way that concentric circles are impressed on the lens surface. A Fresnel lens is often used for the condenser lens in overhead projectors, in rear-screen projection systems, and in studio spotlights.

front-screen projection   A system that employs a light-reflecting screen for use when the image will be projected from a source in front of the screen.

f-stop   Also called *f-number*, the ratio of focal length to the effective diameter of a lens. It represents how much light is able to pass through the lens.

fundamental frequency   Known as *pure tone*, the lowest frequency in a harmonic series.

gain   1. Electronic signal amplification. 2. The ability of a projection screen to concentrate light.

gate   An audio processor that allows only signals above a certain setting or threshold to pass.

gated automatic mixer   An audio mixer that turns microphone channels either on or off automatically.

gateway   The highest router in the hierarchy of routers. It connects a local network to an outside network, and all traffic must travel through it. A gateway will pass traffic to the routers below, and the routers below look to the gateway to find names (DNS addresses) that are not found on the local network.

gauge   A thickness or diameter of a wire.

genlock   To lock the synchronization signals of multiple devices to a single source.

GFCI   See *ground-fault circuit interrupter*.

glass-bead screen   A screen covered with tiny glass beads, each of which provides a spherically reflective surface.

good viewing area   The good area from which to view a screen. It is typically defined as any point within 45 degrees to the left or right from on-axis. The total good viewing area is 90 degrees.

graphic equalizer   An equalizer with an interface that has a graph comparing amplitude on the vertical with frequency on the horizontal.

graphics adapter   Commonly referred to as a *video card*, outputs computer signals.

ground   1. The earth. 2. In the context of an electrical circuit, the earth or some conductive body that extends the ground (earth) connection. 3. In the context of electronics, the 0V (zero volt) circuit reference point. This electronic circuit reference point may or may not have a connection to earth.

ground fault   1. An unintentional, electrically conducting connection between an ungrounded conductor of an electrical circuit and a normally noncurrent-carrying conductor, metallic enclosure, metallic raceway, metallic equipment, or earth. 2. The electrical connection between any ungrounded conductor of the electrical system and any noncurrent-carrying metal object.

ground lift   1. Interruption of a cable shield connection by means of a switch or by simple omission in an attempt to solve a hum or buzz problem from current flowing on a cable shield due to a pin 1 problem, detrimental ground loop, and so on. 2. Interruption of the connection between the chassis ground and signal ground, usually by means of a switch. 3. Incorrect term used for a 3-pin-to-2-pin AC adapter. (See also *grounding adapter*.)

ground loop   An electrically conductive loop that has two or more ground reference connections. The loop can be detrimental when the reference connections are at different potentials, which causes current flow within the loop.

ground plane   A continuous conductive area. The fundamental property of a ground plane is that every point on its surface is at the same potential (low impedance) at all frequencies of concern.

ground potential   A point of no potential in a circuit.

ground reference   The 0V (zero volt) reference point for a circuit.

grounded   See *grounding*.

grounded conductor   A system or circuit conductor that is intentionally grounded.

ground-fault circuit interrupter (GFCI)   A safety device that deenergizes a circuit (or a portion of that circuit) within an established period of time when a current to ground exceeds the values established for a Class A device. Class A GFCIs trip when the current to ground is 6 mA or higher; they do not trip when the current to ground is less than 4 mA.

ground-fault current path   An electrically conductive path from the point of a ground fault on a wiring system through normally noncurrent-carrying conductors, equipment, or earth to the electrical supply source.

grounding   Connecting to ground or to a conductive body that extends the ground connection. The connected connection is referred to as *grounded*. (See also *equipment grounding* and *system grounding*.)

grounding adapter A 3-pin-to-2-pin electrical adapter, the design of which is defined by National Electrical Code 406.9. The rigid tab or lug is to be used for equipment grounding and not left "floating."

grounding conductor A conductor used to connect equipment or the grounded circuit of a wiring system to a grounding electrode or electrodes. (See also *equipment grounding conductor.*)

grounding electrode A conducting object through which a direct connection to earth is established.

grounding electrode conductor The conductor used to connect the system grounded conductor or the equipment to a grounding electrode or to a point on the grounding electrode system.

harmonics Higher-frequency sound waves that blend with the fundamental frequency.

HDTV High-definition television.

headend The equipment located at the start of a cable distribution system where the signals are processed and combined prior to distribution.

headroom The difference in dB SPL between peak and average level performance of an audio system. For a speech-only system, this value is 10 dB.

heat sink A device that absorbs and dissipates heat produced by an electrical component.

hemispheric polar pattern The dome shape of the region within which some microphones will be most sensitive to sound. This pattern is used for boundary microphones.

hertz (Hz) Cycles per second of an electrical signal.

High-Definition Multimedia Interface (HDMI) A point-to-point connection between video devices that is becoming the standard for high-quality, all-digital video and audio in the consumer marketplace. HDMI signals include audio, control, and digital asset rights management information. It is a "plug-and-play" standard that is fully compatible with the Digital Visual Interface (DVI).

hiss Broadband higher-frequency noise typically associated with poor audio system gain structure.

horn A loudspeaker that reproduces mid- to high-level frequencies.

hot spot The part of a displayed image that is unevenly illuminated, usually a bright area in the center.

hum Undesirable 50 to 60 Hz noise emanating from a sound system or evidenced by a rolling hum bar on a display.

IEEE   The Institute of Electrical and Electronics Engineers.

IG   See *isolated ground*.

imager   A light-sensitive electronic chip behind a video camera's lens made up of thousands of sensors, called pixels, which convert the light input into an electrical output. In normal operation, an imager will output a frame of captured video at the frame rate of the video standard.

impedance (Z)   The total opposition to current flow in an AC circuit. Like a DC circuit, it contains resistance, but it also includes forces that oppose changes in the current (inductive reactance) and voltage (capacitive reactance). Impedance takes into account all three of these factors. It is measured in ohms, and its symbol is Z.

inductance (L)   Opposition to the starting, stopping, or changing of current flow in a coil of wire. An inductor's ability to resist the changes in current is represented by the symbol L, but it is measured in henries (H).

induction   The influence exerted on a conductor by the movement of a magnetic field. An example of this is a magnet moving through a coil of wire.

inductive reactance (X_L)   Opposition to the current flow offered by the inductance of a circuit. It is dependent on frequency and inductance. Its symbol is X_L.

infrared (IR)   A frequency range of light used to send information. Remote controls and other wireless devices use IR.

insulation   Also known as the *dielectric*, material applied to a conductor that is used to isolate the flow of electric current between conductors and to provide protection to the conductor.

interlaced scanning   The scanning process that combines odd and even fields of video to produce a full frame of video signal.

Internet Protocol (IP)   A standard networking protocol, or method, that enables data to be sent from one computer or device to another over the Internet.

inverse square law   The law of physics stating that some physical quantity or strength is inversely proportional to the square of the distance from the source of that physical quantity. For AV systems, this law is applied to light and sound.

IP   See *Internet Protocol*.

IR   See *infrared*.

IRE unit   An Institute of Radio Engineers (IRE) unit used as a reference to measure video signal levels.

isolated ground (IG)   A commonly misused term to describe a requirement for a dedicated equipment ground that terminates only at the main panelboard at the service entrance. IG is an equipment grounding method permitted by the National

Electrical Code (NEC) for the reduction of electrical noise (electromagnetic interference) on the grounding circuit. Equipment grounding for isolated receptacles and circuits is accomplished via insulated equipment grounding conductors, and run with the circuit conductors.

isolated grounding circuit   A circuit that allows an equipment enclosure to be isolated from the raceway containing circuits, supplying only that equipment by one or more listed nonmetallic raceway fittings. The equipment is grounded via an insulated grounding conductor. See National Electrical Code 250.96 (B) for additional information.

isolated receptacle   A receptacle in which the grounding terminal is purposely insulated from the receptacle mounting means. Isolated receptacles are identified by a triangle engraved on the face and are available in standard colors. The receptacle (and so the equipment plugged into the receptacle) is grounded via an insulated grounding conductor. See National Electrical Code 2008 250.146 (D) for additional information.

isolated star ground   InfoComm experts consider this term ambiguous. See *isolated ground*.

jacket   Outside covering used to protect cable wires and their shielding.

junction box   1. A portable set terminal for power cables. 2. Generally, metal or plastic boxes where wire and/or cable terminates, combines, or splits. It is used to protect the conductors.

keystone error   The trapezoidal distortion of a square-cornered image due to the optical effect of the projection device being located in an improper position with respect to the screen.

latency   Response time of the network. It is expressed as the amount of time in milliseconds between a data packet's transmission from the source application and its presentation to the destination application.

lavalier   A small microphone designed to be worn either around the neck or clipped to apparel.

LCD   Liquid crystal display. A video display that uses liquid crystals to produce an image. These devices do not emit light directly.

LCoS   Liquid crystal on silicon. A reflective, fixed-resolution LCD imaging technology. LCoS panels resemble LCD panels in size and function. A liquid crystal layer is applied inside an LCoS panel to a reflective complementary metal-oxide semiconductor (CMOS) mirror substrate. The LCoS chip has a fixed matrix of pixels, each backed by a mirrored surface.

least favored viewer (LFV)   The farthest usable seat from the image. The LFV depends on the viewing angle toward the screen, image size, and content being displayed.

LED   Light-emitting diode.

lenticular A screen surface characterized by silvered or aluminized embossing, designed to reflect maximum light over wide horizontal and narrow vertical angles. The device must be held very flat to avoid hot spots.

limiter An audio signal processor that functions like a compressor except that signals exceeding the threshold level are reduced at ratios of 10:1 or greater.

line level The strength of an audio signal. Line levels perform signal routing and processing between audio components, such as loudspeakers.

listed Equipment, materials, or services included in a list published by a Nationally Recognized Testing Laboratory (NRTL), such as Underwriters Laboratories (UL), that is acceptable to the authority having jurisdiction (AHJ). The NRTL list is concerned with the evaluation of products and services. The NRTL maintains periodic inspections of the production of the listed equipment or materials and periodic evaluations of the listed services. The listing states that the equipment items, materials, or services meet appropriate designated standards or have been tested and found suitable for a specified purpose.

load center An electrical industry term used to identify a lighting and appliance panelboard designed for use in residential and light-commercial applications.

local-area network (LAN) A computer network limited to the immediate area, usually the same building or floor of a building. A LAN connects devices within a small geographical area, such as a building or campus. LANs are generally owned and/or operated by the end user.

local monitor A device used to monitor the output of a signal from a system or other device in the local vicinity.

logarithm The exponent of base 10 that equals the value of a number.

loudspeaker A transducer that converts electrical energy into acoustical energy. A loudspeaker is basically a driver within an enclosure.

low voltage An ambiguous term. It may mean less than 70V AC to an AV contractor, while an electrician may use the same term to describe circuits less than 600V AC. The term may also be determined by the authority having jurisdiction (AHJ).

lumen A measure of the light quantity emitted from a constant light source across 1 square meter.

luminance (Y) Also called *luma*, part of a bandwidth-limited video signal combining synchronization information and brightness information. Its symbol is Y.

lux A contraction of the words *luminance* and *flux*. 10.7 lux is equal to 1 footcandle.

MAC (media access control) address The actual hardware address, or number, of a device that has a network interface (NIC). Each device has a globally unique MAC address to identify its connection on the network.

matrix decoder   A decoder that produces red, green, and blue from Y, R-Y, and B-Y.

matrix switcher   An electronic device with multiple inputs and outputs. The matrix allows any input to be connected to any one, several, or all of the outputs.

matte-white screen   A screen that evenly disperses light 180 degrees uniformly, both horizontally and vertically, creating a wide viewing cone and wide viewing angle.

MATV   Master antenna television system. In this type of television system, broadcast programs are received via a master antenna array, and then distributed to users over coaxial or fiber-optic cable.

media retrieval system   A system that allows for remote requests of content to be delivered from a headend location in a facility.

mic level   A very low line level signal. It creates only a few millivolts of electrical energy.

microphone sensitivity   A specification that indicates the electrical output of a microphone when it is subjected to a known sound pressure level.

midrange   A loudspeaker that reproduces midrange frequencies, typically 300 to 8,000 Hz.

mixer   A device for blending multiple audio sources.

modular connector   A connector used with four, six, or eight pins. Common modular connectors are RJ-11 and RJ-45 (8P8C).

modulator   A device that converts composite or S-Video signals, along with corresponding audio signals, into modulated signals on a carrier channel.

monophonic   Uses input from all microphones and relays them from the electronic control system to the loudspeakers using a single path or channel.

MPEG-2   A Moving Pictures Expert Group (MPEG) compression scheme that reduces the number of bits needed to code the video image.

multimeter   A multipurpose test instrument with a number of different ranges for measuring current, voltage, and resistance.

multiplexing   The process used by the combiner to put together a number of modulated signals.

multipoint   Also called *continuous presence*, videoconferencing that links many sites to a common gateway service, allowing all sites to see, hear, and interact at the same time. Multipoint requires a bridge or bridging service.

native resolution   The number of rows of horizontal and vertical pixels that create the picture. The native resolution describes the actual resolution of the imaging device, not the resolution of the delivery signal.

near-field   Sound that has not been colored by room reflections. This is also known as *direct sound*.

network address translation (NAT)   Any method of altering IP address information in IP packet headers as the packet traverses a routing device. NAT is typically implemented as part of a firewall strategy. The most common form of NAT is port address translation (PAT).

network bridge   A device that allows you to connect two networks. It translates one network protocol to another protocol. An example of a bridge is a computer modem. A cable modem converts, or bridges, the Ethernet protocol to a cable television protocol.

network interface card (NIC)   An interface that allows you to connect a device to a network. Many NICs are now integrated into the device's main circuitry.

neutral conductor   See *grounded conductor*.

nit   The metric unit for screen or surface brightness.

noise   Any electrical signal present in a circuit other than the desired signal.

noisy ground   An electrical connection to a ground point that produces or injects spurious voltages into the computer system through the connection to ground (IEEE Standard 142-1991).

notch filter   A filter that notches out, or eliminates, a specific band of frequencies.

octave   A band, or group, of frequencies. The relationship of the frequencies is such that the lowest frequency is half the highest. 200 Hz to 400 Hz is an octave, 4,000 Hz to 8,000 Hz is an octave, and so on.

Ohm's law   A law that defines the relationship between current, voltage, and resistance in an electrical circuit as proportional to applied voltage and inversely proportional to resistance. The formula is $I=V/R$, where I is the current (in amps), V is the voltage (in volts), and R is the resistance (in ohms).

omnidirectional   Describes the shape of the area for microphones that have equal sensitivity to sound from nearly all directions.

on-axis   The center point of a screen, perpendicular to the viewing area for a displayed image. This is considered to be the best location for viewing.

operating system (OS)   Computer platform that enables software applications to communicate with the CPU.

oscilloscope   A test device that allows measurement of electronic signals by displaying the waveform on a CRT.

OSI model   A reference model developed by the International Organization for Standardization (ISO) in 1984. The OSI model serves as a conceptual framework of standards for network communication across different equipment and applications by different vendors. Network communication protocols fall into seven categories, or *layers*: Physical, Data Link, Network, Transport, Session, Presentation, and Application.

overcurrent   Any current in excess of the rated current of equipment or the ampacity of a conductor. It may result from overload, a short circuit, or a ground fault.

overcurrent protection device   A safety device designed to open a circuit if the current reaches a value that causes excessive or dangerous temperatures in conductors or conductor insulation. Examples are circuit breakers and fuses.

overhead projector   A device that produces an image on a screen by transmitting light through transparent acetate placed on the stage of the projector.

packet filtering   A firewall technique that uses rules to determine whether a data packet will be allowed to pass through a firewall. Rules are configured by the network administrator, and they are implemented based on the protocol header of each packet.

panelboard   A single panel or group of panel units designed for assembly in the form of a single panel, including buses and automatic overcurrent devices. A panelboard may be equipped with switches for the control of light, heat, or power circuits. It is designed to be placed in a cabinet or cutout box placed in or against a wall, partition, or other support, and accessible only from the front.

parametric equalizer   Allows discrete selection of a center frequency and adjustment of the width of the frequency range that will be affected. This can allow for precise manipulation with minimal impact of adjacent frequencies.

PDP   See *plasma display panel*.

PDU   See *power distribution unit*.

peak   The highest level of signal strength, determined by the height of the signal's waveform.

peaking   An adjustment method that allows compensation for high-frequency loss in cables.

phantom power   A DC power source available in various voltages.

phase   A particular value of time for any periodic function. For a point on a sine wave, it is a measure of that point's distance from the most recent positive-going zero crossing of the waveform. It is measured in degrees; 0 to 360 degrees is a complete cycle.

Phoenix   A molded, plastic, captive screw connector. Termination requires a wire to be stripped and slid directly into a hole on the connector (compression termination).

phone connector   An audio connector used as a loudspeaker connector. Common types are 1/4 inch and 1/8 inch.

phono   The European name for an RCA connector.

phosphor The substance that glows when struck by an electron beam, providing the image in a CRT. The higher the quality of the phosphor, the brighter and more vivid the image.

pink noise A sound that has equal energy (constant power) in each 1/3-octave band.

pixel A combination of two words, "picture" and "element." The smallest element used to build a digital image.

plasma display panel (PDP) A direct-view display made up of an array of cells, known as *pixels*, which are composed of three subpixels, corresponding to the colors red, green, and blue. Gas in the plasma state is used to react with phosphors in each subpixel to produce colored light (red, green, or blue) from a phosphor in each subpixel.

point source A sound system that has a central location for the loudspeaker(s), mounted high above, intended to cover a large area. This type of sound system is typically used in a performance venue or a large house of worship.

point-to-point Conferencing where two sites are directly linked.

polar pattern Also known as *pickup pattern*, the shape of the area within which a microphone will be most sensitive to sound.

port address translation (PAT) A method of network address translation (NAT) whereby devices with private, unregistered IP addresses can access the Internet through a device with a registered IP address. Unregistered clients send datagrams to a NAT server with a globally routable address (typically a firewall). The NAT server forwards the data to its destination and relays responses back to the original client.

power distribution unit (PDU) A rack-mountable or portable electrical enclosure that is connected by a cord or cable to a branch circuit for distribution of power to multiple electronic devices. A PDU may contain switches, overcurrent protection, control connections, and receptacles.

preamplifier A device that boosts the electronic signal captured by the microphone before it is sent to other equipment.

primary optic The lens that focuses the image onto the screen.

prism A beam splitter that filters the light into its red, green, and blue components.

progressive scanning Scanning that traces the image's scan lines sequentially, such as with an analog computer monitor.

pulling tension The maximum amount of tension that can be applied to a cable or conductor before it is damaged.

pure tone See *fundamental frequency*.

quality of service (QoS) Any method of managing data traffic to preserve system usefulness and provide the best possible user experience. Typically, QoS refers to some combination of bandwidth allocation and data prioritization.

quiet ground   A point on a ground system that does not inject spurious voltages into the computer system. There are no standards to measure how quiet a quiet ground is.

raceway   An enclosed channel of metal or nonmetallic materials designed for holding wires, cables, or busbars, with additional functions. Raceways include, but are not limited to, rigid metal conduit, rigid nonmetallic conduit, intermediate metal conduit, liquid-tight flexible conduit, flexible metallic tubing, flexible metal conduit, electrical nonmetallic tubing, electrical metallic tubing, underfloor raceways, cellular concrete floor raceways, cellular metal floor raceways, surface raceways, wireways, and busways.

rack   See *equipment rack*.

rack unit (RU)   A unit of measurement of the vertical space in a rack. One RU equals 1.75 inches (44.5 mm).

radio frequency (RF)   The portion of the electromagnetic spectrum that is suitable for radio communications. Generally, this is considered to be from 10 kHz up to 300 MHz. This range extends to 300 GHz if the microwave portion of the spectrum is included.

radio frequency interference (RFI)   Radiated electromagnetic energy that interferes with or disturbs an electrical circuit.

RAM   Random-access memory. The most common type of computer memory used by programs to perform tasks while the computer is on. An integrated circuit memory chip allows information to be stored or accessed in any order, and all storage locations are equally accessible.

rarefaction   The action of molecules moving apart.

ratio   A mathematical expression that represents the relationship between the quantities of numbers of the same kind. AV examples include image aspect ratio, which describes the relationship of image width to height; signal-to-noise ratio, which describes the amount of signal to noise; and contrast ratio, which describes the difference in brightness between white and black levels of a display. A ratio is typically written as X:Y or X/Y.

RCA connector   Also known as a *phono connector*, a connector most often used with line-level audio signals.

reactance (X)   Opposition to alternating current resulting from capacitance and inductance in the circuit.

rear-screen projection   A system in which the image is projected toward the audience through a translucent screen material, for viewing from the opposite side. This is opposed to *front-screen projection*.

reference point   The point of no potential used as the 0V (zero volt) reference for a circuit.

reflection   Light or sound energy that has been redirected by a surface.

reflective technology   Any display device that reflects light to create an image.

refraction   The bending or changing of the direction of a light ray when passing through a material, such as water or glass. How much light refracts, meaning how great the angle of refraction, is called the *refractive index*.

relocatable power tap   A cord-connected product rated 250V AC or less and 20 A or less with multiple receptacles. This tap is intended only for indoor use and plugged directly into a branch circuit. It is not intended to be connected to another relocatable power tap.

reserve DHCP   A hybrid approach to IP addressing. Using reserve DHCP, a block of statically configured addresses can be set aside for devices whose IP addresses must always remain the same. The remaining addresses in the subnet will be assigned dynamically. The total pool of dynamic addresses is reduced by the number of reserved addresses.

resistance   The property of a material to impede the flow of electrical current, expressed in ohms.

resolution   1. The amount of detail in an image. 2. The number of picture elements (pixels) in a display.

retrace time   The time it takes for the electron beam to turn off, travel to its next starting point, and then turn back on to begin scanning again.

retro unit   A self-contained rear-projection system.

reverberant sound   Sound waves that bounce off multiple surfaces before reaching the listener, but arrive at the listeners' ears quite a bit later than early reflected sound.

reverberation   Numerous, persistent reflections of sound energy.

RF   See *radio frequency*.

RFI   See *radio frequency interference*.

RF system   A closed-circuit system with the composite video and audio signals modulated at a certain frequency, called a *channel*. RF systems require a display device (such as a television) with a tuner set to a selected channel to display the information modulated onto that frequency.

RGBHV signal   A high-bandwidth video signal with separate conductors for the red signal, green signal, blue signal, horizontal sync, and vertical sync.

RGBS signal   A four-component signal composed of a red signal, a green signal, a blue signal, and a composite sync signal.

RGSB   A three-component signal composed of a red signal, a green signal with composite sync added to the green channel, and a blue signal. It is often called "sync on green."

ring   A network topology that connects terminals, computers, or nodes in a continuous loop.

ROM   Read-only memory. Memory whose contents can be accessed and read but cannot be changed. ROM is permanent memory that can be entered only once, normally by a manufacturer. ROM may not be altered or removed.

router   A device that works on the OSI model layer above the Network and Transport layers. A router knows the IP address of sent packets, and it can send them to specific locations on the network. The IT manager can use a router to change how the network works. A router also allows for redundancy in the network.

RS-232   The interface between data terminal equipment and data circuit-terminating equipment employing serial binary data interchange. RS-232 supports a single-ended mode of operation with one driver and one receiver. It supports a maximum cable length of 50 feet (15 meters) with a data rate of 20 kbps.

RS-422   A standard that provides the electrical characteristics of balanced voltage digital interface circuits. It specifies a balanced signal with one driver and ten receivers with multidrop capability. The maximum cable length for RS-422 is 4,000 feet (1,220 meters), with a data rate of 10 Mbps.

RsGsBs   Red, green, and blue signals with composite sync added to each color channel. This requires three cables to carry the entire signal. It is often referred to as "RGB sync on all three."

RU   See *rack unit*.

safety ground   See *equipment grounding conductor*.

sampling rate   Describes the number of samples taken per unit of time (typically seconds) when converting a continuous (analog) signal to a discrete signal (typically a digital signal). The sample is described by its bit depth.

satellite television   Entertainment or business video and audio transmitted via a satellite.

scaler   A feature in a display device that changes the size of an image without changing its shape. Scaling may be required when the image size does not fit the display device.

scan rate   The frequency of a display drawing one line of information.

scattering   When light hits a textured surface, the incoming light waves are reflected in multiple angles because the surface is uneven.

screen gain   Describes the distribution of light reflected off a projection screen. The amount of gain is compared to a matte-white screen, which reradiates light and distributes it with perfect uniformity.

SDTV   Standard-definition television.

sensitivity specification   A way to determine a device's ability to convert one form of energy into another form of energy. It is used to define the device's efficiency in converting from one form to another.

separately derived system   A premises wiring system whose power is derived from a source of electric energy or equipment other than a service. These systems do not have any direct electrical connections, such as a solidly connected grounded circuit conductor, to supply conductors originating in another system. Examples of separately derived systems are generators, batteries, converter windings, transformers, and solar photovoltaic systems.

service level agreement (SLA)   A contract used to document agreements between an IT service provider and a customer. An SLA describes the services to be provided, documents service level targets, and specifies the roles and responsibilities of the service provider(s) and customer(s).

shield   A metallic partition placed between two regions of space. A shield is used to control the propagation of electric and magnetic fields from one of the regions to the other. It contains electric and magnetic fields at the source or to protect the receiver from electric and magnetic fields. A shield can be the chassis (metallic box) that houses an electronic device or the metallic enclosure (aluminum foil or copper braid) that surrounds a wire or cable.

short circuit   The electrical connection between any two conductors of the electrical system from line-to-line or from line-to-neutral (*Basic Electrical Theory*, by Mike Holt). Note that a short circuit is not the same as a ground fault.

shotgun microphone   A long, cylindrical, highly sensitive, unidirectional microphone used to pick up sound from a great distance.

signal generator   A test equipment instrument that produces calibrated electronic signals intended for the testing or alignment of electronic circuits or systems.

signal ground   1. 0V (zero volt) point of no potential that serves as the circuit reference. 2. A low-impedance path for the current to return to the source.

signal-to-noise (S/N) ratio   The ratio, measured in decibels, between the audio or video signal, and the noise accompanying the signal. The higher this ratio, the better the quality of the sound or picture.

single-phase power   Alternating current electrical power supplied by two current-carrying conductors. This type of power is used for residential and some light-commercial applications.

single-point ground (SPG)   In the context of IEEE Standard 1100, refers to the implementation of an isolated equipment grounding configuration for the purposes of minimizing problems caused by circulating current in ground loops.

SMATV   Satellite and master antenna television system. In this television system, satellite and broadcast programs are received via a master antenna array and distributed to users over coaxial or fiber-optic cable.

Speakon   A specialized connector used to hook up loudspeakers without causing a short circuit. It allows connection of a loudspeaker while it's working, or "hot."

specular reflection   A mirror-like reflection of energy, in which most of the energy is reflected back in a single direction.

SPG   See *single-point ground.*

spherical aberration   Light passing through the edges of the lenses that have focal lengths different from those passing through the center.

splitter   An electronic device that divides a signal into different pieces to route to different devices.

star   A network topology in which all network devices are connected to a central network device, usually a hub or a switch.

star ground   1. A conductor connection by which separate electrical circuits or equipment are connected to earth at one point (IEEE Standard 1100-2005). 2. A grounding configuration where grounds from different circuits are insulated from one another and referenced (connected) to a single point.

static IP address   A manually assigned permanent IP address.

stereophonic   Commonly shortened to "stereo," describes when input from all microphones is split into at least two channels before driving the signal through the loudspeakers.

streaming video/streaming audio   Sequence of moving images or sounds sent in a continuous, compressed stream over the Internet and received by the viewer as they arrive. With streaming video or audio, a web user does not need to wait to download a large file before seeing the video or hearing the sound.

subnet   A logical group of hosts within a local-area network (LAN). A LAN may consist of a single subnet, or it may be divided into several subnets. Additional subnets may be created by modifying the subnet mask on the network devices and hosts.

subnet mask   A binary number whose bits correspond to IP addresses on a network. Bits equal to 1 in a subnet mask indicate that the corresponding bits in the IP address identify the network. Bits equal to 0 in a subnet mask indicate that that corresponding bits in the IP address identify the host. IP addresses with the same network identifier bits as identified by the subnet mask are on the same subnet.

subwoofer   A loudspeaker that reproduces lower frequencies, typically 20 to 200 Hz.

supercardioid polar pattern   The exaggerated heart shape of the area within which a highly directional microphone is most sensitive to sound.

surface-mount microphone   Also called a *boundary microphone,* a microphone placed on a table to pick up sound. This type of microphone is often used in boardrooms and other environments where a number of talkers must be picked up and the microphone needs to remain unobtrusive.

surround-sound system   A stereo playback system that uses from two to five channels for realistic sound production, producing an experience where the sound seems to surround listeners. This is best achieved using surround-encoded material, a receiver, and surround loudspeakers.

S-Video   A video signal, also known as Y/C. Y is the luminance, and C is the chrominance. Y and C are transmitted on separate conductors.

switcher   A peripheral or sometimes integrated device used to select one of a group of signals.

sync   Synchronization. The timing information that keeps images displaying properly.

system   In the AV industry, a compilation of multiple individual AV components and subsystems interconnected to achieve a communication goal.

system grounding   The intentional grounding of one of the current-carrying conductors in a manner that will limit the voltage imposed by lightning, line surges, or unintentional contact with higher-voltage lines, and that will stabilize that voltage to earth during normal operation.

tap   A connection to a transformer winding that allows you to select a different power level from the transformer.

technical ground   An ambiguous term that refers to a branch circuit with an isolated equipment ground that serves an AV system exclusively.

technical power   An ambiguous term used to refer to a separately derived power system that is 120V line to line and 60V to ground. In the AV industry, technical power refers to power that serves an AV system exclusively.

tensile strength   The maximum force that a material can withstand before deforming or stretching.

three-phase power   Alternating current electrical power supplied by three current-carrying conductors, each offset by 120 degrees from one another. A fourth conductor, a neutral, is used as the return conductor. This type of power is used for commercial and industrial applications.

throw distance   The length of the projection beam necessary for a particular projector to produce an image of a specified size.

time code   A method of numbering video frames according to Society of Motion Picture and Television Engineers (SMPTE) standards. The code is the eight-digit address representing the hour, minute, second, and frame recorded on the videotape's control track.

transformer A passive electromagnetic device commonly consisting of at least two coils of wire (inductors) with no physical connection between them. Most often, these coils share an iron-based alloy core. This common core aids in concentrating the magnetic lines of force created by the current flow in one coil (primary), thereby inducing a voltage into the other coil (secondary).

transient disturbance A momentary variation in power, such as a surge, spike, sag, blackout, or noise.

transmission Passing of sound energy through partitions or structure-borne vibrations.

transmissive technology Any display device that creates images by allowing or preventing light to pass.

TRS Tip, ring, sleeve. A three-conductor design of a phone connector that can be terminated as balanced or unbalanced.

TS Tip, sleeve. A two-conductor design of a phone connector used for an unbalanced circuit.

tweeter A loudspeaker that is designed to reproduce frequencies above 3,000 Hz.

twisted-pair Any number of wires that are paired together and twisted around each other. The wires can be shielded or unshielded.

unbalanced circuit A two-conductor circuit in which one conductor carries the signal and the other conductor carries the return. The return conductor is usually the cable shield and is a low-impedance connection, as it is connected to the signal ground and possibly also to the earth ground. The impedance of the signal circuitry is quite different from the return circuitry, hence the impedances of the two conductors are quite different—the impedances are unbalanced with respect to one another.

unity gain Derived from the number 1, refers to no change in gain.

UTP cable Unshielded twisted-pair cable, typically used for data transfer. UTP cable contains multiple two-conductor pairs twisted at regular intervals, employing no external shielding.

V See *volt*.

vectorscope A specialized oscilloscope used in video systems to measure chrominance accuracy and levels.

viewing angle A viewing-area measurement that determines how far off-axis (screen centerline) a viewer can sit and still see a good-quality image. This is no greater than 45 degrees off the projection axis.

viewing cone The best viewing area for the audience. The term *cone* is used because there is width, height, and depth to the best viewing area, and this area emanates from the center of the screen.

virtual local-area network (VLAN)   A network that connects separate LANs to form a logical group. For instance, the LANs at each branch of a large company could be combined into one company-wide VLAN.

virtual private network (VPN)   A network that uses the Internet to create a tunnel between two or more local-area networks (LANs). VPNs are used to create virtual wide-area networks (WANs), and for remote monitoring, troubleshooting, and control. VPNs are typically controlled and configured by the enterprise network administrator. Each host requires the proper software, access rights, and password to log in to the client network.

volt (V)   The basic international unit of potential difference or electromotive force.

voltage   The electrical potential to create current flow in a circuit.

waveform monitor   A specialized oscilloscope used to display and analyze the video signals synchronization, luminance, and chroma levels.

wavelength   The distance between two corresponding points of two consecutive cycles measured in meters.

white noise   A sound that has the same energy level at all frequencies.

wide-area network (WAN)   A data communications system that uses telecommunication circuits to link local-area networks (LANs) that are distributed over large distances. A WAN covers a large geographical area, such as a state or country. The Internet, which covers the entire world, is an example of a WAN.

wire   A single conductive element intended to carry a voltage or electronic signals.

wireless local-area network (WLAN)   A network that shares information by radio frequency (RF) transmissions.

woofer   A loudspeaker that has low frequencies, typically 20 to 200 Hz.

XLR connector   A popular type of audio connector featuring three leads: two for the signal and one for overall system grounding. This is a secure connector often found on high-quality audio and video equipment. It is sometimes called a *cannon connector*.

Y/C   A video signal, also known as S-Video. Y is the luminance, and C is the chrominance. Y and C are transmitted on separate synchronized conductors.

zero reference   InfoComm experts consider this term ambiguous. See *reference point*.

zoom lens   A lens that allows the operator to adjust the focal length for sizing or distance.

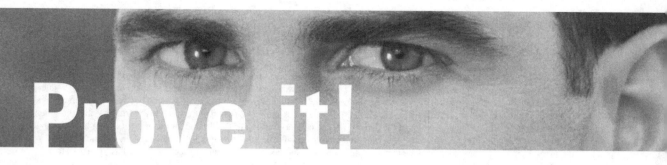